D1570294

THE DARK ABYSS OF TIME

THE DARK ABYSS OF TIME

THE HISTORY OF THE EARTH & THE HISTORY OF NATIONS FROM HOOKE TO VICO

PAOLO ROSSI

Translated by

LYDIA G. COCHRANE

THE UNIVERSITY OF CHICAGO PRESS
CHICAGO & LONDON

Paolo Rossi is professor of the history of philosophy at the University of Florence. His previous books in English include *Francis Bacon: From Magic into Science*, also published by the University of Chicago Press, and *Philosophy, Technology, and the Arts in the Early Modern Era*.

This work was originally published as *I segni del tempo: Storia della terra e storia delle nazioni da Hooke a Vico*, © Giangiacomo Feltrinelli Editore, Milano, 1979.

The University of Chicago Press gratefully acknowledges a grant from the Italian Ministry of Foreign Affairs, which supported in part the costs of translation of this work.

D
16.8
.R6813
1984

The University of Chicago Press, Chicago 60637
The University of Chicago Press, Ltd., London

© 1984 by The University of Chicago
All rights reserved. Published 1984
Printed in the United States of America

93 92 91 90 89 88 87 86 85 84 12345

Library of Congress Cataloging in Publication Data

Rossi, Paolo, 1923–
The dark abyss of time.

Translation of: I segni del tempo.
Bibliography: p.
Includes index.
1. History—Philosophy—History. 2. Science—Philosophy—History. 3. Vico, Giambattista, 1668–1744.
I. Title.
D16.8.R6813 1984 901 84-8481
ISBN 0-226-72835-8

Contents

Preface

IT often happens that historians of science—and philosophers of science—flatter themselves that the discipline they study has always existed. They draw their subject matter from a variety of texts belonging to different epochs and to heterogeneous fields, and they plot the development of an imaginary object. The case of geology is an excellent example of this. In the two or three works that have become classics in the field we find summarized arguments on the structure of the earth, on fossils, on the formation of mountains, or on springs, and we find them grouped together in a presumed uniformity and continuity in the object studied. Figures of primary importance—those whom the textbooks classify as great philosophers or great scientists—who denied that the process of the formation of the Earth could be the object of rational investigation, or who identified the "shells" found in the Alps as the garbage of travelers' meals, find absolutely no place in this sort of "history." Everything becomes simple and easy in them, and reality places no obstacles before the omnipotence of epistemology: after the observation of things, theories are developed, observation is ever more refined, and the theories become ever more all-inclusive and satisfactory. New facts contradict old theories, and the new theories incorporate new facts. Superstitions, over-audacious hypotheses, the philosophies, the romances of physics, and the images of science all find no *droit de cité* in these histories. They are only mentioned in a page here or a note there, like a bothersome and irrelevant presence. But how is it that the same *thing* could seem to some people a remnant, or a document in a history of nature, and to others, during the same years, as simply one of the many objects or the many forms that abound in nature? What were the effects on the image of man and the investigation of nature of the adoption of a chronological scale enormously vaster than the traditional one? What is the significance of the strong resistance to the idea of a lengthened time—apart, of course, from the instances obviously due to preoccupations of a religious nature? Why is it that only when the metaphysical hypotheses underlying Burnet's "romance" were accepted could the books on modern "geology" be written? What interrelationships are there between the "philosophical romances" and the texts of "scientific" geology? Questions of this sort made no sense in those histories.

The truth is, as Canguilhem noted, that the object of the history of science does not in the least coincide with the object of science. Hélène Metzger's work on the history of crystallography does not speak of crystals in the same way as does a current treatise on crystallography. She considers arguments that do not coincide in the slightest with those in the terms of which crystals have subsequently become objects in the science of crystallography—that is, now that science has constructed its own specific object on the basis of a specific definition/theory.

The very existence of the object of a science (of any science) presupposes definitions and theories. It requires the consensus of a scientific community, the existence of divergences and of conflicting programs within a "convergent thinking," or a terrain that is recognized as common. Thomas Kuhn was certainly right about that. It is true that the textbooks are continually being rewritten, but it is also true that the existence of the textbooks—that is, of codified forms of knowledge, transmitted as such—is not something marginal, but something intrinsic, something essential to those particular forms of creative knowledge which we qualify as "scientific." But even consensus, common ground, or "convergent thinking" are not givens, always present; they are constructed laboriously through the course of time. They are presented as results, not as presuppositions, at least as historians view them. As Kuhn himself has remarked, this consensus is evident in mathematics and astronomy from ancient times; in optics, geometry, and part of physiology, we can trace it back to the classical world; we can see that it is recent as far as chemistry, the study of magnetism, electricity, and heat are concerned; and we have to wait for the middle of the nineteenth century to find it in geology and the nontaxonomic sectors of biology. We sense this consensus as extremely recent in sectors that have to do with the mind and with human and animal behaviors, and for other forms of understanding we view it as a goal to be reached (Kuhn 1977, 231–32). As the shrewder historians of science realized, it makes no sense to speak of seventeenth-century astronomy and the chemistry of the same century on the same plane or within one global argument. Astronomy had for some time possessed a highly organized theoretical structure, it made use of sophisticated techniques, and in 1543 it underwent a radical turning point, which was followed, and not preceded, by an enormous job of acquisition of factual material. Chemistry in the seventeenth century had no organized structure as a science, it possessed no coherent theory of changes and reactions, and it had no definite tradition behind it: craftsmen and artisans had at their disposal a much more ample array of chemical knowledge than the natural philosophers (Hall, in Clagett 1962, 19).

Whatever its influence and its extent might be, the "convergent thinking" of which Kuhn speaks was in reality constructed extremely laboriously. In the hundred and twenty years that separate Robert Hooke's work on earthquakes from James Hutton's theory of the earth, what was written regarding the earth and its history was shaped by radical alternatives. Not only do the discussions concern different models of the history of the earth; they also differ concerning the very possibility of making that history the object of a "scientific" investigation. If physics and natural philosophy were concerned with the world *that is*—as it has been put in motion by God—then it made no sense, nor was it in any way legitimate, in a context of physics or of natural philosophy, to pose the problem of the *formation* of the world. In the Newtonian view of science the line of demarcation between science and religion left that problem out of the realm of science, relegating it to the "nonscientific" terrain of the explication of the Genesis narration or confining it within the sphere of gratuitous hypotheses, fables, or the "romances" of physics—or, as we would say today, science fiction. Once the legitimacy of a *historical* investigation of nature was established, choices opened up between markedly divergent theoretical models: between a history made of slow processes, of uniform and imperceptible shifts, and a history punctuated by violent catastrophes, made of qualitative leaps and of revolutions. Within the catastrophist hypotheses there were opposing theses that took as their principal agents floods, earthquakes, or volcanic eruptions, the action of water or the action of fire. These distinctions, as is almost always the case, were not absolute: they gave rise to curious intertwinings of different theses, to compromises, and to eclecticisms.

The "scientific metaphysics" of Descartes, of Newton, and of Leibniz was at work at the deepest level in these discussions, proposing different ends to research, orienting it and limiting it in differing ways. In these debates, great questions arose again and again: the relationship of science with theology, how to regard the Lucretian, materialistic tradition, a consideration of the relationship of man to nature, the comparison with the Hermetic and Neoplatonic tradition, the choice between an anthropomorphic and a naturalistic view. Images of science act on theories and on observation at the deepest level. The discussion concerning fossils, between the seventeenth and the eighteenth centuries, seems to confirm Norwood R. Hanson's thesis in *Patterns of Discovery* (Cambridge 1958): epistemological vision does not coincide with the retina's vision. Where Scilla saw sharks' teeth petrified by time, Robinet saw intermediate forms, halfway between the mineral and the animal worlds, documents of the "apprenticeship" that nature went through on its way toward the formation of man.

A great many pages have been written on the Copernican revolution, taking it as symbolic of a radical change in the position of man, who was thus removed from the center of the universe and relegated to its edges. There has been no such insistence, in the same way and with the same intensity, on other, no less decisive changes. Men in Hooke's times had a past of six thousand years; those of Kant's times were conscious of a past of millions of years. The difference lies not only between living at the center or at the margins of the universe, but also between living in a present relatively close to the origins (and having at hand, what is more, a text that narrates the *entire* history of the world) or living instead in a present behind which stretches the "dark abyss" (the term is Buffon's) of an almost infinite time. Similarly, it is a different matter to live on an earth that is to this day as it was shaped by the benevolent hands of God and is populated by the plants, animals, and men that He created, or, on the other hand, to be conscious of the mutability of the forms of nature and the forms of life and conscious that the earth on which we find ourselves walking hides within it (in Hutton's expression) a "succession of worlds."

The infinity of the universe in space, which Bruno celebrated, filled Kepler with a sense of bewilderment and "hidden horror." The near infinity of time, the image of a history conceived according to an enormously broad chronological scale, also provoked different reactions. Reflection on the length of human history, on the boundless antiquity of "nations," took place during these same years, and, in many cases, appeared closely intertwined with the discussions on development within nature and on the history of the earth.

In the course of that long process that Stephen Toulmin and June Goodfield have called the "discovery of time," closely comparable (at times astonishingly similar) theses were used for very different purposes. When placed in different philosophical and ideological contexts, the same affirmations could serve those intent on subverting tradition or defending orthodoxy. Attacks on what were termed "hypotheses" and "physical romances" came *both* from the most obtuse champions of tradition and from those who rejected a type of science that seemed on the verge of losing contact with experience. The same sort of thing occurred in discussions of the history of "nations" and of "barbarism" or "savages." Combating the libertines and furnishing arguments to the libertines and to the deists were two apparently contradictory undertakings that often seemed paradoxically to merge into one undertaking. Vico and Warburton, and many other writers as well, could be attacked ferociously as potential destroyers of the faith and, during the very same years, praised as its alert defenders.

I have no intention of anticipating the matter of this work, but I do want to warn that this is *also* a book on Vico and a renewed attempt to examine, in a European context, some of the problems that Vico took on and some of the solutions that he worked out. The great, direct interlocutors of Vico, the men with whom he engaged in dialogue in the writing of the *Scienza Nuova*—when he managed to free himself briefly from the "stimulus to further lamentation against adverse fortune and to more inveighing against the corrupt fashion of letters" (Vico 1911, 175)—did not live in Naples. They were those scholars and philosophers of European reputation who, with the exception of Bayle, whom Vico perhaps never read directly, had published their works between 1600 and 1680: Lapeyrère and Hobbes, Marsham and Spencer, Huet and Grotius, Selden and Petau, Bochart and Gerhard Vossius, Bacon and Pufendorf, Descartes, Horn, and Spinoza. Many Vico scholars, even some of the most illustrious, have never shown more than a slight familiarity with some of these authors. Precisely because of their relation to Vico, they ought to be taken up again and studied anew. I have tried in this book to fulfill this need and to treat these authors—some of them in more limited fashion—more fully than in the past.

I am also more and more persuaded that it makes little sense to shirk the task of reading the many texts of European origin that Vico had read—and which he noted—while continuing to speak of texts that he *should have* read or that we would like him to have read—even when he never mentions them or when they were written in languages inaccessible to him. Sir Isaiah Berlin, who is the author of a fundamental work on Vico, recently affirmed the decisive importance, for Vico, of the work of many French historians and grammarians. He states that the absence of the least reference to these sources in Vico proves nothing because "ideas travel without labels," as demonstrated by the fact that many writers have been influenced by Marx or Freud (to the point of proclaiming themselves Marxist or Freudian or even the two together) but who have never read even one page of *Capital* or the *Interpretation of Dreams*. Precisely because it is true that ideas travel without labels, it is important not to establish nonexistent relationships, but to document possible correspondences or similitudes in problems and solutions. *Independently* of Vico, and on cultural terrains very different from his, doctrines and theories came to be elaborated, between the middle of the seventeenth and the middle of the eighteenth centuries, that Italian scholars of Vico designate as "Viconian." There is little sense in trying to avoid the task of reading the texts of the European scholars to whom Vico does refer, however, and in continuing to call him a "solitary precursor" or speak of his "divinatory innovations." As far as I am concerned, I will continue to believe *both* in Vico's "isolation" in his times *and* in the presence in his thought of many great and decisive "innovations."

Vico thought that many of the luxurious books published in his time were like spoiled dishes in which the bad odor of the meat or the fish is masked with spicy sauces. He thought that his contemporaries were frail-hearted readers, too easily swayed by fashion (his century was "dilicato e vistoso") and he claimed that he had not read a contemporary book in twenty years. He held his times to be quite like those on which Tacitus meditated when he declared "corrumpere et corrumpi saeculum vocatur." Where is it written that all great philosophers must feel themselves contemporary to the men who live in their own times? Or, what is worse, that they must be "up to date"?

Like my other books and like my preceding book, *Philosophy, Technology and the Arts in the Early Modern Era* (New York, 1970), this volume is born of the conviction that a historiography attuned simultaneously to scientific theories, to philosophies, and to currents of ideas has a precise function. This sort of research, which strictly speaking belongs neither to the history of philosophy nor to the history of science, has a totally respectable tradition and seems to be the object of renewed interest in recent years, especially on the part of historians of science. This may be in large part a result of the extreme (and probably irreversible) crisis that is taking place in that type of "epistemological" history of science that conceives of historical reconstruction as little more than a sort of collection of "comforting examples" in support of already solidified epistemological points of view.

Thomas Kuhn attacks with extraordinary efficacy the "simplifications" of the philosophers of science when they present "standard examples." He insists on the fact that the choice between paradigms in not resolvable *by proof*, he emphasizes the importance of the choices made by the scientific communities, choices which, in some cases, take on the function of *criteria*, and he states that the listing of the canons in light of which the scientific enterprise can be defined as such *is not sufficient* to determine the complex of the choices made, either in the decisions of single scientists or in those of the scientific community. He also insists on the noncumulative character of the growth of science, on the fallaciousness of the belief in a neutral language of observation, and on the fact that theories are never "dislodged" from experience, but only from other theories. He emphasizes that changes of paradigm, which reinterpret all descriptive propositions and give rise to new rules, occur according to the model of the "Gestalt switch theory," in which the world itself changes when a paradigm changes, and he stresses the fact that a common language among paradigms does not exist.

These theses have kindled widespread discussion (not always free of "scholastic" aspects, however), important contributions to which have been made, among others, by P. K. Feyerabend, I. Lakatos, J. P. Sneed, W. Stegmüller, J. Watkins, S. Toulmin, and, in Italy, E. Bellone and G. Giorello. I do not for one minute intend to enter into this discussion—at least not here. I would like, however, to linger briefly over three points that are more closely related to some of my own long-term interests.

1. The discussion of Kuhn's theses cannot avoid also involving, it seems to me, a careful reconsideration (on the part of epistemologists and the so-called internist historians of science) of the relations that link the history of science and the other so-called specialized histories. "The similarities of science and art," Kuhn writes, "came as a revelation. . . . Gombrich's work, which tends in many of the same directions, has been a source of great encouragement to me, as Hafner's essay" (Kuhn 1977, 340, 343). Such a reconsideration has been lacking up to now.

2. Two possible lines of development seem to emerge from among many others in the discussion as it has taken place thus far. The first—which is the direction Feyerabend takes—leads to a radical anarchism in methodology (the notion that "anything goes") and to an absurd cancellation of all the differences between science and art, religion, or magic. The second—the one Kuhn has proposed in more recent years—aims at the identification of differences (which are no longer those codified by tradition) and considers as an essential and constitutive part of any scientific discourse a *search for the historical and cultural contexts* within which the individual scientist and the scientific community exercized choices and made decisions and within which the changes in paradigm took place. The history of ideas seems to have to be connected, much more closely than it has been in the past, with the history of science.

3. Kuhn has now moved farther in this direction, proving the error, in substance, of the criticisms that too hastily equated his position with Feyerabend's. (For my point of view, see Rossi 1977, 172–73.) But it also should be said that Kuhn now speaks of real "error" committed at the time of the writing of *The Structure of Scientific Revolutions*. He there identifies and differentiates the various scientific communities on the basis of the questions or the "objects" they studied: optics, physics, heat, electricity, and so forth (Kuhn 1977, xv–xvi). In the eyes of a historian of ideas, this is truly an unpardonable error.

Important consequences derived from the correction of this "error," and Kuhn's 1976 essay, "Mathematical versus Experimental Traditions" (see Kuhn 1977, 31–

65) abounds in new perspectives. As distinguished from A. Koyré, K. Popper, and many others who qualified the "Baconian" movement as a sort of gigantic fraud or, in any event, as a movement extraneous to modern science, Kuhn here clearly realizes that the scientific revolution embraced both a mathematical tradition and the "Baconian" or experimental tradition out of which the new sciences of magnetism, the study of heat, chemistry, etc., emerged. Kuhn says of the mid-seventeenth century: "The Baconian sciences were then in gestation, the classical being radically transformed. Together with the concomitant changes in the life sciences, these two sets of events constitute what has come to be called the Scientific Revolution" (ibid, 52). The scientific revolution cannot be explained without determining "new ways of life, and new values combined to promote the status of a group formerly classified as artisans and craftsmen [artist/engineers] . . . whose expertise included painting, sculpture, architecture, fortifications, water supply, and the design of engines of war and construction" (ibid, 55). Kuhn concludes with the most important point: "The major contribution of Hermeticism to the Baconian sciences and perhaps to the entire Scientific Revolution was the Faustian figure of the magus, concerned to manipulate and control nature. . . . Recognizing Francis Bacon as a transition figure between the magus Paracelsus and the experimental philosopher Robert Boyle has done more than anything else in recent years to transform historical understanding of the manner in which the new experimental sciences were born" (ibid, 54).

The completely imaginary Bacon of Sir Karl Popper and of many of his followers has up to now played the ungrateful role of Turk's head in discussions of the logic of discovery and the scientific revolution. But Kuhn's affirmations are not important because they give the Lord Chancellor his proper historical place, and even less so because they recognize the significance of research carried on in a now remote past (the first edition of my own Francis Bacon dates back to 1957). They are important for another, more general reason: they show that *today*, as an integral part of the most productive and significant discussions on science, debate on the cultural contexts, on the connections between ideas and theories, and on the influence that the images of science exert at the very heart of theories, is seen as essential.

Several provisory conclusions can perhaps be drawn from these considerations: (1) The discussion between philosophers of science and historians of science is particularly lively today, but the time for *theorizing* a total detachment from historical considerations in the construction of scientific knowledge seems to be waning. (2) The territory of "science" as an object of interest for philosophers of science and for historians of science does not necessarily coincide with the territory in which theories can be reduced to axioms. (3) The history of science takes as the object of its study not only the theories that can be reduced to axioms, and not only "complete" theories, but also the attempts to construct theories. (4) A close analysis of these attempts shows that the listing of "criteria" operates in many cases with excessive ridigity. As J. B. North commented during a meeting at Pisa in 1978, we must not let ourselves be hypnotized by the delights of the axiomatization of limited theories in physics.

We can hope that new tensions will open up between philosophers and historians in this partially renewed field. Because the point is not to arrange marriages. Competing programs are preferable to normative methodologies, both in the individual sectors of research and in the relationships among the various sectors of research.

Earth science and social science were undeniably emerging forms of knowledge during the period under consideration in this book. As I have said, the themes here treated do not principally concern the "discoveries" made in geology and, even less, in the search for the laws of history. I have instead tried to show by what paths (and they are often twisted paths) things or phenomena—or objects like shells and fossils, monuments, languages, writing systems, and the traces of the most ancient civilizations—came to be a part of a historical consideration of nature and of those forms of understandings of the past which Vico proposed to call new science.

Perhaps it is true, as Sir Hermann Bondi once wrote, that so-called scientific progress often consists not so much in a progress *within* science as it does in taking something that formerly was not science and making it part of science itself. This was not a matter of unconnected processes, however, but of paths that interwove at more than one point: this happened not only because the men of those times spoke of monuments when they spoke of the history of the earth and of documents when they were considering fossils, but because during those years natural history came to be the history of nature, and the awareness of a new relationship between natural history and human history was born.

One last observation: in discussing the contents of this book with its original editor, Giampiero Brega, I realized that a good part of it treats not so much the "successes" that took place during the process that led to the "discovery of time" as the resistances put into play in various ways to exorcize that discovery. I have only one argument in my own defense: in order to understand the life of individuals and that of communities, the study of defense mechanisms is no less relevant, and certainly no less interesting to study, than that of achievements and discoveries of the truth.

I began to gather the material for this book in 1970 at University Library, Cambridge. Some conclusions that I had reached in 1969 in relation to Burnet and Vico constitute an indispensable preliminary to the present work. Thirty pages of my work on Vico (Rossi 1969, 133–64) and twenty-four pages on Burnet (Rossi 1971, 267–91) are incorporated, with some variants, into this work. The rest is new material. Among the persons to whom I take pleasure in expressing my gratitude must be numbered, aside from my friends at Wolfson College, Cambridge, Mary B. Hesse, Georg Buchdal, William Shea. For suggestions and indications received on specific problems I thank Ferdinando Abbri, Enrico Bellone, Salvatore

Califano, Gianfranco Cantelli, Alessandro Dini, Chiara Giuntini, Sergio Landucci, Massimo Mugnai, Stefano Poggi. I owe a particular debt to the studies of J. C. Greene, M. J. S. Rudwick, S. Toulmin and J. Goodfield and, regarding Vico, of I. Berlin and A. Momigliano. But I also owe much to the conversations I have had during these years with J. Agassi, M. Boas, I. B. Cohen, Y. Elkana, A. R. Hall, M. C. Jacob, and R. K. Merton. Finally, I am in debt for several citations to Silvia Betocchi, Maria A. Cocchi, Chiara De Matteis, Serena Esposito, Francesca Paoletti, Grazia Santini, and Eleonora Sorbi, who have worked with me on questions involving natural history and the philosophy of history between the seventeenth and eighteenth centuries.

I dedicate this book to young Martino, in hope that I will be lucky enough to remain in his memory as a grandfather who tells stories in an acceptable manner.

P. R.

Note

FOR both original texts and secondary studies, the notes refer back to the bibliography, where the year of the editions used is indicated. The date of the *first edition* of the works mentioned is given in the text in parentheses. The abbreviations *SNP* and *SN* refer to the *Scienza nuova prima* of Vico and to the *Scienza nuova* of 1744; the number immediately following indicates the paragraph numbers assigned by Fausto Nicolini. The abbreviation *DU*, followed by a page number, refer to the *Diritto universale* in the Laterza edition.

Translator's Note

TEXTS originally published in English are cited from the original; English language editions of standard works have been substituted for the Italian editions when feasible and are indicated in both footnotes and bibliography, following the author's system. Quotations from Vico are given in the translations of Leon Pompa *(SNP)* or Thomas Goddard Bergin and Max Harold Fisch *(SN; Autobiography)* when available.

1
THE EARTH, TIME, AND SHELLS

1·Fossils: An Alternative

AT the beginning of the seventeenth century "the great question, now so much controverted in the world," a question over which many a scholar of the sixteenth and seventeenth centuries had labored, was summarized with great clarity by Robert Plot, the first director of the Ashmolean Museum and author of *The Natural History of Oxford-Shire* (1705). Plot asked himself whether the stones that had the form of shells were

> *Lapides sui generis*, naturally produced by some extraordinary plastic virtue, latent in the Earth or Quarries where they are found? Or whether they rather owe their Form and Figuration to the Shells of the Fishes they represent, brought to the places where they are now found by a Deluge, Earth-quake, or some other means . . . ?

Those strange natural objects, he notes, still conserved the forms, the lines, the sutures, the bulges, the cavities, and the orifices of shells. Are we to conclude that they are shells and fish that were "filled with Mud, Clay, or petrifying Juices [that] have in tract of time been turned into Stones"? Or were they rather special, singular, and strange stones? Plot subscribes to the second hypothesis:

> I must confess I am inclined rather to the opinion of Mr. Lister [*Historia conchiliorum*, 1685–92], that they are *Lapides sui generis;* than that they are formed in an Animal Mould. The latter Opinion appearing at present to be pressed with far more, and more insuperable difficulties than the former.[1]

Martin Lister took an authoritative stand in 1671 in opposition to the hypothesis Nicolaus Steno had advanced on the organic origin of fossils. But the choice, as Plot formulates it, was not based entirely on the difference between the Mediterranean shells on which Steno had worked and the "English" ones—much more difficult to interpret—that Lister had examined.[2] Hidden behind a technical quarrel differing premises of a philosophical sort were also at work, and very different cultural traditions. And, in any event, grave difficulties emerged from within *each* of the two different solutions. The "thesis of organic origin" drew attention to notable differences between living species and the fossil animals. The importance of these differences led necessarily to the conclusion that some animal species had become extinct. But did not the admission of the extinction of living species imply an inadmissible break in the "fullness" of natural reality and in the "great chain of being"? Was it not equivalent to the recognition of incompletion and imperfection in the work of the Creator? John Ray was to construct a way out of these difficulties and still retain the thesis of organic origin: fossils are organic of origin but nevertheless do not derive from extinct animal species, but rather from species that still exist even now, even though unknown to men, in some remote part of the globe. On the other hand, did not acceptance of the thesis of the plastic virtues imply the existence of forms that exist in nature that have no function and that completely fortuitous similarities exist in nature? How could one admit to such similarities in a universe conceived of as a perfect succession of forms? Robert Hooke, who defended organic origin, used just such an argument: given that nature does nothing in vain, is it not "contrary to that great Wisdom of Nature, that these prettily shap'd bodies should have all those curious Figures and contrivances . . . generated or wrought by a *Plastick virtue*, for no higher end than onely to exhibit such a form?" Could not those objects, John Ray asked himself in 1692, if they are conceived of as produced by chance rather than as the "effects of a Counsel or Design" become an arm in the hands of devotees of atheism?[3]

The hypotheses concerning the Flood and the effect of great catastrophes in the transformation of the earth and in the formation of fossils generated other difficult choices. To sustain, as for example Woodward was to do, the genuinely "miraculous" character of the Flood was to invite accusations of skepticism and implicit atheism from those who affirmed instead—backed up by a venerable tradition—that it was legitimate and possible to reconcile the biblical account with the findings of natural philosophy. On the other hand, interpreting the Flood simply as a historical and natural event meant inviting the accusation of determinism and lack of respect for Holy Writ.

But the alternatives, on closer examination, presented an even more radical choice: that between one image of nature as a series of immutable forms and as an ordering of permanent structures or, on the other hand, an image of nature as a process that takes place in time, as a set of structures that were only apparently constant. In the first case fossils come to be seen as stones and natural objects stranger than other stones or objects existing in nature. In the second case they can be seen as *documents* or *vestiges* of the past, as traces of processes that have taken place. In the first case they are observed, in the second they are observed and read, as a document is read. Nature, through them, is no longer opposed, as the reign of the immutable, to history, which is the reign of becoming and change. *Nature itself has a history* and these "shells" are among the documents of that history.

When the discussion of fossils is placed in a much broader, more general perspective than the one typical of post-Darwinian geological science, it is evident that between the mid-seventeenth century and the mid-eighteenth century it took into consideration "facts" of an extremely heterogeneous nature. It encountered and intermingled with the Genesis story, with cosmological themes of the formation and destruction of the universe, with millenarianism and catastrophism, with the theological principles of the plenitude of nature and the great chain of being, and with problems relating to the Flood and to the existence of the first men on earth. The parallel between the history of the earth and the history of civilization which arose on the level of analogy and metaphor, gave rise to a methodology and an epistemology of a "historical" type, which in turn interacted with the historiographic constructions concerning the most ancient civilizations. In the midst of an extremely rich harvest of theories (which reached its peak between 1680 and 1705), problems of a general nature arose, philosophical and religious assumptions were debated, and situations of great theoretical complexity were generated—situations that placed natural philosophers and theologians before difficult alternatives.

In order to have a clear idea of some of these alternatives and of the multiplicity and variety of the theories, it is helpful to return to Anton Lazzaro Moro's lucid summary (1740) of the "opinions on marine-mountainous bodies" that had been elaborated during the course of many centuries.[4] All those who have considered the problem, Moro says, can be divided into two classes: those who affirm that marine-mountainous bodies come from the sea, and those who seek elsewhere for their origin.

Within the first group various explanations had been advanced:

1. Winds transported the shells to the mountain tops.
2. Fishermen carried snails and crustaceans to the mountains, ate the animals, and threw away the shells, which became petrified and calcinated.
3. There is no one reason for their presence: some of these bodies are tricks of nature, some are marine products flung up on the mountains by subterranean fires or by earthquakes, or by violent inundations or gushing waters, some are generated on the mountains by the same "virtue" that generates them in the sea.
4. The marine bodies remained "trapped" by the earth when, after the Creation, the waters retired from the dry land and gathered in the seas.
5. The marine bodies were transported on the mountains by "great overflowings and particular inundations" of the sea.
6. The seas were at one time much more extensive than now, and many mountains were once covered by the sea. The shifting of the axis of the earth and the consequent shift of its center of gravity provoked a gigantic displacement of the waters on the surface of the globe.
7. The marine bodies (this was one of the most widespread explanations) were carried to the mountain tops "on the occasion of the universal Deluge."

The principal explanations elaborated by the second group of scholars, those who saw the marine-mountainous bodies as coming from sources other than the sea, were the following:

1. The eggs and extremely light sperm of fish and crustaceans were carried up by vapors from the subterranean waters present in the cavities and the internal layers of the earth; the fish and the marine shells were born in the water that resulted from the condensation of those vapors. Once the water had retired or evaporated, they "remained high and dry, wrapped up in the earth," and when the ground became petrified, the marine bodies were "petrified" as well.
2. Many plants and animals are present in the sea that are similar to the plants and animals of the land: the land, "trying to rival the sea in fertility," produced plants and animals similar to the marine ones.

3. The marine-mountainous bodies were produced by the land, but it produced only their "shells" as the land is not able to produce the animals they contained, which are produced only in the seas.
4. The marine-mountainous bodies are tricks of nature. They are merely "stones which bizarre Nature has figured in this manner," in imitation now of a mollusk, now a fish, now an insect, now a crab.
5. These bodies take their origin "from the generative powers of putrefaction."

Anyone intending to draw connections between the discussions of "geology" and the "epochs of nature" and the "discovery of time" should, however, begin by acknowledging some initial premises. George Agricola, in his *De natura fossilium* (1546), worked out a classification which included gemstones, gypsum, metals, amber, rocks, "hardened fluids" (like amber and pitch), and also the "stones" that included what we now call fossils. Two things were necessary in order to abandon a classification founded on identifying as a "fossil" anything found under the Earth's surface that has the common property of "stoniness" and to arrive at the modern definition of fossils as the remains or impressions of organisms that once lived on the earth: it was necessary not only to "discriminate between the organic and the inorganic within a broad spectrum of 'objects dug up,'"[5] but also—and this is what it is important to emphasize—to accept the premise that those curious objects might be explained by appealing to their origin, by interpreting them, precisely, as vestiges or prints and as documents. It had long been possible, on the basis of theoretical premises that were closely attached to more general Aristotelian or Neoplatonic philosophies, to explain their presence without the slightest need to see in them documents of the history of the earth or the results of the work of time.

2·The Force of Tradition: The Paracelsian Heritage and Mosaic Philosophy

In the Aristotelian view some of the simpler organisms can be formed by spontaneous generation, even out of nonliving matter, and this can happen, although it does not happen very frequently, under the earth's surface. The "seed" of a fish, endowed with its specific form, can grow from the material "stone" and generate a fishlike object in the rock, the "matter" of which, unlike a living fish, is, precisely, rock. In other terms, the form of a genuine living organism is combined in fossils with the stonelike matter that is typical of all other "fossils" (or objects-that-lie beneath-the-ground). The combination is caused by spontaneous generation, or else by the presence of specific "seeds" which have penetrated underneath the earth's surface. In this view, if should be noted, even stones are "generated." As we see in the *De ortu lapidum* (1665) of J. C. Schweigger: "Stones are generated as plants are—like produces like—gold gives birth to gold, gems to gems, stones to stones. By virtue of their seminary power, each species reproduces and multiplies itself and it preserves its own species intact and perfect."

Schweigger admits that Sennert had stated that the seed or seminal principle cannot be isolated in stones, but this was no reason to deny its existence: "Just as the willow, although it resembles other plants, has its seminal principle dispersed through the entire body of the tree so that a new willow tree can be propagated by planting any slip cut off from it, so the reproductive principle exists in all parts of the stone."[1]

The soul of stones, as Alsted wrote in 1649, "is latent and close to the vegetative soul"; when one states that minerals are inanimate bodies *hoc accipiendum est non absolute, sed comparate*. According to Thomas Shirley (1672), stones, including "the Stone in the Kidneys and Bladders of Men," are generated from "meer water; which by the power of proper seeds is coagulated, condensed, and brought into various forms." Under the earth's crust—according to a theory defended by George Agricola, among others—there circulates, in liquid form, a *succus lapidificus* or *lapidescens* (or "petrifying juice") or, in gaseous form, an *aurea petrifica* or *lapidifica*, that transform various substances into stone. Again, in 1730, Giacinto Gimma draws a parallel between this essence and the power of the Gorgon's head.

In the texts that refer more closely to Hermetic or Neoplatonic views (but it is common knowledge that the various "philosophies" were continually and persistently mingled), the similarity between some fossils and some living organisms was instead traced to the secret relationships of similitude that penetrate and give form to every part of the universe. The "modeling" or "plastic virtue" that governs the growth of organisms can operate within the earth as well, and it brings into being those affinities and those "images" that constitute the universe as a unified reality. Tommaso Campanella, in his *De sensu rerum* (1604, published 1620), conceives of the earth as *mater*, as a living animal: within it the stones vegetate and crystals develop.[2] In his *Mundus subterraneus* (1644), Athanasius Kircher refers to the *spiritus plasticus* diffused throughout the entire geocosm (analogous to the microcosm), and he considers the waters, which circulate perpetually within the body of the earth, as analogous to the blood, which moves within the complicated network of the arteries and the veins. Kircher found in rocks geometrical figures, letters of the Greek and Roman alphabets, depictions of celestial bodies, and the outlines of trees, animals, and men; mysterious symbols springing from profound religious meanings, each one of which could offer a way to the revelation of the divine significance that permeates the world. The goal of his voluminous and much-read work on the subterranean world was the search for the *abscondita latentis naturae sacramenta*. This expression should suffice to give an idea of Kircher's attempt to be encyclopedic.

One theme that emerges clearly in Kircher's work—a theme that was to have a singularly successful career—was that of a world maintained by God in a perennial and harmonious equilibrium. The *virtus lapidifica* has the power of coagulating and hardening, of reducing "subjected matter" to one form or another. If this force had not been placed by God within the earth, "the entire mass of the Geocosm, through the innumerable injuries of time, of storms, of earthquakes, would by now be consumed, the mountains would have disappeared, and the sea would have absorbed the continents." The *vis lapidifica* also has the task of reconstructing what time has destroyed in nature.[3] At the beginning of time the world was like a *seminarium*—an image of the world which is founded on two assumptions: universal magnetism (the attraction of likes and the rejection of unlikes) and the structure of "insensible corpuscles" which characterizes vapors and exhalations. The principle of solidity and consistence is identified with the *sale* of chemical philosophy. This principle was present in the original chaos, after the division of the waters: it gives rise to the solid rocks and the friable soil; in combination with niter, alum, and vitriol it gives rise to the different varieties of stones and gems. Processes of generation and corruption alternate on the earth. The story of Moses tells us that God created a "chaotic matter" out of nothing. Aside from that matter and the human soul, the Divine Architect, according to the text of the sacred pages, created nothing. Out of that chaotic matter, which was "an underlying matter fertilized by the *incubitus* of the divine spirit," spring the heavens and the elements, out of which, in their turn, come the plants and the animals. The *panspermia* or *vis seminalis* that produced physical reality was a material spirit "composed both of more subtle celestial air and of portions of the elements." It was a "spiritous sulfureo-salino-mercurial vapor" that was "the universal seed of things, created with the elements, and the origin of all of the things in the world."[4]

Buffon gives a beautifully clear summary of the positions of Burnet, Whiston, and Woodward in his *Proofs of the Theory of the Earth*, and he is to be thanked (or blamed) if their names have become a sort of triad that is inevitably discussed by all historians who are in some way or for some reason interested in the problem of the history of the earth and in the cosmologies of the latter part of the seventeenth century. The last decade of that century was undeniably decisive from that point of view. But reflections on the formation of the earth and on cosmology, not to mention the explanation of fossils, were not the exclusive patrimony of the scholars whom Buffon discussed and who have since become a part of the official histories of geology and cosmology. Those same reflections were present, and in those same years, in thinkers who were far from sharing in the atmosphere of the scientific revolution, some of whom came out of the Hermetic tradition, some from the Paracelsian tradition, and who sustained the superiority of Mosaic philosophy, identifying it with Hermetic wisdom.

Robert Fludd's principal works were published between 1616 and 1638. His *Philosophia Moysaica* (1638) offers a mystico-chemical interpretation of Genesis: divine light is the active agent in creation and water and the original darkness interact with it. In the *Ortus medicinae* of Jean Baptiste van Helmont (1648), a description of the elements is connected to the biblical narration. In Joachim Becher's *Physica subterranea* (1669), which appeared only a few years after Kircher's work, the great themes of chemical philosophy, as it was called, come to the foreground: the alchemistic interpretation of the creation; the creation as a cyclical process; and the thesis that animal and human life emerged spontaneously from the earth. All three of these themes had already been fully discussed in the *Philosophia ad Athenienses* (1564) of Paracelsus, and Thomas Erastus had already in 1572 judged the Paracelsian solutions to these problems impious and the sacrilegious fruit of a sick mind.[5]

The thesis of the spontaneous generation of life, present in Lucretius and Avicenna and repeated by Paracelsus, was also restated and defended in various ways by Pomponazzi, Cardanus, Johann Boemus, and Giordano Bruno.[6] It is only marginally connected to the reflections concerning the history of the earth. However, it is important to emphasize that chemical philosophy aimed at reducing to one process both nature and the formation and origin of nature. The "chemical" interpretation of the creation and the equating of the original genesis to a process of separation—both of which elements are present in the *Liber de natura physica ex Genesi desumpta* (1583) of Gerhard Dorn and in the *Traicté de la matière* (1626) of Quercetanus (Joseph Duchesne)—ended up dissolving the specificity of animal and human life into a global vitalism of forms and principles. At the very moment in which matter revealed itself capable of progress based on "digestion," life tended to be resolved in processes of "combustion" and the latter tended to be equated with "respiration." The spirit of God that "moves over the face of the waters" is indeed the source of life, but it is also identified as an "aerial niter." Metals "grow" within the earth as a human fetus grows in its mother's womb, and Mother Earth has been impregnated by the Heavens. It is against the background of these ideas and these visions that Kircher develops his explanation of the origin of springs, and Gabriel Plattes (1639) explains the existence of mineral veins in mines. The idea of the earth as a living organism underlies as well their many references to physical forces and their parallels with man-made machines.

At the end of the seventeenth century this type of literature was still alive and thriving. A quick look at Allen Debus's works gives a good idea of this vitality. We can see many of these themes in one author of the 1690s who escaped Debus's attention: Thomas Robinson published, in 1694, a short work entitled *The Anatomy of the Earth*, and in 1696 a longer work, *New Observations on the Natural History of This World of Matter and This World of Life*. The themes he treats are typical of the histories of the earth and of the cosmologies of the late seventeenth century: the Mosaic system, the Flood, the final conflagration. Robinson's adherence to the alchemistic and Paracelsian theme of the superiority of practice appears particularly evident in the second of these two works. The "gentlemen miners" are wrong to complain if some of their "subterranean projects" fail: they can always remedy their errors by means of their experimental knowledge of "those hidden regions of matter." Philosophers, on the other hand, can give no account of them, "unless by way of hypothesis and conjectures." Robinson was well informed concerning those hypotheses and conjectures. He had read the works of Burnet and Woodward and had found them ill-founded, inconsistent with the senses and with experience, and in disagreement with both reason and the Mosaic account. Burnet was wrong to believe that the antediluvian earth was the same as the one we know, and Woodward—who, however, contradicts himself—was right to refuse the hypothesis that the mountains have their origin in the diluvian catastrophe.[7]

Robinson had a decidedly higher opinion of those who study the internal structure of the earth than of those who construct great cosmogonic hypotheses. He cites with respect the classics of the geology of his time: Nicolaus Steno and Agostino Scilla, Robert Hooke and John Ray, Martin Lister and Fabio Colonna. In particular, he greatly admires "the accurate Discourse of Bernardini [*sic*] Ramazzini printed four years ago" (the *De fontium scaturigine* of 1691). He reproposes the image—much debated, as we will see—of a Moses who had learned culture and knowledge at Pharaoh's court. He repeats the doctrine of universal sympathies and antipathies and the explanation of fossils based on plastic virtues. The plastic powers are "the first principles of life in this natural world." The plastic spirit is a "subtle saline volatile which (whilst matter was in a fluid substance) diffus'd it self through all the lax strata and consistences of it." The formation of fossils is a process similar to digestion: as the acido-salinic humor, together with vital heat, separates in the stomach of the animals "the more pure and spirituous parts of the nourishment from the crasser and more excrementitious parts," the plastic spirit, operating together with "that subterranean flame" that pervades the veins and hollow parts of the earth, "did separate, collect and coagulate the more simple, pure and homogeneous parts of matter" and gave rise to rocks and gems.[8]

The rotation of the earth is "natural" because the earth is a great animal and a living body capable of movement. The cosmogony outlined in the *Anatomy of the Earth* is a characteristic mixture of Cartesian and atomistic themes and Paracelsian and Neoplatonic themes:

> The first part that the Almighty created of this sublunary world was matter; which consisted of innumerable particles of divers figures and different qualities, running a reel in a dark confusion. In this state they continued, bound up and hamper'd with a vast thick Fogg or Mist of waterish substance, extending it self as far as the Moon's Vortex, until the *Spirit of God moved upon the face of the Waters* . . . and, by vital heats and incubations, did digest it into an orderly World: by whose infusing into it, and every part of it, a vital Spirit, it became a great Animal; having skin, flesh, blood, bones, nerves etc.[9]

The surface or covering of the earth is a skin. Cooked and "digested" by the rays of the sun, and with the help of the plastic power infused in it by the *spiritus*, it produces grasses and trees. When this process reached full force and vigor, it produced birds, animals, and snakes spontaneously, just as even now, Robinson argues, various species of insects are born spontaneously. Through the influence of the *spiritus naturae* the seminal forms of fishes are generated in matter, followed by those of the birds, the land animals, and finally that of man, made in the likeness of God. Adam and Eve are the first individuals of this nobler species, but Robinson prefers to leave to "those who pretend to knowledge in mystical Philosophy" the task of establishing whether Adam and Eve were really single persons or "a whole generation of men and women, which stock'd all the world at once," or if by Adam we should understand only "the rational and masculine faculties of the soul" and by Eve the "feminine and subservient."[10] The Moses account is an *ad hominem* account, and the origin of all mankind from one common couple presents us with a choice between two theses: either we must admit that the Ethiopians and the Europeans do not descend from one identical couple or, as it is more probable, "the different soyls, or various modifications of matter in several parts of the world produced men of different colours and complexions." For this same reason,

> some kinds of animals are, to this day, peculiar to their own several countries: whence Africa breeds so many venemous creatures, Ireland none; Athens owls, Crete none . . . and thousands of strange beasts and birds are proper to America; such as no Greek, Roman, or Hebrew author has ever mentioned, and of which some have thought it reasonable to doubt whether any of their great variety of species ever came into Noah's Ark.[11]

The themes of the Hermetic tradition and of Mosaic philosophy, often solidly joined to alchemistic ideas and to those of the Egyptian wisdom cult, penetrated broadly, as every reader of Toland knows, into seventeenth-century culture. The *Physica vetus et vera: Sive tractatus de naturali veritate Hexaemeri Mosaici* of the Oxonian doctor of medicine and alchemist Edmund Dickinson, published in 1702, is a further example. God revealed directly to Moses a corpuscular or Democritan philosophy that the Hebrews did not understand. An atomistic philosophy is hidden within the Genesis account, which does not regard the genesis of the entire universe, but only that of "this material world." The divine spirit set the particles of matter into motion, and from them were born the elements, drops of water, and aeriform *bracteae*. Then the abyss, or the total mass of matter, took on a rotary motion. The heavier particles were pushed by this motion toward the center. A globe was formed. The *bractaea* penetrated through cracks in this globe, kindled the central fires, and helped in the generation of metals. The animals that live underground were generated from the subterranean heat produced by the central fire. A great reservoir of subterranean water lay next to the central fire. The two poles of the earth were connected by a subterranean passageway. A great many canals ran through the center of the earth, putting the great reservoir in contact with the seas, lakes, and springs, and permitting a continuous internal circulation of water. When the force of the waters weakened, some of these canals filled with earth, and this explains why shells, animal bones, and the debris of ancient ships or tools is found under the ground. In the fifth and sixth days of creation God infused animal spirit into the eggs of the various animal species and distributed them over the earth, after which the animals developed, taking their nourishment from the soil and the water, with the help of heat.[12]

3·Changes and History: Robert Hooke

FRANCIS Bacon distinguished three types of natural history: the history of *nature in course* or in its *constancy*, that of nature *erring or varying*, and that of nature *altered or wrought*, modified by the presence of man. The first type gave rise to the history of *creatures*, the second to the history of *marvels*, the third to the history of *arts*. The history of nature in course (in both of its aspects, regularity and the violation of regularity) must be accompanied by the history of nature modified by the hand of man. Bacon's projected natural and experimental history was intended to weld into one history both the traditional description of nature and a new history of the dominion of man over nature.

The dimension of time is present in the history of man's arts—the history of human intervention in nature—but it is completely absent in the other two types of natural histories planned (and in part realized) by Bacon. From this point of view these histories are natural histories in the traditional sense. They are freed, or mean to be freed, from imposture, from insufficient verification by fact, from myth and legend, and from carefully detailed descriptions of imaginary beings, but they are still—just like the histories of Aristotle and Pliny—descriptions of nature *as it is*.[1]

Robert Hooke, writing in the middle of the seventeenth century, returns to Bacon's polemic. The natural histories of Aristotle and Pliny are "uncertain and superficial, taking notice only of some slight and obvious things, and those so unaccurately, as makes them signify but little." They include no "Dissections, Experiments, or Mechanical Tryals." They consider together things true, probable, and false, and they concentrate only on the "outward Shape and Figure" or the beauty of objects, or else on the magical or superstitious effects that can be gotten from them, even then "seeming to aim at creating Pleasure, and Divertissement or Admiration and Wonder, and not of such a Knowledge of Bodies as might tend to Practice."[2] These are nearly the same words that Bacon uses in the *Novum Organum* and in the *Parasceve*. But Hooke has a broader and different concept of natural history. Natural history had aimed at describing and classifying nature; it had not studied the alterations and transformations that nature undergoes during the course of time. When he confronts the question of the "shells," Hooke states that science should investigate "how they came to be disposed, placed, or made in those parts where they are, or have been found." He admits that "it is very difficult to read them, and to raise a *Chronology* out of them," but he states that the chronology of nature constitutes a problem that needs a solution. The modifications of nature are not only the result of the intervention of man: they are also the result of the work of nature itself.[3]

When Hooke elaborates on some of the points he discusses in the *Micrographia* (1665) in the *Lectures and Discourses on Earthquakes* (1668), he confronts the problem of fossils, which "have hitherto much puzzled all the Natural Historians and Philosophers." It seemed to him difficult to think that nature forms these curious bodies only "to play the Mimick in the Mineral Kingdom, and only to imitate what she had done for some more noble End, and in a greater Perfection in the Vegetable and Animal Kingdoms." Fossils, which contain mixtures of the most varied sorts of substances, show no sign of any internal form of substance, but only external surfaces and their features. Why, Hooke asks, should one search for a plastic nature and think that nature followed a path different from the one she follows with all other bodies? And if it is difficult to believe in the plastic natures, it is just as difficult to believe that those bodies have been transported, either by man or by other ordinary means, to where they are now found. This is true, Hooke adds, whether those bodies are considered true shells, the bodies of fish or other animals, or "impressions" left by those bodies on other substances.[4]

Hooke keeps a careful distance from both Aristotelian theses and Neoplatonic ones. He also frees himself from the very start from the theses that make the Flood responsible for the existence of fossils. The diluvian thesis is for Hooke "improbable," documented neither by Holy Scripture nor by profane history. How could the Flood possibly have carried fossils to the summits of the highest mountains? How can one reconcile the brevity of the Noachian Deluge with the radical nature of the changes that it was supposed to have provoked?[5]

Several objections had been raised to the thesis of the organic origin of fossils, and Hooke affronts them squarely. He first asks, *How* were those bodies transported to the places where they were found? Second, *Why* were many of the bodies found composed of substances different from the original substance; why were the shells not made of the same material as common shells, but of clay, chalk, marble, soft and hard stone, flint, and marcasite? Those curiously shaped bodies were, for Hooke, animal or vegetal substances the pores of which had been filled with some sort of "petrifying liquid substance" that cemented their parts together, or else they were "impressions" left by those animal substances on matter, as it happens when a seal is pressed into wax. Or else, finally, they arose when matter hardened or "subsided" after, "by some kind of fluidity," it had perfectly filled the plant or animal and given rise to a sort of mold, as in the casting of plaster statues.[6]

There is one point on which Hooke has no doubts: matter does not coagulate into stone by itself, and ordinary or common causes are insufficient to explain fossils. They do not result from exhalations from subterranean eruptions or from earthquakes, from saline or sulphurous substances that work by crystallization or precipitation, from bituminous materials capable of gluing sandy bodies together, nor from long periods of intense cold or compression. All of these are only "concurrent Causes." Some sort of "extraordinary Cause" is needed to explain fossils. This is true, principally, because it is impossible to produce fossil objects experimentally and because it is impossible to imagine that the natural principle of universal decay or the universal putrefaction of organic substances be violated. Animals and plants "after they have left off to vegetate, do soon decay, and by divers ways of Putrefaction and Rotting, loose their Forms and return into Dust."[7]

For Hooke, the earth and the forms of life on the earth have a history. A series of "natural powers" and of physical causes—earthquakes, inundations, eruptions, torrential rains—have altered the earth and life on it. From the time of creation a great part of the surface of the earth has been transformed and altered in its nature, in particular, "many Parts which have been sea are now Land, and divers other Parts are now Sea which were once a firm Land, Mountains have been turned into Plains and Plains into Mountains and the like." At its beginnings, the earth consisted of fluid substances. These, "in tract of time," crystallized and solidified little by little, giving rise to the solid bodies that surround us. In this sense, the earth is transformed and grows almost in the same way as animals and plants. The changes may have been so profound as to have altered the earth's axis, shifting the original position of the poles and the equator and reversing the earth's magnetic poles. In preceding ages there may have been "whole countries either swallowed up into the Earth or sunk so low as to be drown'd by the coming in of the Sea." Thus it was for Plato's Atlantis, while other regions—England, for example—emerged out of the waters or from the depths of the globe.[8]

To explain the existence of fossil shells and fossil fish belonging to no known species, Hooke abandoned completely the idea of immutable and eternal species and formulated the hypothesis of the destruction and the disappearance of living species. He makes the existence of varieties within a single species depend on changes that take place in habitat, climate, or feeding habits:

> In former times of the World [there may have been] divers Species of creatures, that are now quite lost, no more of them surviving upon any part of the Earth. . . . For since we find that there are some kinds of Animals and Vegetables peculiar to certain places, and not to be found elsewhere; if such a place have been swallowed up, 'tis not improbable but those Animal Beings may have been destroyed with them. . . . There may have been divers new varieties generated of the same Species, and that by the change of the Soil on which it was produced; for since we find that the alteration of the Climate, Soil, and Nourishment doth often produce a very great alteration in those

> Bodies that suffer it; 'tis not to be doubted but that alterations also of this Nature may cause a very great change in the shape, and other accidents of an animated Body. And this I imagine to be the reason of that great variety of Creatures that do properly belong to one Species.[9]

In "former Ages," therefore, there probably existed "Species of Creatures . . . of which we can find none at present." Today, on the other hand, "'tis not unlikely . . . but that there may be divers new kinds now, which have not been from the beginning." Not only that: the species existent in remote times may have been of a form so different from the one familiar to us that "we should not have judged them of the same Species." The changes and modifications that have altered the earth have also profoundly changed living forms. To study those changes is to write a history in which those shells and other fossils appear as "the medals, urnes, or monuments of nature." Indeed, they are "the greatest and most lasting monuments of antiquity, which, in all probability, will far antedate all the most ancient monuments of the world, even the very pyramids, obelisks, mummys, hieroglyphics, and coins, and will afford more information in natural history, than those others put altogether will in civil."[10]

Reading these monuments is a difficult task, and it is particularly difficult to "raise a *Chronology* out of them, and to state the intervalls of the times wherein such, or such Catastrophies and Mutations have happened." Nature's history is interwoven with changes. The greatest among them took place in those "Uncertain or fabulous Times," as Varro termed them, and now, in "the Historical time"—the time familiar to men and close to the present—it is difficult to find reliable aids to their reconstruction. Slow, uniform changes, unforeseen catastrophes, the destruction and emergence of continents, the mutations within species, all spoke to Hooke of a history that is like a river of unknown course, which might even contain, in its past, treasures of wisdom now lost forever. Together with plants and animals, entire cultures might have been destroyed or might have disappeared. The Hermetic myth of a semidivine wisdom that has been canceled from history, leaving only tenuous traces, reemerges cautiously—in a context totally different from the traditional one—in the pages of the *Discourse on Earthquakes:*

> 'Tis not impossible but that there may have been a preceding learned Age wherein possibly as many things may have been known as are now, and perhaps many more, all the Arts cultivated and brought to the greatest Perfection, Mathematicks, Mechanicks, Literature, Musick, Opticks, etc. reduced to their highest pitch, and all those annihilated, destroyed and lost by succeeding Devastations. Atomical Philosophy seems to have been better understood in some preceding time, as also the Astronomy evinc'd by Copernicus.[11]

It is undeniable, as Haber and, later, Toulmin and Martin Rudwick have emphasized, that Robert Hooke inserts his history made of catastrophes and continual changes into the brief span of sacred history. And it is true that he intended neither to reject traditional chronology nor cast doubt on the "accordance" between nature and Scripture.[12] Hooke, like Steno and many other writers after him, tends to compress the periods of the history of the earth and the history of man within the chronological scale permitted by orthodoxy. In Hooke, however—the writers mentioned above notwithstanding—this does not take place on the level of the acceptance of the obvious. Hooke was a calm scholar, cautious and meditative, and little given to polemics. He loves repetitions and clarifications. He makes wide use of dubitative terms. But he was aware of the fact that the problem had been posed, and that there existed alternatives to the traditional time scale. The histories of Egypt and China, he says, "tell us of many thousand Years more than ever we in Europe heard of by our Writings, if their Chronology may be granted, which indeed there is great reason to question."[13]

The "natural powers" have provoked catastrophes and altered the face of the earth. Has the history of nature left some sort of trace in *human* history? Men, according to Hooke, have felt the need to *hide* those catastrophes. The "Mythologick Stories of the Poets" represented catastrophes as the "monstrous and seemingly impossible representations of humane Powers." The "fabulous stories" of the Egyptians, the Chaldeans, and the Greeks were constructed both to communicate and to hide; to transmit the truth and at the same time to prevent full understanding of it. Those "fabulous stories" contained a major part of the history of the most ancient ages of the world, yet they also contain a deliberately concealed wisdom; they served to communicate to the few the "real history" and the "Philosophical or Physical hypothesis" on which that knowledge was based. It is clear that Hesiod's *Theogony* contains the wisdom the Greeks had received from the Egyptians and the Phoenicians, but the majority of the texts and the myths are difficult to interpret. On the truth of fables, Hooke, who had undoubtedly read carefully Bacon's *De sapientia veterum,* assumes a very cautious position:

> The Aegyptians, Greeks, and other Nations converted their true Histories into these Romantick Fables. Not that I do here undertake for the truth of History in every Fable, for I conceive that there are as various kinds of Fables as there are of Histories. Some are repeated and believed Fables which are true Histories, others are believed true, but are really Fables. Some are believed Fables and are really so, and others are believed true and really are so. But of this fourth Head I fear is the smallest number.[14]

What was true of the fables was true for the other remains and documents of human history: medals, monuments, and inscriptions. In all fields Hooke saw evidence of the truth and conterfeits of the truth. One certainly should not state, because counterfeits exist, that all documents are without truth. The criteria of verification that are valid for human history can thus be applied to fossils as well, since they are the "Medals, Inscriptions, or Monuments of Nature's own stamping." Even happy coincidences are not to be excluded: it is possible that some fables that contain "the truth of History" conceal within them the account of events that the history of fossils might document. For example, the fables of Perseus, of Atlas, of Andromeda and Medusa "have relation to the *Herculean Columns* and to the *Atlantis*, or to those parts of *Lybia* which were near it." It might be true that Plato's account is not sufficient to document the reality of Atlantis. But it surely is valid documentation that Plato believed in the existence of profound changes and great catastrophes that had occurred in the preceding ages of the world.[15]

The Baconian spirit and scientific caution are intermingled in Hooke with the repetition of traditional themes. In the very same years during which he was theorizing a science founded on ascertainable evidence and on the cautious construction of testable hypotheses, Hooke also fell back on the image of a golden age of mankind and on the myth of a wisdom canceled by time. He associated the image of a nature that "does nothing in vain" with the idea of a Providence that operates through physical causes; to the image of an open and collaborative pursuit of knowledge he juxtaposed the project—shared by many English intellectuals of his day—of an invisible "college." He noted in his diary, under the date of 1 January 1675: "We now began our New Philosophical Clubb and resolv'd upon ingaging ourselves not to be speak of any thing that was then revealed *sub sigillo* to any one, nor to declare that we had such a meeting at all."[16]

4·The Modes of Production of a Natural Object

TOWARD the middle of the seventeenth century, the problem of the *interpretatio naturae* tended to exceed the exclusively spatial and structural dimensions on which it had been based. It turned out to be connected with the dimension of time. Analyzing and interpreting a substance came to mean more than simply breaking it down, reducing it to particle motion, and studying it in its geometrical aspects. Other questions began to acquire meaning: How has a natural object come to be formed through time? How can we read it and interpret it from this point of view? How has nature produced a particular object during the course of time? In 1669 Nicolaus Steno stated the terms of the new "theorem" concerning fossils with Cartesian clarity: "Given a substance possessed of a certain figure, and produced according to the laws of nature, to find, in the substance itself the evidences disclosing the place and manner of its production." The *Prodromus* (the Forerunner) of 1669, as is known, shows clear signs of the influence of Galileo and Descartes. Steno uses the corpuscular theory of matter in this work to draw a clear distinction between "crystals" and "shells and other sea objects." His famous theory of "strata" (which he formulated on the basis of Tuscan rock formations but took as generally valid) explained the presence of the fossils "enclosed" within the stratalike foreign bodies. It also represented a coherent attempt to reconstruct the sequence of geological events.[1] The Universal Deluge can be explained, Steno argues, providing that the earth's central fire was surrounded by a sphere of water, or at least that there were "huge reservoirs" out of which the waters could have surged. Even without taking into consideration the center of the earth, the shape of the earth, or the principle of gravity, the Flood could have been caused by the concomitant action of a series of natural causes if we accept five hypotheses: (1) that fragments of collapsed strata obstructed the passages through which the sea sends water to springs; (2) that the water contained in the bowels of the earth was forced out by the subterranean fire, in part through the springs, in part into the atmosphere through the pores of the earth where it was not already covered with water; (3) that the sea bottom rose as a result of the ballooning of subterranean caverns; (4) that the other cavities under the earth's surface were filled with soil that washed down with the rains; (5) that since the earth was "nearer to its birth," its surface was less irregular than it is now.[2]

> Neither Scripture nor Nature tell what happened on the surface of the Earth when it was covered by the waters. On the basis of Nature, we can only state that the deep valleys were produced at that time: because, due to the force of subterranean fires, the enlarged cavities offered room for more extensive ruins; because a way was opened for the waters to return to the deepest parts of the Earth; finally, because today, in places far from the sea, deep valleys are found, full of various sorts of marine sediments.[3]

"Fere ubique de Antiquorum dubio altum silentium est": The theme of the silence of earliest history—natural as well as human—is present in Steno. The writers of classical antiquity had no sense of time: they spoke of memorable sequences of years, but *obiter tantum et quasi aliud agendo*.[4] As far as nature was concerned, Steno argues, men now are confronted with two histories: the one narrated by fossils and strata, which tells of the six successive aspects of *facies* of the earth, and the one given in the Bible. Disagreement between the two is impossible, and the scriptural narration can be interpreted in light of knowledge about nature. Indeed, what we know about nature finds confirmation in Holy Writ. Anyone who investigates patiently "concerning the time of the universal deluge of Sacred History will find no disagreement with profane history." The ancient cities of Etruria had their birth more than three thousand years ago, and some of them were built on hills produced by the sea. The history of Lydia goes back nearly four thousand years, and we can see from that fact that "the time when the sea abandoned the land fits with the time mentioned in Scripture."[5] Where Scrpture is silent, it is possible and permissible to advance hypotheses based exclusively upon nature. The greatest difficulties arise with those periods in the history of the earth for which hypotheses can be formulated but cannot be supported by Scripture or corroborated by any other source in Gentile history. The fifth and the sixth phases of the history of the earth were characterized by great and decisive transformations, but Scripture says nothing about them and the *historia gentium,* which refers to the centuries immediately after the Flood, "is considered dubious and full of fables." Postdiluvian histories are confused, and for the most part they celebrate the great actions of heroes rather than narrating natural events. This does not authorize us, however, to interpret as meaningless or erroneous all the descriptions they contain of an earth unlike the present one. These texts describe a world that is often different from the one we know. We must take each instance separately and distinguish fantasies and fables from real history. Above all, Steno insists, we must pay attention when we interpret these texts not only to the changes that have occurred in human history, but also to those that characterized natural history:

> I would not like to lend faith too easily to the fabulous narrations of the ancients, but they contain many things to which I would not deny credence. I take from them a series of affirmations the falsity of which seems to me dubious, not the veracity. For example: the separation of the Mediterranean Sea from the Western Ocean; the existence of a passage from the Mediterranean Sea to the Red Sea; the submersion of the island of Atlantis; the accurate description of many places in the voyages of Bacchus, Triptolemus, Ulysses, Aeneas, and others. And I do so even if these descriptions do not correspond to things as they are now.[6]

During the year that followed the publication of Steno's *Prodromus* (1670), Agostino Scilla, a painter and a member of the Academy of the Fucina (under the academic name of *lo Scolorito*—the faded one), published his *La vana speculazione disingannata dal senso. Lettera responsiva circa i corpi marini che petrificati si truovano in vari luoghi terrestri* ("Vain speculation undeceived by sense. Answering letter concerning the marine bodies that are found petrified in various terrestrial locations"). To the "vain speculation" that interpreted fossils as "grown" within rocks, Scilla, who was unaware of Steno's work, opposed the thesis of their organic origin:

> Shells, echinoids, sea urchins, teeth (which they call *glossopetre*), vertebrae, corals, sponges, crabs, spatangoida, turbinidae, and so many other innumerable bodies that some have judged to be generations of pure rock and tricks of nature, used to be not only animals and bodies of that species, but bodies and animals quite appropriate to the sea, which arrived by some accident within the earth with the matter they contained. . . . I claim that all of the empty shells that we find petrified within the earth were real and animate . . . and I intend to oppose the broken, dismembered and, to speak clearly, dreamed-up generation [proposed by] those who would have nature play tricks, just like their fantasticating brains.[7]

The material that encloses the fossils now appears to us in the form of hills and mountains, Scilla continues. Many men, fooled by the familiarity of the landscape and by their remembrance of it as immutable, are unable to *see* the history that is written in nature. As Scilla says—and he speaks at length of vision, of eyesight, and of his painter's craft—"they are unable to grasp with their eyes the true history that the Omnipotent clearly registered in every place and offers to us."[8]

Scilla—who refers to and makes use of, in one way or another, Cardano, Gassendi, Sennert, Peiresc, Aldrovandi, Francesco Calzolari, Cesalpino, Redi, Fabio Colonna, and Kircher—firmly maintains that fossils "were real animals, and certainly not tricks of nature simply generated of a rocky substance." He is very decided—often ironic and cutting—in his systematic demolition of traditional theses. He does not believe in the "intrinsic growth" of metals in the mines and rejects energetically the opinion of those who would have "the quarries fill up all by themselves with fresh, burgeoning metal." He argues that the walls of mines abandoned more than forty years earlier "show clear, fresh signs of pick blows" and that in long-abandoned mines metal does not seep out of the veins, nor heap up in "prodigal and precious pyramids." The opinion, "supported by valiant men," of the vegetation of rocks and of the "production of various bodies, resembling those of the sea, out of pure stone within rocks" seemed to him the fruit of pure fantasy.[9] He attacks the interpretation of fossils as symbols and the thesis of correspondences and "analogies" in forceful passages:

> I will leave to others the liberty to believe as they will, conceding to them that the marine Ball, the Hermodactylus, the Phallus, the "Boratmets" resemble the chestnut more than a grape stalk, the hand more than the knee, the god of the orchards, Priapus, more than the human breast, and, to finish, a lamb more than a snake, but certainly not that they are these very same things in their form. It would take much to persuade me that they could be produced from a same seed or from one single formative virtue. Let it be clearly understood that whoever follows such a way of philosophizing is doing his utmost to be estranged from knowledge of the truth.[10]

Scilla continues: to understand the dentition of an anglefish or a dogfish we need only look in its mouth. A tooth from the left side of the jaw will not fit on the right side, and the "full part" of each tooth is turned toward the throat. Every time we pick up a tooth "loose and apart from its site" we can state confidently and "with no suspicion of error" that "this is a tooth from the right side, this other one is from the left." The objection that the *glossopetre* could not be animal teeth because they were so varied was not a valid argument because the animals involved were "of very different species, whose mouths Nature armed with unlimited quantities of very different teeth." The argument of their being found apart from the jawbone was no more valid, since—as with human bones—the skeletal bones fall to dust, leaving the teeth intact.[11] Scilla found particularly invalid the objection to the organic origin of fossils that argued that if this were the case, living or recently dead animals ought to be trapped inside the rock as well:

> Not so. All we need is to imagine them alive in their lairs in the ground and that somehow they remained shut in there, dead and petrified, [and they will have] satisfied every requirement: their only task is to give evidence of their misfortune—that is, to having been seized by some accident that managed to agglutinate and amass that clay and that silt, together with them, into a rock mass.[12]

Scilla then insists that the thesis of those who saw in fossils a "generation of particular members" or of isolated parts of organisms was equally false and untenable. Who could ever accept the thesis that a particular seed was needed for every part of an animal—the nose, eye, ear, mouth, etc.? A formative virtue or, as Sennert called it, "a soul diffused by the generant" had to be united to the seed. Even if one accepted the thesis that seeds are scattered through the universe, this is not to say that there are scattered seeds capable of generating a single part of an organism. The Democritic *minima* were an acceptable thesis,

> but that there be some among these same *minima* that both within the whole and apart from it can, all by themselves, produce one leaf of a given tree, one member of the human body, one animal tooth, one vertebra, one piece of bark, one bone, must be taken as a fantastical opinion.[13]

Scilla believed in the Flood and in the Mosaic account: "As for me, I believe and will continue to believe in the universal inundation, just as Moses told it, and also that the waters covered all." But he insists on the fact that the opinions of the Church Fathers and the theologians "are most varied in determining the manner God chose for drowning this world." He declares that he does not for one minute believe that all of the beetles and fish "died out brusquely" and that he knows that "the world is ancient" and that many authors have spoken of "many particular inundations." He absolutely does not believe that, as Kircher would have it, the island of Malta was created by God "in the form in which we see it today." A series of floods and of "most terrible waves" convulsed the earth. No one remained alive to tell posterity of the precise hour at which that tragedy struck, but that hypothesis alone can account for the facts.[14] Beyond his rejection of philosophies, beyond his diffidence toward the great cosmological hypotheses, the "path our eyes show us" led Scilla to catastrophism:

> I have no idea how the sea could reach so far into the land; I do not know whether this happened during the universal Deluge or during other special floods. Neither do I know whether this beast of a World (as it seems to some who hold it so and who have observed everything in it down to the movement of its entrails) at some particular time, tired of lying on one side, might have turned over and exposed its other side, which had been under the water and was full of so many bits of

> the refuse of the sea, to the rays of the Sun. I do not know, neither do I know the way to find out. Nor do I care. What I do know is that the corals, the shells, the sharks' teeth, the dogfish teeth, the echinoids, etc. are real corals, real shells, real teeth; shells and bones that have indeed been petrified, but that are not formed from stone. The composition of the soil persuades me of this despite myself, and it seems to me impossible to arrive at any sort of knowledge of the truth if I abandon the path my eyes show me.[15]

Scilla refers continually to his painter's profession; he claims that he is a "simple man," that he has "little inclination for high philosophizing," that he "carefully shies away from speculations" and trusts only "observation of things." He also holds, having read Descartes, that not only is it important to question everything, but that a "great sublimity of intellect" is not needed for the discovery of truth. He declares that he has "fixed fast in his mind that doubting things is the best and the only means to know them, or at least to come closer to knowing them and to increase the probability of doing so." He has faith in the senses and cites Lucretius: "Invenies primis ab sensibus esse creatam/Notitiem veri, neque sensus posse refelli" ("You will find that the concept of the true is begotten from the senses at first, and that the senses cannot be gainsaid" [4.478–79]; Cyril Bailey trans.). If sense "has tricked me," he says, "to whom could I turn?" He considers speculative philosophy a vain juggling of concepts, a complex of "caprices and beautiful ways to explain what we can in no way understand." He holds that the ancient philosophers "did not have the certainty of truth in their opinions" and that modern philosophers have always "sought only pasture for their fine wit," subjugating everything, rightly or wrongly, to their genius. Given that the things "which we do not know or we have not seen" are infinitely more numerous than those seen and understood,[16] he claims that he has no reason to be ashamed of his uncertainties, all the more so because "the hypotheses on the great machine of the Universe" had proven to be inconsistent:

> I am not ashamed of my perplexities, and I say less and less as I reflect on the hypotheses of the great machine of the Universe. One among these was presented more forcefully by Ptolemy, who with such clear and valuable proofs distributed its parts, both stable and revolving; others, with no less clarity of demonstration collapsed it all, unhinged the earth, and stopped celestial motion itself with no regard to the eyes of every living person. Nor would ways be lacking to human genius, through philosophizing, to deny both of these systems and to preach many others, should the thought of innovation, and not the obligation to track down the truth, become the goal of its speculations.[17]

When sense experience tricks us, this depends not on the senses, but on "some principle supposed," and it is better to have faith in an observation, even "very brief and confused," than in the "murky abstractions of the metaphysicians." Science should be purely a description of reality: "Halfway blushing for my simplicity, I would like things that are subjugated to sense to be established uniquely by what sense determines." Scilla knew that this was impossible. He would like it if at least "some little particle of history were embraced by philosophy" "and that in things that have no need for long, drawn-out speculations we not take flight with the intellect to distant and spacious fields of the possible, as some of the noblest minds of today, who scorn mere history in all matters, are wont to do."[18]

Scilla presents his theses as the simple result of chance, the fruit of a mind "totally unpreoccupied by opinion and not impelled by the authority of any master." Undeniably, he worked on the relatively "easy" Cenozoic fossils observed in Malta, in Calabria, and in his native Messina.[19] But it is also true that his professions of philosophical ignorance should not be taken at the letter. Scilla declares his intention "to live and to die under the dictates of the Holy Roman Church," but underlying his discussion of fossils—beyond his insistent display of his simplicity and lack of preparation—there are clear signs not only of the major contemporary texts in geology, but also of Descartes, Gassendi, and especially Epicurus:

> I must still praise certain philosophers, and most particularly Epicurus, inasmuch as I do not perceive him—as the vulgar defame him—as an evil glutton [*uno sciagurato crapulone*] but as one of the most self-possessed of the ancient philosophers, just as the ultra-moral Seneca, the highly erudite Gassendi, and a hundred others of the greatest literary figures declare him to be.[20]

Scilla's reference to the work of Giovanni Ciampoli seems particularly significant. Behind Scilla's strongly sense-oriented and "skeptical" arguments, the teachings of Galileo—who is never named—are undoubtedly present. Only one philosophy seems to him in any way acceptable: the one "that recognizes the great disparity that exists between what men think and what Nature has been able to do."

> I would like it if Nature's methods of petrifying things were never ascertained, inasmuch as she must have thousands of ways to manage her affairs that we do not know. . . . The methods used by those interested in investigating natural things are truly different from the methods of those who plead causes before a court of law. In the latter, the texts carry full authority because they are full of laws that are either good or are negotiated by common consent. But in philosophizing there is no argument—no matter how authoritative it might be—sufficient to contradict the evidence that our eyes furnish.[21]

Scilla's book did not remain confined to Italy. When Leibniz, in his *Protogaea*, attacked the fantasies of Kircher and Becher, who thought they saw Christ and Moses on the walls of the Baumann Grotto, who discerned Apollo and the Muses in the veinings of an agate, and who recognized the pope and Luther in the rocks at Eisleben, he called on "the testimony of a learned painter," according to whom the more those stones were examined, the less one saw those pretended resemblances. A full summary of Scilla's work can be found in the *Telliamed* of Benoît de Maillet,[22] and other texts could be cited. It is appropriate, however, to note the presence of Scilla in the circles and during the years in which the debate about fossils was most lively and impassioned. In January and February of 1696, William Wotton presented an abstract of Scilla's work before the Royal Society. The following year, Wotton published in London, in the form of a letter to John Arbuthnot, a short work attacking the celebrated Woodward and entitled *A Vindication of an Abstract of an Italian Book concerning Marine Bodies*.[23] Wotton was an illustrious figure: in 1694 he had published *Reflections upon Ancient and Modern Learning* and he was to publish in 1701 a history of Rome from the death of Antonius Pius to the death of Alexander Severus. He later attacked Toland in his *Letter to Eusebia* (1704) and Tindal in 1706. In 1713 he published a study on the confusion of languages at Babel and, in 1718, one on the traditions, uses, and customs of the scribes and the Pharisees. As in the case of Woodward, Wotton's geological interests were firmly linked to his activities as a historian of the ancient world and to his theological positions.

Wotton pointed out that John Woodward, in a work entitled *An Essay toward a Natural History of the Earth* (1695), had sustained that the shells in question were "*exuviae* of once-living Animals." "The proving of which propositions," Wotten continues, "being likewise the design of Scilla's Treatise," and since the latter was "a Book then very little known in *England,* I thought a large Account of it would not be unacceptable to the Society." Wotton's attitude toward Woodward is not particularly friendly. He does not go so far as to accuse Woodward of plagiarism, but limits himself to stating that *if* he had read Scilla, it would have proved most useful to him.[24] An explicit accusation of plagiarism was advanced during those years, however, by Thomas Robinson, and John Harris, responding to Robinson and springing to Woodward's defense, accuses poor Scilla of "easiness and credulity."[25] Wotton defended Scilla from these accusations with conviction and passion, but he added nothing new on the level of theoretical divergence. The dispute was by this time confused and prevalently theological. Nevertheless, Wotton's pages serve to document the presence in England of this "Galileian," a figure a good deal more widely studied in the England of Wotton's time than in that of our own contemporary, Martin J. S. Rudwick.

5·The Four Impious Hypotheses of the Origin of the Universe

IN the *Origines sacrae* of Edward Stillingfleet (1662) there is a chapter entitled "Of the Origins of the Universe." In it the author discusses at length the doctrines concerning origins formulated in different historical epochs which oppose the Mosaic thesis of a world created *ex nihilo* by the "Omnipotent Will and Word of God," offering no possibility whatsoever of compromise or conciliation. For Stillingfleet, those doctrines, or "Hypotheses," can be reduced to four:

1. The doctrine that asserts that the world has existed from all eternity (Aristotle).
2. The doctrine that attributes the formation of the world to God, but nevertheless sustains the preexistence and the eternity of matter (the Stoics).
3. The doctrine that denies the eternity of the world and explains the origins of the physical universe by means of the chance combination of atoms (the Epicureans).
4. The doctrine, finally, that attempts to explain the origin of the universe and of all natural phenomena exclusively by means of "the Mechanical Laws of the motions of Matter" (Descartes).[1]

Stillingfleet declares his intention to discuss each of these four hypotheses "in point of Reason." But he also feels the need to express a moral condemnation. A correct understanding of the problem of the origins of the world has, in fact, "a great influence upon our Belief of all that follows in the world of GOD." If, for example, it is true that the world is eternal, then "the whole Religion of Moses is overthrown, all his Miracles are but Impostures, all the Hopes which are grounded on the Promises of God, are vain and fruitless." If the world exists necessarily, God is not a free agent. If that is so, "then all instituted Religion is to no purpose; nor can there be any expectation of Reward, or fear of Punishment." But if the Aristotelian thesis of the origin of the world destroys all possible religion, so do those of Epicurus and Lucretius. If, following Epicurus, we assert the existence of a divinity but we then conceive of the world as the result of chance, "all that part of Religion which lies in obedience to the Will of GOD is unavoidably destroy'd." What remains is the veneration of a supreme being who, however, has no control over life. But how can the world still depend on a God who does not interfere in or manifest himself in the world?[2]

The first hypothesis, Stillingfleet explains, was unknown to the earliest Greek philosophers and to Plato. After the advent of Christianity, it became widespread among pagan philosophers (even among the Platonists) as an anti-Christian position. The conclusion that the world is eternal was reached, according to Stillingfleet, starting from an acceptance of the notion that it was not generated in the same way that animals and plants are generated. But this, he continues, was not a legitimate conclusion. The Aristotelians admit the existence of a Cause of the world and nevertheless they hold the world coeternal to God. They admit a priority in the ordering of causes, but they deny priority to time and duration. They draw a parallel to the sun, from which light flows, even though the sun cannot be conceived of without light. Since *omne quod movetur ab alio movetur*, Aristotle affirms the existence of a Prime Mover. But his assertion is founded on the presupposition of the necessity of motion, which is what is instead denied when one affirms that God in his infinite power can give motion to what was initially without motion. Time, furthermore, "denotes nothing real in it self existing, but only our manner of Conception, of the Duration of Things, and as it is conceived to belong to Motion, and so can argue nothing as to the real existence of Things from all Eternity." In addition, the comparison with the sun does not hold, because even if God is necessarily and essentially good, still "the communications of his Goodness are the effects of his Will, and not merely of his Nature."[3]

Many have attributed the second hypothesis to Plato, Stillingfleet continues, but there is no reason why the primal matter of which Plato speaks could not instead be identified with the Chaos that preceded the ordering of the world. The Stoics, on the other hand, held as necessary the existence of a passive principle coeternal with God, or the active principle. But the assertion of the necessary existence of matter is in contradiction with the power of God for two reasons. If a preexistent matter is necessary to the production of things, then the power of producing something out of nothing no longer has any meaning: "that Being which hath a power to produce something out of nothing, hath only a power to produce something out of something, which is a plain contradiction." In the second place, if God has the power to produce something out of nothing, "either this power doth extend to the production of this Matter, or not." If it extends to the production of matter, "then it depends on him; if not, his power is not Infinite."[4]

Stillingfleet distinguishes very carefully between a legitimate and an illegitimate use of the third hypothesis—the atomistic or Epicurean hypothesis. It is proper to make use of it in the context of investigations of nature. It is improper to derive from it general hypotheses on the formation and the origin of the world. It is one question, in other words, to accept an explanation of natural phenomena founded on concepts of matter and motion rather than on substantial forms and on occult qualities; it is quite another question to affirm that *only* matter and motion exist in the world, or that the origin of the universe can be explained *only* as "a casual concourse of Atoms." Men's minds, he says, are often prone "to be intoxicated with those Opinions they are once in love with." Some refuse to subordinate the atomistic hypothesis to God and Providence, "but think these *Atoms* have no force at all in them unless they can *extrude* a *Deity* quite out of the World." Moreover, Stillingfleet argues, the Epicurean hypothesis is destined to fail as a general hypothesis because it encounters three problems that it seems impossible to resolve within its own terms: (1) it seems impossible to reduce the great variety of natural phenomena to simple differences of dimension, shape, and particle motion; (2) the order and the beauty of the world cannot possibly be conceived as the result of chance; (3) the problem of the origins of the human species certainly cannot be resolved, as the atomists would have it, by a recourse to the "fable" of spontaneous generation out of mother Earth.

Stillingfleet insists with particular energy on the second of these points, which by the end of the century had come to acquire increasing importance and later came to be the central theme of the "Newtonians'" thought:

> Not only the variety, but the *exact Order and Beauty of the World* is a thing unaccountable by the *Atomical Hypothesis*. . . . That there should be nothing else but a blind *impetus* of *Atoms* to produce those vast and most regular Motions of the heavenly Bodies, to order the passage of the *Sun* for so great conveniency of Nature, and for the alternate succession of the Seasons of the Year . . . to believe, I say . . . these things . . . is the most prodigious piece of Credulity and Folly that human Nature is subject to.[5]

The fourth hypothesis, which Stillingfleet notes was the one recently advanced by the "late Famous *French* Philosopher Mr. *Des-Cartes*," is analyzed and explained on the basis of *Principia* 3:46. Stillingfleet, who makes full use of Henry More's letters, shows a certain caution in dealing with Descartes, but his condemnation is none the less rigorous:

> For altho' there be as much Reason as Charity to believe that he never intended his *Hypothesis* as a foundation of *Atheism*, having made it so much his business to assert the Existence of a DEITY, and Immateriality of the Soul; yet because it is apt to be abus'd to that end by Persons *Atheistically* dispos'd, because of his ascribing so much to the power of Matter; we shall therefore so far consider it, as it undertakes to give an Account of the Origin of the Universe without a DEITY.[6]

The Cartesian doctrine of the three elements claims to give an account, with the first, of the sun, the fixed stars, and the vortices; with the second, of the heavens; and with the third, of the formation of the earth, the planets, and the comets. There are, for Stillingfleet, two essential weaknesses in the Cartesian position: the Cartesian argument presupposes the existence of matter, but matter is conceived of as producing motion and motion is conceived of as essential to matter. Stillingfleet agreed with Henry More on this point: if motion pertains necessarily to matter, there can be no sun, no stars, no earth, no man. In fact, matter being uniform, all of its particles will show equal motion and each particle will be in movement to the limit of its capacities. No particle can be superior to any other particle, either by its size or by its motion, and consequently "Universal *Matter*, to whom Motion is so essential and natural, will be ineffectual for the producing of any variety of appearances in Nature." Without a God who sets matter into motion and regulates that motion, the variety of the forms found in the world is inexplicable.[7]

The attack on the concept of the eternity of the world—found in many other writers from the sixteenth century to the eighteenth as well as in Stillingfleet—nevertheless left a good deal of room for a "historical" treatment of nature. The world had had its beginnings in time, and time was created with the world. Even if the world is not eternal, it *has not always been, nor will it always be, as it now appears*. For Stillingfleet, the thesis of the eternity of the world was founded on a false premise, which states that "*it is inconceivable that things shou'd ever have been any otherwise than they are*." The concept of *production* implies that of *mutation*, and it is not legitimate to project the present order back into eternity:

> For supposing a Production of the World, several things must of necessity be suppos'd in it, different from what the present order of the World is; and it is an unreasonable thing to argue from a thing when it is in its greatest perfection, to what must always have been in the same thing; for by this means we must condemn many things for falsities which are apparently true, and believe many others to be true which are apparently false.[8]

Stillingfleet, who had read Maimonides carefully, draws on the attack on the Aristotelian thesis of the eternity of the world in the *Guide to the Perplexed* (2:17). How could a baby abandoned on a deserted island, deprived of any contact with men and women or with male and female animals, ever believe that men come into the world generated in the mother's womb? How could such a person possibly be persuaded that it is possible to live within the body of another person without eating, drinking, or breathing? How could that child possibly be persuaded that men are produced as indeed they are produced?[9]

The relationships between the Christian tradition and the idea of time have been studied in full detail. Henri Charles Puech has seen in the adoption of the Mosaic account, in eschatological themes, and in the idea of a progressive realization of universal redemption the three pillars fundamental to the idea of a rectilinear time and to the consequent double rejection of the idea of the eternity of the world and of the "Platonic" idea of time as an infinite succession of cycles, as a "mobile image of an immobile eternity." In this view, God has even transmitted to men a *history* of the formation of the world. The Mosaic account excludes an explanation founded on a combination or a recombination of principles. It is possible to try to reconcile the biblical narrative with a series of hypotheses concerning the formation of the world. One can attempt to narrate the history of those six days. Descartes, as is known, was to take great pains to avoid every possible attempt at reconciliation, and was to present his rigidly mechanistic theory as a fiction, a dream, and a fable. But even he was seduced by this possibility. In a fragment of a letter written in 1630 Descartes writes:

> I have come to the description of the birth of the world, in which I hope to include the greater part of Physics. And I will tell you that during the last four or five days, rereading the first chapter of Genesis,

I discovered as if by miracle that it could be explained according to my thoughts much better, it seems to me, than in all the ways in which the interpreters have explained it, something that I had never hoped for until now. But now I propose, after having explained my new Philosophy, to show clearly that it agrees with the verities of faith much better than Aristotle's.[10]

6·Accidental Variation in the Time Process

DESCRIBING things that once were—the works of nature and the works of man—on the basis of what now is. Reading the past in historical texts and in the ancient fables, in mountains and in fossils. Constructing hypotheses on the formation of rock strata, on changes in the land and seas, on the origins of civilizations, on the great migrations of peoples and animals, on the first forms of man's dominion over nature. Searching for a link between these hypotheses and the sacred history of the Hebrew people told in the Bible. Consideration and debate of these themes could be reached by many paths, depend on different traditions and embrace different cultural heritages. One could start from Epicurus or Lucretius, or from Descartes and the origin of his imaginary world, or even from an energetic opposition to the view of the world as the result of a chance meeting of atoms or to the impious thesis of the eternity of the world.

Sir Matthew Hale, as his biographer, bishop Burnet, informs us, dedicated his life to combating atheism. Aside from his monumental work on the origins of mankind, which was part of an even more vast project, he wrote on physics, morality, and theology. He left thirty-nine unpublished manuscripts, one of which, *Concerning Religion,* stretches over five folio volumes.[1] The aims of his major work, *The Primitive Origination of Mankind* (1667), are made clear in his preface "To the Reader." He intends, in the first place, to demonstrate that the thesis of the eternity of the world is false and that the world had a beginning in time. In the second place, he will document his charge that all the philosophers, ancient and modern, who have advanced hypotheses contrary to the Mosaic account are in error. He discusses Empedocles, Plato, Aristotle, Epicurus, and Avicenna, and, among the moderns, Cardano, Cesalpino, and Berigardo. What these authors assert concerning the eternity of the world and the formation of animal and human life out of raw matter can in no way be reconciled with the truths contained in the "Mosaical system," which, even when it is "abstractively considered"—that is, without taking into consideration the divine inspiration of its author—is "highly consonant to Reason."[2]

The greatest stumbling-block for the champions of biblical orthodoxy was the discovery of America. This discovery, Hale declares, "hath occasioned some difficulty and dispute touching the Traduction of all Mankind from the two common Parents supposed of all Mankind, namely, *Adam* and *Eve*." The American continent is contiguous to no part of Asia, Europe, or Africa, and is populated by men and all sorts of animals. These men are different from Asiatics, Europeans, and Africans in their aspect, their language, and their customs, and those animals are different from the ones that populate the other three continents. The difficulties this creates have given new vigor to the impious thesis of the eternity of the world and the origin of life out of matter. Some have tried to explain the difference of the Americans, for example, by saying they "derive not their Original either from *Adam*, or, at least not from Noah, but . . . they were Aborigenes, and multiplied from other common Stocks than what the Mosaical History imports." Whereas one could suppose that a few men, starting from Asia, Africa, or Europe, might have crossed the seas, either *ex industria* or "by accident," still "it is not easily conceptible how Beasts, especially of prey" could have been "conducted over the Seas to be preserved in the Ark, and after be transported again thither." Some have concluded, Hale continues, that the biblical flood was not universal, or that a new creation of animals took place in America after the Flood. Isaac Lapeyrère, for example, had advanced a series of hypotheses on the local or purely Palestinian character of the Flood, on particular floods, and on the antiquity of peoples. These hypotheses, Hale states, may very well solve the difficulties at hand, but they are in contradiction with the Mosaic system, they cast doubt on the authority and infallibility of Holy Writ, and, finally, they give more credit "to the Fabulous Traditions of the *Egyptian* and *Babylonian* Antiquity . . . than to the Holy Scriptures."[3]

In order to shore up the thesis that all mankind sprang from Adam, to reassert the universality of the Flood, to reconstruct a past in conformity with the Mosaic system, and, finally, to resolve the difficulties connected with the discovery of America, Hale puts forth four solidly interconnected theses to explain the presence of man in America: (1) the Americans descend from inhabitants of Europe, Asia, and Africa who transmigrated to that continent; (2) those migrations were not of one people, but of several different peoples; (3) they did not take place at one and the same time, but at different times; (4) it is not possible to determine the epochs of the individual migrations, but they all took place during the four thousand years that separate our age from the Flood. Hale cites Tyrian and Phoenician navigation; he insists on the perfection of the Ark and on progress in shipbuilding; he quite specifically supports the thesis of a sea passage: "It seems not only possible, but very probable that either by casualty and tempest, or by intention and design . . . several parts of this great continent at several times have been planted with inhabitants." On a more general level, however, he is not disposed to defend the thesis of *any sort* of passage or migration, which is why he supports the "transformistic" thesis that emerged from the new geology. It is possible, he states, that during the four thousand years that separate us from the Flood, "there hath been great alterations in the situations of the Sea and Earth" and that "possibly there might be anciently Necks of Land that maintained passage and communications by Land between the two Continents." Many cities that are now ports were once far from the sea. John Ray, Hale notes, had documented the existence of a great number of subterranean forests, buried ten or twenty meters underground, and had given a full description of petrified shells found in great quantities in the middle of the continent and at a great distance from the sea. When faced with a choice between a hypothesis that interpreted fossil shells as residue left by the sea or as the effects of a plastic power of the earth, Hale clearly leans toward the former. His "digression" on fossils enables him to demonstrate

> that we can by no means reasonably suppose the Face, Figure, Position and Disposition of the Sea and dry Land to be the same anciently as now, but there might then be Sea where there is now dry Land, and dry land where there is now Sea; and that there might have been in former times Necks of Land, whereby communication between the parts of the Earth, and mutual passage and re-passage for Men and Animals might have been, which in long process of time within a period of 4000 Years may have been since altered; that those parts of *Asia* and *America* which are now dis-joyned by the interluency of the Sea, might have been formerly in some Age of the World contiguous to each other.[4]

In order to explain the existence of men in America, Hale accepted the hypothesis of great geological changes. In order to explain the existence of American animals he accepted the thesis of biological transformation. In the *Historia natural y moral de las Indias* 4:36 (1589, translated into English in 1604), Jose de Acosta had written that there exist in the Americas thousands of different species of birds and animals that are unknown, both in their form and by name, in other parts of the world. Hale found the problem of the animals even more difficult that that of man: How could one think that animals and plants could have been transported by land? Or that bears, lions, tigers, and foxes could have been transported by sea? How could the American animals get themselves all the way to the Ark and then get back again? And if this indeed happened, why are species common in America not found in other places? At this point, Hale found it easier to think that the variation between the animals and birds of the Old and New Worlds "may happen by Mixtures of several *Species* in Generation, which gives an anomalous Production," as, for example, when pheasants and ordinary hens are crossed and produce animals both similar to and different from the parents, or as it is said to occur among African animals, where "couplings of Males and Females of several Species" take place. As Harvey asserts in *De generatione animalium*, Hale argues, the conformity of offspring to their parents, both in man and in the animals, depends on a "specifical operative Idea in the phantasie or Imagination of Animals, fixed and radicated in them, and conformable to their *Species*." Anomalous or monstrous beings result from a "disturbance or discomposure of that specifical *Idea,* by some other inordinate *Idea.*" Hale holds it necessary, though, to insist on a relation between variation and environment:

> Variety of Soils and Climates makes admirable and almost specifical Variations even of the same *Species* of Vegetables, Animals and Men. In Vegetables, a fruitful Soil or Climate improves in Beauty, Bigness, and Virtue: a barren Soil or Climate impairs them: among Animals, the *Indian* Elephants are larger than the *African;* the *English* Mastiff degenerates in his courage and fierceness, at least in the first succession by generation, when brought into *France;* the *Barbary* horse . . . degenerates in a great measure in the first or second generation when removed from *Barbary.* Nay let us look upon Men in several Climates, though in the same Continent, we shall see a strange variety among them in Colour, Figure, Stature, Complexion, Humour. . . . It is an evidence that this ariseth from the Climate, because long continuance in these various Climates assimilate those that are of Foreign extraction to the Complexions and constitutions of the Natives after the succession of a few generations.[5]

According to Hale, the American animals mentioned by Acosta and by other writers were insufficient to refute sacred history for three reasons: (1) there may possibly be animals similar to the American ones in Africa or in parts of Asia that Acosta had never seen; (2) even if this proved untrue, American animals could have originated in "an anomalous mixture of species"; (3) finally, it was possible that those animals could be "of the same species with the Primitives, but received some accidental variations in process of time,"

> as the various kinds of Dogs here in *England,* Mastiffs, Spaniels, Hounds, Greyhounds, etc. might in their Primitives be of one *Species;* the like may be said of various kinds of Apes, Baboons, Monkies; of Elks, Buffalo's and Cows; the like of several sorts of Parrots, which primitively might be but one *Species* and receive accidental Variations in process of generations . . . and so possibly might the Sheep of Peru, called by Acosta *Pacos* and *Guanacos,* be primitively Sheep, but differenced by their long abode in successive generations in Peru.[6]

All of this, however—the recognition of accidental variations, intraspecies combinations and environmental influences in the transformation of species, as well as the abandonment of the concept of fixed species—was in no way intended to question the biblical account. It was instead intended to confirm the validity of the "Mosaical system" against the grave objections raised by the problem of the "Americans." For Hale, the differences between the animals of America and those of Asia, where the Ark was built, "is no argument" against the original form of the animals who were preserved from the waters by Noah in the Ark.

In order to confirm the Mosaic account as the only acceptable source for an explanation of the origins of mankind and to prove the fantastic character of Epicurean and Aristotelian hypotheses in comparison with the biblical truth of creation within time, Hale devotes many a page to an analytic refutation of the claim that life appeared on earth by chance or through the combination of the elements in reaction to celestial forces. But before clarifying his position on spontaneous generation, Hale develops an argument, which he draws from demographic density, to support the thesis that the world is not eternal. The earth's human (and animal) population increases with great rapidity. To prevent it from growing too rapidly, nature has introduced a series of checks, which Hale examines in his ninth chapter, "The Correctives of the Excess of Mankind." Hale identifies the "extraordinary" correctives introduced by Providence (extraordinary in the sense that they are not everyday or habitual) as: (1) "Plagues and Epidemical Diseases"; (2) famines; (3) "Wars and Internecions"; (4) floods and inundations; (5) conflagrations. He takes an example from the *Bills of Mortality* to show deaths by epidemic in London, which were:

11,503 in 1592
10,662 in 1593
30,562 in 1603
35,400 in 1625
10,400 in 1636
65,596 in 1665

If the population still continues to augment in spite of these "correctives" to an excessive growth, Hale argues, how can one claim that the world is eternal? How many men would there now be on earth if the past behind us had no limit?[7]

7·The Sacred Theory of the Earth

THE *Telluris theoria sacra* of Thomas Burnet, a disciple of Cudworth's at Christ College, Cambridge, and a correspondent of Newton's, was published in London in 1680. The first edition, in only two volumes, was translated into English in 1684. The new Latin edition, in four volumes, appeared in 1684 and it, too, was translated into English in 1690–91.[1] Burnet's theory of the history of the earth was "sacred," as he explains in his preface, inasmuch as it was not limited to a consideration of the "common physiology" of the earth, but included an examination of those *maiores vicissitudines* related by Scripture and on which Divine Providence seems to turn. The past states of the earth are the paradise and the Flood; the future states are the universal conflagration, followed by a new paradise. Burnet, like Hale, explicitly rejects the thesis of the eternity of the world, and he takes care to keep within traditional chronology. When faced with the problem of the Deluge, he takes a "scientific" attitude—that is, he seeks for a rationally satisfactory solution. Thus he cites Descartes and points out the similarity between his own hypothesis and the Cartesian: "An eminent Philosopher of this Age, Monsieur des Cartes, hath made use of the like *Hypothesis* to explain the irregular form of the present Earth." Descartes, however, never speaks of the Flood, and he conceived of the surface of the earth as a transitory crust and not as an inhabitable world with a past of more than six thousand years.[2]

Belief in the Flood involved, from the point of view of physics, a series of serious difficulties. How is it possible, Burnet asks, to believe in the Flood if, as the world now stands, there is not enough water on earth to explain such a phenomenon? The waters of six or seven oceans would be needed to cover the highest mountains. From what source, from what reservoirs could we imagine that enormous amount of water to have flowed? Must we admit that God created new matter? Or did God first create and then destroy that new matter? Burnet is well aware of the existence of an antireligious and libertine tradition that had tried to avoid these difficulties—or that had made use of them—to deny the universality of the Flood and reduce it to an episode in local history involving only the Hebrew people. "Some modern Authors observing what straits they have been put to in all Ages, to find out water enough for Noah's Flood, have ventur'd upon an expedient more brisk and bold, than any of the Ancients durst venture upon: They say *Noah's* Flood was not Universal."[3] Burnet rejects these theses and insists on the universality of the Flood. But in order to explain that catastrophe, it seems to him necessary to accept the Cartesian hypothesis—to admit, that is, that the form of the earth in the antediluvian age was different from the one we know—and to assert, as both sacred and profane authorities agreed, that the world originated out of Chaos. By Chaos, Burnet means "the matter of the Earth and Heavens, without form or order; reduc'd into a fluid mass, wherein are the materials and ingredients of all bodies, but mingled in confusion one with another." The divine word transformed that chaos into a world; the heavier parts precipitated toward the center in diminishing order of specific gravity, where, pressed on one another, they formed the innermost parts of the earth. The rest of the mass of matter subdivided, following the same principle of gravity, into a liquid body and an aerial or volatile body.

In this manner, a "fluid Mass, or a Mass of all sorts of little parts and particles of matter, mixt together, and floating in confusion, one with another" gradually became organized, until it gave rise to a gigantic body of egglike structure. The central fire formed its "yolk," and around it lay a great liquid mass, "the *great Abyss*," or "the Sea, or Subterraneous waters hid in the bowels of the Earth," the outer surface of which was made up of an oily layer. Before the process of the formation of the earth had become fully stable, the surrounding air was still full of particles of dust or soillike material. Falling in toward the center, these particles mixed with the oily matter that floated on the surface of the water. In this way, a level crust came into being, completely free of irregularities, and it gradually hardened by the action of the sun. This smooth, perfect surface, over which the air was perennially calm, serene, and free of those disordinate movements that are caused by winds and by the existence of mountains, coincides—all of it—with the terrestrial paradise. This world was an inhabited world, for antediluvian humanity lived in it in simplicity, purity, and innocence.

A universal and terrible catastrophe transformed that spherical paradise into the world we know—a world that is irregular, wrinkled, and contorted, made of great areas of water, or ragged-coasted continents, and of islands emerging irregularly out of the sea. Heat and the consequent expansion of the vapors generated in the liquid interior slowly brought on a greater and greater dilation of the crust. Crevaces opened up and a gigantic earthquake split open the earth's surface. As the crust broke apart, the internal vapors escaped, condensed at the poles, and precipitated near the equator as torrential rains. The mountains, the ocean deeps, and caverns were formed, and the earth's axis shifted.[4]

This naturalistic and mechanical explanation, based from beginning to end on the recognition of a chain of causes and effects, raised a problem: Was it not in contradiction with Scripture? Did it not lead to a denial of the thesis that the Flood was brought on by divine fiat to punish the sinful behavior of mankind? Burnet's answer is characteristic: the coincidence between the series of natural events, determined by mechanical causes, and the series of moral events constitutes in itself a proof of divine wisdom. God, to use a modern term, *synchronized* the two worlds, human and natural, intellectual and material, the events of human history and the chain of causes that produced the Flood:

> This is no reason to doubt that the Flood was ordained as a punishment for mankind and that Providence presided over all its movements. Divine wisdom shines at its brightest in just this: that it adapts the natural world to the moral one in such a way that the order and disposition of the first always correspond to the tendencies of the second, and, the movements of each being in equilibrium, the times and the vicissitudes of the one coincide with the other and are connected. This seems to me to be the great Art of Divine Providence, so to adjust the two Worlds, Humane and Natural, Material and Intellectual, as seeing through the possibilities and futuritions of each, according to the first state and circumstances he puts them under, they should all along correspond and fit one another, and especially in their great Crises and Periods.[5]

This thesis of a providential synchrony was not destined to be developed very fully in Burnet's work. His rigidly Cartesian and naturalistic orientation ended up by provoking a crisis in the ancient idea of the absolute stability and perfection of the fundamental structure of the natural world—stars and mountains, oceans and living species—even more than did the works of Nicolaus Steno and Robert Hooke.

Agostino Scilla had said that many men cannot manage to *see* natural objects as *documents* of a history of nature. Burnet certainly did not belong to this category of persons: "Since I was first inclin'd to the Contemplation of Nature, and took Pleasure to trace out the Causes of Effects, and the Dependance of one thing upon another in the visible Creation, I had always, methought, a particular Curiosity to look back into the Sources and ORIGINAL of Things."[6]

This curiosity about nature's past coincided in Burnet with the ability to *mutare theatrum,* to imagine a world different from the one that is familiar to us, to abandon the idea of eternal structures, and to attempt to determine the characteristics of a universe different from the one in which we now live:

> The form and the situation of the Earth in its current state are well known. Our globe, bipartite in water and land, is rocky, mountainous, irregular, cavernous, almost as if it were a broken and shattered bulk. Such is the state of our abode, and here we conduct our brief and laborious existence. Let us now try to change our theater and imagine, if you can, scenes and parts that are the contrary of the present state: the terrestrial orb whole, continuous, without seas or marine depths, smooth and everywhere of level surface, without mountains, reefs, caverns, or cavities. . . . This is the primigenial Earth, or, better, its image.[7]

Mountains and rocks, islands and grottos appeared to Burnet—as they had to Robert Hooke—to be the ruins of an extremely ancient past that called for reconstruction and that deserved the same respectful attitude that the historian accords the remains of ancient civilizations. The mountains of the earth are "the ruins of a broken world." They show "a certain magnificence in Nature, as from old Temples and broken Amphitheaters of the *Romans* we collect the greatness of that people."[8]

The "discovery of time" in nature also led, in many instances, to a difficulty concerning the traditional image of the perfection of the universe. The order, beauty, and proportion that lie at the root of the world and that reflect the work of God were not denied but, so to speak, *shifted back in time;* projected onto a world that was different and opposed to the one in which man leads his life and has realized his history. The postcatastrophic world was seen as without order or proportion—a sort of new chaos. For this reason, Burnet says, "it seems reasonable to believe that it was not the work of Nature according to her first intention, or according to the first Model . . . but a secondary work, and the best that could be made of broken materials." Both are "the image or picture of a great Ruine, and have the true aspect of a World lying in its rubbish."[9]

This image of rubble and grandiose ruins becomes a sort of metaphysical leitmotif that gives a singular fascination to Burnet's work as it involves nature, but also man and his history, in the announcement of a future catastrophe. All that we have admired and venerated as great and magnificent is destined to vanish into nothing. Where, Burnet asks, will the great empires of the world be, and the great imperial cities with their glorious monuments? Even eternal Rome, ruler of the world, will disappear from the face of the Earth into perpetual oblivion. Two themes, destined to have a wide resonance in Baroque and neo-gothic culture, are found in full force in Burnet's pages: the theme of the funereal and austere beauty of ruins, which, rather than generating horror or fear, gives a sense of the transitory and is transformed into a meditation on the fragility of man's destiny; and the theme of the awesome and majestic spectacle of the great mountain chains, risen out of the first gigantic earthquakes but destined to disappear one day.[10] The image of the harmony and the beauty of the universe was gradually being substituted, on the one hand, by the ambiguous sensation of a history that flows inexorably toward its total consumption and, on the other, by the idea of a nature that offers a spectacle that is too "grand" for us to understand, a nature that subverts the mind with its "excesses" and that has characteristics of the "sublime":

> The greatest objects of Nature are, methinks, the most pleasing to behold; and next to the great Concave of the Heavens, and those boundless Regions where the Stars inhabit, there is nothing that I look upon with more pleasure than the wide Sea and the Mountains of the Earth. There is something august and stately in the Air of these things, that inspires the mind with great thoughts and passions; We do naturally, upon such occasions, think of God and his greatness: and whatsoever hath but the shadow and appearance of INFINITE, as all things have that are too big for our comprehension, they fill and over-bear the mind with their Excess, and cast it into a pleasing kind of stupor and admiration.[11]

As Marjorie Nicolson has so perceptively noted, Burnet conceived his *Sacred Theory of the Earth* as a contribution to the science of nature and to theology, and he never would have dreamed that his work would also occupy a major place in the history of aesthetics. The theme of ruins, however, which was of central importance in the culture of the seventeenth century and is connected to the idea of the aging and slow corruption of the world, of a deterioration in nature, finds in the *Telluris theoria sacra* a precise and important benchmark. The birth of a new geology coincided with the birth of a new aesthetics: the theme of the "reduction to nothing" of the variety of nature, of works of art, and of man's labors, and the motif of a distant catastrophe joined with the image of a world which is not the best of all possible worlds, not the mirror of divine wisdom, but is a sort of "Paradise spoiled," the fruit of a slow but inexorable process of degradation and decay.[12]

Burnet draws a characteristic distinction between the way orators and philosophers consider nature. Orators represent nature with all of its graces and its ornaments and say nothing about all the things that do not agree with their harmonious image; philosophers consider nature with an impartial eye, free from prejudice. They give a free accounting of it and are far from praising the current perfections of the universe.[13]

The second part of Burnet's *Theoria* was published after the Glorious Revolution, which many of his contemporaries viewed in apocalyptic terms. The mixture of millenarianism and science that Margaret C. Jacob has studied with such great perspicacity has important bearings on the discussions of time and the earth as well. In 1675 Comenius's *Lux in tenebris* was published in Amsterdam. Between 1680 and 1689 the *Apocalypsis Apocalypseos* of Henry More, John Evelyn's treatise on the millennium, and *The Judgement of God* of Drue Cressner were all published. Among the Huguenot refugees, Pierre Jurieu denounced the king of France as the incarnation of the Beast (*A New System of the Apocalypse* was published in English in 1688). The political millenarianism within the Anglican Church had a profound influence on Burnet's works. As he says in a passage that has been removed from the English version of the *Sacred Theory:*

> How is it possible to confide in a transient world, which will be reduced to cinders and smoke in the space of a century or two? . . . The age in which we live is indicated in the Holy Scripture with the term *tempora novissima* and that seems to indicate the last age of the world. . . . But the Millennium or the Reign of Christ, as it is described by the prophets, is an age so new that it will not arrive before the end of the world. . . . It is reasonable to conclude that there are no more than fifteen hundred years to go till the end of the world.[14]

Burnet states theses in his *Archeologia philosophica sive doctrina antiqua de rerum originibus* (1692) that are also interesting. He does not limit himself in his work to reworking the problem of the relation between the biblical text and his hypotheses concerning natural history: he poses an entire series of problems regarding the "obscure age" preceding the Flood, the "fabulous age" between the Flood and the Olympiads, and the "historical age." This permits him to treat at some length the origins and characteristics of Indian and Assyrian, Chaldean and Persian, Hebrew and Orphic philosophies. The problem of the origin of these philosophic traditions seems to him difficult to resolve, and he rejects as unsatisfactory the various solutions others had offered.

> It is not easy to explain from where the Egyptians, the Ethiopians, the Chaldeans, the Phoenicians, the Arabs, the Indians, and the other Orientals drew their philosophies and their very ancient credences. There are two opinions, both widespread, concerning the origin of barbaric philosophy. Some affirm that these philosophies were discovered by those peoples themselves, and resulted from their own genius; others sustain that they were derived and taken from the Hebrews—that is to say, from Moses and Abraham, whom these consider the founders of all that is knowable. I find it difficult to declare one or the other of these opinions correct.[15]

When Burnet refused to trace the whole of human culture back to Abraham and Moses and rejected the thesis of a number of autonomous sources of cultural and civil life, there remained the hypothesis of one older common source: Noah and antediluvian culture:

> When we abandon these prejudices we must go back farther to search for the origin of barbaric philosophy. Farther than Moses and farther than Abraham: to the Flood and to Noah, the common father of the Jews and the Gentiles. . . . Why not believe that from this source, from this original man have descended to posterity, that is to say, to postdiluvian man, those principles of theology and philosophy that can be found among antico-barbarian peoples?[16]

There was, according to this hypothesis, a common Adamic-Noachian culture that was transmitted through Noah's descendants to all of the peoples of the postdiluvian age. The culture of the Gentiles and Jews thus had a common root and a common source. Truth and morality are in some way present in all cultures and in all peoples, independently of the historical forms in which they are embodied. Burnet took several occasions to attack the remains of "superstition" in the Christian religion and, defining religion as a natural instinct present in different degrees among both barbarian and civilized peoples, he went so far, in *De fide et oficiis Christianorum,* as to reject the doctrine of original sin and the attribution of a "magical" value to the sacraments. He reached explicitly deistic conclusions: "Every true religion, whether of divine institution or of human institution, has its roots in natural religion." The latter, he continues, is universal and immutable, "and this cannot be said of any positive religion."[17]

Burnet's discussions of the creation of man, the terrestrial paradise, sin, and divine malediction have a characteristic libertine flavor and Enlightenment tone. This is particularly evident in the pages of his *Archeologia philosophica.* It is not easy to know, we read in this work, nor is it very important to know what material God used for the creation of the first human being, or whether it was male or female. Why should Eve have been created out of Adam's rib, given that God could use wood, stone, or any other material? Did Adam have one rib too many? If not, God deprived Adam of an essential part of his body, given that it is inconceivable that anything superfluous have been present in the original human body. Was he some sort of monster, like men with three hands or three feet? Burnet declares that he has no interest in resolving problems of this sort, but he really would like to know ("id me magis anxium habet") how, from that one rib, a woman's body was created. And how can we explain the "marvelous story" of the talking serpent? Furthermore, how much time did it take to complete the events of the Paradise story? What things—and how many things—had to be accomplished during the sixth day of creation! God created all the animals, including the wild animals and the reptiles. Then he created Adam and led an example of each species before him so that Adam could name them. We do not know what language Adam spoke, but, given the large number of animal species, the operation must have taken a certain amount of time. Then God put Adam to sleep, extracted the rib, and made Eve. During the same day, the two "coniugium ineunt sine sponsalibus." The new bride, wandering, who knows why, among the trees, meets the serpent, who "begins a Discourse with her." Then "they argue on one side and t'other about a certain Tree and eating, or not eating a certain fruit." Eve eats the fruit and offers it to her husband. After this meal, they both discover they are naked, they blush, and they make themselves some clothing: all of this must have taken another hour or two. At dusk, God descends into the garden, examines their respective guilt, and establishes their respective punishments. Chased from Paradise, after God has set up angels at all of the entrances to the garden, Adam and Eve wander about alone in the wilderness and spend the night among the wild animals. "All which things we read to have been done within the small space of one day. . . . In the Morning God said all things were good; and in the Evening of the same Day, all things are accursed."[18]

Burnet saw no fewer difficulties, contradictions, and absurdities in sacred history when he turned to the question of the origin of the universe. *Theoria telluris* and Mosaic cosmogony had only one point in common: both hold that chaos makes up the material substratum of the physical world, and that the world was at one time without life and later became populated with animals. For all the rest, "they do not a little differ." The astrophysical theses found in the Mosaic account start from the premise that the sun was created only on the fourth day of creation, therefore putting the earth at the center of the universe as its base and foundation. They view the sun and the stars as "created meerly for the use of the Earth, and in a manner but as so many servile Bodies." Between the two world systems—the one of the scholars, which puts the sun at the center of the universe and conceives of the stars as "very noble Bodies," and the one that is closer to common sense and could be called *plebeium*, Moses chose the second. He "followed the popular System; that which most pleases the People, which most flatters our Senses, is believed and comprehended, at least seems to be comprehended, by the greater numbers."[19]

In Moses' times the Hebrew people were no different from other peoples, nor were they wiser: they were poorly gifted for the comprehension of things both natural and divine, and were incapable of grasping what is intellectual and abstracted from the senses. In order to describe the creation, Moses made use of rough and approximate images, well adapted to the vulgar mind. Rather than following the order and truth of nature, he followed the order that proved to be more comprehensible to the simple people. In doing this, "he rightly consulted the public Safety; when neglecting Philosophy, he adhered to more serious Counsels and Reasons of greater weight."[20]

The Mosaic story arises, then, from an abandonment of philosophy, and it operates on the level of a popular narrative motivated by "political" reasons. It aims not at the enunciation of the truth, but at distracting the Jews from accepting the decadent cosmogonies of the Egyptians, the Assyrians, and the Chaldeans, which teemed with demons and occult forces. But if Moses spoke less than the truth he did so for an important reason. If he had exposed to his people "the first Originals of Things," if he had spoken of the seasons as caused by the incredibly rapid motion of the earth around the sun, if he had referred to the immobility of the sun, he would have been listened to with irritation and greeted with laughter.[21] Moses "did not philosophize or astronomize." He does not offer a description of the real world. The attempt to make the biblical account and the truth of natural philosophy agree had no sense to it whatsoever. The biblical account is the report of a political operation. Not only is it without scientific truth; it cannot even be translated into science. If we want to reach "pure naked Truth," "alia tela texenda est" ("we must go quite upon another Foundation" [1698]).

8·The Origin of Things and the Course of Nature: Boyle, Newton, Descartes

IN a letter written to Thomas Burnet in January, 1681, Newton "read" the real process of the formation of the physical world, which lies beyond the popular, divulgative exposition Moses gave of it in the Scriptures. Chaos, Newton explains, common to the planets and to the sun, separated into various fragments (one for each planet) by the action of God's Spirit. Even before it had assumed the form of a compact, well-defined body, the sun began to shine, and darkness and light, projected from the solar chaos onto the chaos of each of the planets, made the evening and the morning of the first day of which Moses speaks. That day (and the same is true for the second day) can be understood as being as long as we want because, until the moment at which the earth's globe makes 365 annual rotations, the day can have been extraordinarily long. Under the effects of some force, the Earth's motion gradually accelerated until it reached its present velocity. At this point, its force ceased to act on it, and the earth continues to rotate at the rate of 365 rotations a year. The planets were distributed in a definitive manner *before* the principle of gravity becomes operative. A divine act presides over the constitution of the world: "I must profess I know no sufficient naturall cause of the earth [*sic*] diurnal motion. Where natural causes are at hand, God uses them as instruments in his works, but I doe not think them alone sufficient for ye creation."[1]

The studies that have been made in the last few years on the manuscripts on alchemy, theology, and chronology in the Portsmouth collection have deeply modified the traditional image of Newton. Frank Manuel's research in *Newton Historian* has shown the close connection in Newton's work between the "physical history" of the universe and the "history of nations." In Newton's system, Manuel tells us, a chronological event in the history of monarchies can be translated into an astronomical event, and vice versa, because "there were parallel histories in the heavens and on the earth." Just as "the formation of the planetary masses and the regulation of their movements had a temporal beginning, so the world was destined to have a consummation as prophesied in the Book of the Apocalypse."[2] These statements, which are undeniably just, do not in the slightest weaken the image Milhaud suggested more than a century ago, however, of a Newton who stands apart from the "evolutionistic" tendencies that can be found in Descartes, and later in Burnet and in all the theorists of a history of the universe and of the earth. As Manuel himself recognizes, the *Mathematical Principles of Natural Philosophy* constitutes a description of the universe "in the intervening time period between the two absolute poles of creation and destruction." Nor should we forget, when we mention Newton's name in connection with investigation of the history of the universe during the last twenty years of the seventeenth century, that at that time Newton's writings on chronology and on the prophecies had not yet been published. An abstract of the chronology appeared only in 1715, the *Chronology of Ancient Kingdoms Amended* was published in 1728, and the *Observations upon the Apocalypse of St. John* came out in 1733.[3]

It is not necessary to cite Newton's "secret" and "heretical" manuscripts, however, nor his letter to Burnet. Both in question 31 of the *Opticks* (which was added to the 1717 edition) and in the *General Scholium* (1713) Newton's position is expressed with great clarity: a "blind Fate" would never have been able to make all the planets move "one and the same way in Orbs concentrick": "Such a wonderful Uniformity in the Planetary System must be allowed the Effect of Choice." The point that is decisive to our interests, however, is another: if it is true that "the solid Particles were various associated in the first Creation by the Counsel of an intelligent Agent," and if it is true that "it became him who created them to set them into order," then "it's unphilosophical to seek for any other Origin of the World, or to pretend that it might arise out of a Chaos by the mere Laws of Nature; though being once form'd, it may continue by those Laws for many Ages."[4]

In the *General Scholium* as well, as is known, Newton appealed to final causes. He who ordered the universe placed the fixed stars at immense distances from one another "lest the systems of the fixed stars should, by their gravity, fall on each other." The primary planets continue to move in their regular orbits by the laws of gravity, "but it is not to be conceived that mere mechanical causes could give birth to so many regular motions. . . . This most beautiful system of the sun, planets, and comets, could only proceed from the counsel and domination of an intelligent and powerful Being."[5]

Natural laws begin to operate *only after* the universe is formed (better, after it has been created). The formation of the universe and the establishment of those laws is the work of God and *there is no reason to search for any other origin of the world*. The science of Newton is a rigorous description of the universe taken as it *is*. And it *is*—stable and harmonious—as it is comprehended between the creation that Moses narrates and the final annihilation that Saint John predicts. The remote "past" of nature is not an object of scientific investigation; the "formation" of the universe cannot be made the object of rational study. The laws of nature are of no value to explain the emergence of the universe out of chaos. If an organized universe *cannot* arise out of the laws of matter and motion, what sense can there be in setting oneself to study how that organization came about? A cosmology seen as the history of the universe makes no sense. If God is the legislator of the universe, would not a mechanical interpretation (founded on the laws of nature) extended to the Genesis render the very image of God superfluous? For Newton, the biblical account tells real facts in a "popular" manner. But if these same facts could be explained "mechanically" or "scientifically," would not the biblical account necessarily end up as just an "allegory" or a metaphorical description of a different process of events? Would not that narration come to be dissolved into a different narration? And would not this different narration necessarily have the aspect and the substance of a *theory*?

Newton discusses these matters in works that date from the second decade of the eighteenth century. Already during the course of the preceding century, however, authoritative scholars had sustained and promoted theses that were equivalent in every respect.

The themes—and the difficulties—that we have found in Newton emerge clearly in a work by Robert Boyle *(About the Excellency and Grounds of the Mechanical Hypothesis)* which dates from 1665 and therefore precedes not only Burnet's works, but also those of Robert Hooke. The strict connection between a refusal of the mechanistic and materialistic aspects of the Epicurean and Cartesian position and an affirmation of the impossibility, the uselessness, and the irrelevance of a historical and scientific reconstruction of the formation of the world emerges just as clearly.

While praising the excellence of the corpuscular and mechanical philosophy, Boyle took care to draw two distinctions. The first set his position apart from that of the Epicureans and those who hold that atoms, "meeting together by chance in an infinite vacuum, are able of themselves to produce the world, and all its phaenomena." The second served to set him off from the "modern philosophers," or the Cartesians. For the latter, if it were supposed that God introduced into the total mass of matter an invariable quantity of motion, the "material parts" would be able "by their own unguided motions, to cast themselves into such a system." It would no longer be necessary, or of any consequence, that God should make the world. The corpuscular or mechanical philosophy that Boyle championed was not to be confused either with Epicureanism or with Cartesianism. Indeed, such a philosophy distinguishes

> between the first originals of things, and the subsequent course of nature, teaches, concerning the former, not only that God gave motion to matter, but that in the beginning so guided the various motions

> of the parts of it, as to contrive them into the world he designed they should compose, . . . and established those rules of motion, and that order amongst things corporeal, which we are wont to call laws of nature. And having told as to the former, [the first originals of things,] it may be allowed as to the latter [the successive course of nature] to teach, that the universe being once framed by God, and the laws of motion being settled and all upheld by his incessant concourse and general providence, the phaenomena of the world thus constituted are physically produced by the mechanical affections of the parts of matter, and what they operate upon one another according to mechanical laws."[6]

The line of demarcation between the atheistic mechanism of the Epicureans and the Cartesians and the mechanism of Boyle and Newton coincides with a separation of those who did not from those who did distinguish between the first origin of things and the successive course of nature. On one side were those who elaborated hypotheses or theories on the origin of the universe, and on the other, those who refused to consider matter as a problem and restricted themselves to a reliance on the Mosaic account. For the latter, the rules governing motion, order among things, and the laws of nature were initially established by God and have no history. Only *after* this beginning did the world exist as a system, as a structure that lends itself to investigation and can be known using the instruments of reason.

Anyone who did not draw the necessary distinction between the origin of things (which is not subject to scientific discourse) and the successive course of nature (which is the object of science) was a mechanist in the Epicurean and Cartesian sense of the term. In the eyes of Boyle or of Newton and the Newtonians, the Epicureans and the Cartesians were not only atheistic mechanists; they also represented a type of philosophy that had the presumption to subject the *origin of things* to the methods of rational understanding, to "deduce the world," to "construct hypotheses," and to "construct systems." Why, Turgot was to ask Buffon, do you undertake to explain such phenomena as the origins of the solar system? Why "do you wish to deprive Newtonian science of that simplicity and wise restraint which characterize it? By plunging us back into the obscurity of hypotheses, do you want to give justification to the Cartesians, with their three elements and their formation of the world?"[7]

If it was true, as Condillac would have it, that "we can create true systems only in the instances in which we have made enough observations to grasp the concatenation of phenomena," how could one possibly arrive at acceptable results on the question of the formation of the world? How could the Cartesian claim of "explaining the formation of the world" be justified?

> If a philosopher, deep within his study, should try to move matter, he can do with it what he wishes: nothing resists him. This is because the imagination sees whatever it wishes to, and sees nothing more. But such arbitrary hypotheses throw light on no verity; on the contrary,

> they retard the progress of science and become most dangerous through the errors they lead us to adopt.[8]

Do not imagine, Voltaire was to reiterate, that you can do with a romance what Newton could not do with his mathematics. Discussions of the formation of the universe tended to become identified with a pretention to "deduce the world" or a spurious tendency to construct hypotheses and systems.

What else had Descartes done in the fifth part of his *Discourse on Method* (1637), in his *Principia* (1644), in his short treatise, *Le monde ou le traité de la lumière* (finished in 1633, but published only in 1664 and 1677), if not present an alternative account to that of Genesis? If not describe the birth of the physical world? There are sections of the *Principia*, for example, entitled "How the Sun and the fixed Stars are formed" (3:54); "What light is" (3:55); "Concerning the creation of all the planets" (3:146); "How the Earth was created, according to this hypothesis" (4:2). In a fragment that dates back to 1630, already cited, Descartes affirms that his "thoughts" were capable of explaining the first chapter of Genesis much better than the ways used by other interpreters. He proposed to "show clearly" that his "description of the birth of the world" agrees much better than Aristotle's "with the verities of faith." But he was not to pursue this path by any means, and he later presented his cosmology as "an hypothesis which is perhaps very far from the truth," stating that he did not "wish it to be believed that the bodies of this visible world were ever created in the manner which was described above." He declares that, rather than accepting or refuting the opinions of the learned, he has resolved to "leave all this world [the real world] to their disputes and to speak only of what would happen in a new world"—an imaginary world—adding that he certainly does not intend to imply by what he had said that "this world has been created in the manner which I described." To understand his theories, he says, one must permit thought to "pass for a little while beyond this world, that you may behold another wholly new one," which arises in "imaginary spaces." He claims no interest in explaining "the things actually found in the real world" as other philosophers do. Rather, "since we take the liberty to fashion this matter according to our fancy, we will attribute to it . . . a nature in which there is nothing at all that anyone cannot know as perfectly as possible." To prevent his long discourse from becoming too boring, he says, he will "enclose some of it in the invention of a Fable." There is no room in this "fable" of the formation of an imaginary universe for either God or Moses—precisely because it is fabulous and imaginary. The planets take their origin from the motion of the vortices, from the condensation of igneous matter, which forms a brillant star on which a solid crust then comes to be formed. The Earth emerges through the work of mechanical forces. Geogony is now an integral part of cosmogony.[9]

Descartes certainly was not unaware that his "fable" was dangerous. He did not want it to be known that he was working on *Le monde*, "to continue to have the possibility of disavowing it." After the condemnation of Galileo, as is known,

Descartes was to end up renouncing its publication.[10] There was obviously a certain ambiguity involved in the presentation of this "fable." While asserting that it is quite probable "that at the beginning, God made [the world] such as it was to be," Descartes also declares that the nature of material things "is much easier to understand when we see them coming to pass little by little than were we to consider them as all complete to begin with." From this "false hypothesis," he states in the *Principia*, one can deduce "things . . . entirely in conformity with the phenomena." In that case, the hypothesis "will be as useful to life as if it were true." The same hypothesis can give "totally intelligible and certain" reasons for all that is observed on the earth and, although the world "was immediately created by God," still all the things it contains are "of the same nature" now as if they had been produced according to this hypothesis. In 3:45, Descartes declares that he does not doubt that the physical world was created "with all the perfection which it now possesses," complete with the sun, the moon, the stars, and an earth that contained not only plant seeds but was covered with plant life, and with Adam and Eve, not as children, but as a perfect man and woman. The Christian faith, he says in conclusion, "teaches us this," but natural reason also persuades us of this truth because when we consider the omnipotence of God, we are necessarily led to think that "everything that He created was perfect in every way." Descartes turns the sense of his argument inside out, however, with a procedure typical of him:

> But, nevertheless, just as for an understanding of the nature of plants or men it is better by far to consider how they can gradually grow from seeds than how they were created [entire] by God in the very beginning of the world; so, if we can devise some principles which are very simple and easy to know and by which we can demonstrate that the stars and the Earth, and indeed everything which we perceive in this visible world, could have sprung forth as if from certain seeds (even though we know that things did not happen that way); we shall in that way explain their nature much better than if we were merely to describe them as they are now, or as we believe them to have been created.[11]

When we know how the fetus is formed in its mother's womb, or we know how plants grow out of their seeds, we know something more than we would if we knew a child or a plant *as it is*. The same is true of the universe. Science allows us to know more about the nature of all things if it is able to tell us something not only about how the world is, but also about how it was formed and has become what it now is. Descartes differs radically with the science of Boyle and Newton on this point. The "first originals of things" can be investigated, and that investigation can even tell us something useful to our understanding of the successive "course of nature." Beyond Descartes's infinite precautions a path opened up that was to lead to the substitution of another account for the Genesis account.

In the scintillating pages of Bernard de Fontenelle's *Entretiens sur la pluralité des mondes* (1686), Cartesian physics becomes ideology and a cohesive view of the physical world, and it is transformed—to the utter satisfaction of the Marquise de G . . .—into a new "common Sense." Fontenelle warns, "You, perhaps, expect to hear of Feasting, Parties at play, and Hunting-matches. No, Sir; you will hear of nothing but *Planets*, *Worlds*, and *Tourbillions*." The theory of the vortices is here combined with the libertine themes of extraterrestrial voyages, the plurality of worlds, and inhabitants of other planets. Here a sense of the relativity of the human position and of the long time periods of nature comes together with a rejection of anthropomorphism, with the picture of a universe in mutation, and with the idea that the fossils give evidence to great alterations in the earth.

> And tho' there had never happened any Change in the Heavens to this Day, and tho' they shou'd seem to last for ever, I yet would not believe it, but wou'd wait for a longer Experience; nor ought we to measure the Duration of anything by that of our own scanty Life. . . . Several Mountains at a great distance from the Sea have extensive beds of Shells that necessarily indicate that Water formerly covered them. Often, again quite far from the sea, we find Stones in which Fish are petrified. Who can have put them there, if the Sea has not been there?[12]

There were the fossils, but there were also the fables: they too were "disfigured history" of the most remote past, and in them as well reflections could be found both of the vicissitudes of human history and of alterations in nature:

> That Hercules separated two Mountains with his two hands is not very credible; but that at the time of some Hercules, for there are fifty of them, that the Ocean might have broken into two Mountains weaker than the others, perhaps with the aid of some sort of earthquake, and threw itself in between Europe and Africa, I could believe without much difficulty. This was then one beautiful Spot that the Inhabitants of the Moon saw appear suddenly on our Earth.[13]

In the entry "Ovide" in the *Dictionnaire historique et critique* (1697–98, second edition 1702), Pierre Bayle enumerates and analyzes many of the radically different theories on transition from the original chaos to the present order, or system, of the world found in the culture of his time. If God is not the prime mover of matter, he argues, and if matter is able to organize itself on the basis of mechanical forces, then God becomes a superfluous hypothesis:

> The supposition that God is the prime mover of matter is a principle that leads naturally to this consequence: that God formed the Heavens and the Earth, the Air and the Sea, and that he is the Architect of this great and marvelous edifice that we call the World. But if you take from him this quality of prime mover, if you assert that matter moved independently of him and that it had in itself diversity of forms, that regarding some of its parts its motion tended toward the center

> and regarding others it tended toward the circumference, that it contained corpuscles of fire and corpuscles of water and corpuscles of air and corpuscles of earth; if, I say, you assert all of this, with Ovid, you use God to no end and inappropriately in the construction of the World. Nature could very well do without the ministry of God; it had sufficient powers to separate the particles of the elements and to bring together those that were of the same class.[14]

The objection that had been made against the Epicureans that the world cannot have been produced by chance and that it is impossible to obtain an organized system by chance—just as the *Iliad* cannot come out of a chance alignment of typographical characters—was not a valid objection. Between the two instances—and here Bayle is repeating Lamy's argument—there was an enormous difference. The *Iliad* is a unique result among an infinite number of other possible results, and it could not be formed except by the precise, determinate conjunction of a definite number of characters. To form a world, on the other hand, "it is not necessary that atoms meet and combine in one certain, unique, and determinate manner: for however they join, they will necessarily form assemblages of bodies and, consequently, a World." To bear meaning, the words of the *Iliad* must be "arranged in conformity with human institutions." The system to which the atoms give rise is instead independent of conventions. If the atoms had not formed this world, they would necessarily have formed another: "Whatever their arrangement might be . . . they produce considerable effects, capable of eliciting man's admiration."[15]

"The Cartesians, the Gassendists, and the other modern Philosophers," according to Bayle, had chosen a path that necessarily led to materialism. They "are obliged to maintain that the motion, the situation and the figure of the parts of matter suffice for the production of all natural effects, without even excepting the general arrangement that has placed the Earth, the Air, the Water, and the Stars where we see them."[16]

Bayle tried to attenuate the variance between Descartes and Newton, saying that he was sure that Newton must have been thinking of a mechanical origin of the world. He indicated with extraordinary clarity the direction that Buffon was later to take: an acceptance of the Cartesian proposal of a science capable of going back to the origin of the universe and an explanation of the formation of the world that made use of the science of Newton and of the "small number of mechanical laws" that he had established:

> I know that there are persons who do not approve the fiction that Mr. Des Cartes advances concerning the way in which the world may have been formed. . . . I am well persuaded that Mr. Newton, the most redoutable of all the critics of Mr. Des Cartes, does not doubt in the least that the effective System of the World can be the production of a small number of mechanical laws established by the author of all things.[17]

In reality, Newton had never accepted the idea that the physical world could have been *produced* by the laws of mechanics. In 1763 Immanuel Kant saw clearly that Newtonian science had explicitly renounced all attempts at rational explanation of the origin of the universe. Considering that celestial mechanics, through Newton's efforts, had become a completely clear, certain, and comprehensible science, Kant declares:

> It is not likely that one should then also hit upon the assumption that the state of nature [as it was when] these motions that still persist according to such simple and comprehensible laws were first imparted to it should likewise be easier to perceive and to understand than perhaps most things the origin of which we see in nature? . . . But for this reason, to bypass promptly all mechanical laws [as Newton had done] and by a bold hypothesis to postulate that God himself threw the planets into motion, explaining why they should move in circles in conjunction with their gravity, was too large a step to remain within the bounds of wordly knowledge.[18]

9·Possible Worlds and the History of the Real World

IN Descartes (*Principia* 3:47) there is an affirmation that Leibniz considered highly dangerous ("Je ne croy pas qu'on puisse former une proposition plus perilleuse que celle là").[1] That affirmation, Leibniz continues, "is the *prōton pseudos* and the foundation of Atheistic Philosophy." It does away with physics' task of "nourishing piety," and it draws the philosophy of Descartes closer to that of Hobbes and Spinoza. Descartes states here that the laws of nature are such that "even if we were to suppose the Chaos *of the Poets, that is, a total confusion of all of the parts of the universe,* it could still be demonstrated that, by means of them, this confusion must little by little return to the order that is presently in the world." How and why does this order come into being? The Cartesian thesis that Leibniz holds to be impious (and cites in full) is the following:

> Besides, the way in which I assume this matter to have been arranged in the beginning is of very little importance, since this arrangement must subsequently have been changed, in accordance with the laws of nature. . . . Given that these laws cause matter to assume successively all the forms it is capable of assuming, if we consider these causes in order, we shall finally be able to reach the form which is at present that of this world. . . . Although I speak of suppositions, I

> nevertheless do not make any which, even if known to be false, could give rise to doubts about the truth of the conclusions which will be drawn from them.[2]

If matter, Leibniz comments, can receive all possible forms in succession, it follows that there is nothing that we can imagine—be it absurd, bizarre, or injust—that has not happened or could not one day happen. As Spinoza put it, justice, goodness, and order thus become concepts relative to man, and the perfection of God would consist only "in this plenitude of his operation, so that nothing is possible or conceivable that he does not actually produce." If, as Hobbes would have it, all that is possible *is* in the past, in the present, or in the future, if God "produces all and makes no choices among the possible beings," there will be nothing to expect from Providence. Leibniz says, and repeated on many other occasions:

> For if all is possible, . . . if every fable or fiction has been and will become true history; then there is nought but necessity and there is no choice, nor providence. . . . To sustain also that matter passes successively through all possible forms is to destroy indirectly the wisdom and the justice of God, for if all that is possible exists necessarily in his time, God makes no choice between good and evil, just and injust, and, among an infinity of worlds, there will be some that will be completely upside down, where the good will be punished and the wicked rewarded. . . . Descartes says in several places that matter passes successively through all possible forms: that is, that God does all that is doable and passes, following a necessary and fatal order, through all possible combinations: but for that the necessity of matter alone was sufficient or, rather, his God is no more than this necessity or this principle of necessity acting in matter as it can.[3]

Leibniz, as Jaako Hintakka has clearly shown, had touched on an essential point. In infinite time, Lucretius had declared, all possible sorts of encounters, motions, and combinations of atoms are produced and, finally, the beginnings of the real world. As Hobbes, the materialist, had written: "Every act, which is not impossible, is POSSIBLE, every act, therefore, which is possible, shall at some time be produced." In infinite time no genuine possibility can remain unrealized; rather, what persists always, persists necessarily. There are no eternal accidents; what does not happen is impossible.[4]

Leibniz later cited this last formula of Hobbes's ("ce qui n'arrive point est impossible") in his *Essais de Théodicée* (1710), comparing Hobbes to Spinoza, who taught that "all things exist through the necessity of the divine nature, without any act of choice by God," and who did not believe "that God was determined by his goodness and by his perfection . . . but by the necessity of his nature." Leibniz affirms that his system "is founded on the nature of the possibles, that is, of things that imply no contradiction." Leibniz continues to say that Spinoza's followers, as

Bayle had recognized, subvert a universal and evident maxim, according to which "all that which implies contradiction is impossible, and all that which implies no contradiction is possible." This maxim, Leibniz adds, "is in fact the definition of the *possible* and the *impossible*."[5] His discussion, especially of Descartes, not only attacks the so-called principle of plenitude, but also concerns the order of the world and the passage from chaos to order. It may be helpful to a better grasp of the dichotomies Leibniz draws up to return for a moment to one or two passages in Descartes.

"I am busy unraveling chaos, to make light come out of it," Descartes wrote to Mersenne on 23 December 1630. The laws of nature, he was to say in the sixth chapter of *The World*, "will be enough to ensure that the parts of chaos can unravel themselves unaided, placing themselves in such fine order that they form a perfect world." Again, at the beginning of the following chapter: "I do not want to put off any longer telling you by what means Nature can be herself unravel the confusion of chaos of which I have spoken, and what are the laws that God has imposed on her." When he hypothesizes an imaginary world, in part 5 of the *Discourse on Method*, Descartes states that "if God now created . . . matter sufficient wherewith to form it, and if He agitated in diverse ways, and without any order, the diverse portions of this matter," letting nature act in accordance with the laws he had established, this would result in "a chaos as confused as the poets ever feigned." Following these same laws, "the greatest part of the matter of which this chaos is constituted" would necessarily "dispose and arrange itself" so as to resemble the heavens we know, while "some of its parts" would form "an earth, some planets and comets, and some others a sun and fixed stars." The mountains, the seas, the springs, and the rivers "could naturally be formed" on the earth, "metals come to be in the mines and the plants to grow in the fields." Thus "all bodies, called mixed or composite, might arise."

> In this way, although [God] had not, to begin with, given this world any other form than that of chaos, provided that the laws of nature had once been established and that He had lent His aid in order that its action should be according to its wont [ainsi qu'elle a de coutume], we may well believe, without doing outrage to the miracle of creation, that by this means alone [par cela seul] all things which are purely material might in course of time have become such as we observe them to be at present; and their nature is much easier to understand when we see them coming to pass little by little in this manner, than were we to consider them as all complete to begin with.[6]

Nature in the Cartesian system is not "some sort of Goddess or some other sort of imaginary power." From the very fact that God continues to conserve matter "with all the qualities, globally understood, that I have attributed to it" (and this is the Cartesian definition of *nature*), it follows that there will be a series of changes in the parts of matter—changes that are not to be attributed to immutable divine action but to nature. The laws of this nature are "the rules according to which these changes take place."[7]

The structures of the present world result from matter, the laws of matter, and time. As we have seen, the Cartesians were criticized precisely because they claimed to be able to describe the transition from chaos to nature and from the confusion of origins to the present order of the world. The original order of things, their critics insisted, must be distinguished from the successive course of nature: for Boyle and for Newton the process of the formation of the world did not constitute a possible object of science. For them, that process was narrated in the Bible, and the laws of nature are helpless to explain the world's way out of chaos. The physical world, just as it now exists, between its creation and its final dissolution, is the only object of rational investigation.

The systems of Descartes, Newton, and Leibniz, as it has already been said, operated in the later seventeenth and in the eighteenth centuries as three different "programs" or three "scientific metaphysics" in competition with one another. Although we should avoid overly rigid categories, we need to see that Leibniz's proposals, in the contexts under discussion here and in the preceding section, present some markedly original elements. In what way or according to what perspectives can one construct a history of the earth and of the cosmos? Has this history a meaning and a direction? How can we imagine its beginnings? Does it make sense to speak of chaos and order, or of the transition out of chaos into the order of the natural world?

For Leibniz, God does nothing that is not within order, and everything conforms to universal order. In whatever manner God may have created the world, it was "always regular and within a certain general order." God has chosen the most perfect world, the one that is "the simplest in hypotheses and the richest in phenomena." What seems extraordinary in the universe is so only "in regard to some particular order established among creatures." The existent world is contingent, and there was "an infinity of other equally possible worlds" with equal claims to existence. God, as the intelligent cause of the world, "has had regard for or relation with all of these possible worlds in order to determine one." If this world were not the best *(optimum)* among all possible worlds, "God would not have produced any of them."[8]

Leibniz's concern to combat materialism can be seen in his very definition of the universe:

> To avoid the objection that many worlds could exist in different times and different places, I call world all the series and all of the collection of all created things existing. . . . When I speak of this world, I mean the entire universe of creatures material and immaterial taken together, from the beginning of things.[9]

That "beginning of things" is the expression of an intelligible model and is subordinate to an "uncreated logic." God does not create the logic of the world (in the sense that he does not impose any sort of complex of laws on a passively indifferent matter, as in Descartes). The order created is subjected to the rules of order, and the world that is brought into existence implies that all of the particular laws of that world have been brought into existence.[10] Since every individual substance expresses, in its own way, all of the universe, "in the notion of it all of the events [of the universe] are included with all of their circumstances and all consequent external things." God has not chosen an Adam "the notion of whom is vague and incomplete," but *one certain* Adam, with a certain posterity, and he has chosen him from among all of the possible Adams who would have had a different posterity. Furthermore, God did not take any decision regarding Adam "without considering everything in any way connected with him."[11] In God, a limited number of "free, primitive decrees that can be called laws of the universe" have been joined to the free decree to create Adam. Possible individual notions include some possible free decrees:

> For example, if this world were only possible, the individual notion of any body belonging to this world, which includes certain motions as possible, would also include our laws of motion (which are free decrees of God), but also as only possible. For, as there is an infinity of possible worlds, there is also an infinity of laws, some proper to one world, and some to the other, and every possible individual of any world includes in its notion the laws of its world.[12]

Therefore there is an infinity of possible ways to create the world "according to the different designs that God might form," and every possible world "depends on certain principal designs, or ends, of God . . . that is to say, on some primitive free decrees (conceived *sub ratione possibilitatis*), or Law of the general order of this possible Universe, to which [these laws] are appropriate and the notion of which they determine, as well as the notion of all the individual substances that must enter into this same universe."[13] The present world "is not necessary absolutely or metaphysically. In reality, given things are as they are, it follows that certain things are born as a consequence."

This metaphysics determines the characteristics of the history of the universe:

1. History is the working out of possibilities already implicit in its beginning and already "programmed," as an embryo is programmed. The choice of the "program" is God's: in God, in fact, "the idea of the act always precedes the act and the present state of things was preconceived."
2. It is not chaos that lies at the roots of this history and at its beginnings but the free decrees of God, or the "Laws of the general order of the possible universe" which has been chosen by God to become real.
3. This history is realized through changes or "disorders," which are only apparent and which seem without order to our human eyes because man is incapable of an all-embracing but particular vision of the entire process.

From this point of view, both Leibniz's negation of chaos and the affirmation with which the *Protogaea* begins—"Deus incondita non molitur"—seem totally comprehensible. Leibniz wrote to Louis Bourguet, referring to the *Protogaea,* on 22 March 1714:

> When I affirm that there is no Chaos, I do not at all intend to say that our globe or other bodies have never been in a state of external confusion: for this would be contradicted by experience. The mass Vesuvius expels, for example, is such a chaos; but I mean that anyone who had sense organs penetrating enough to perceive the smallest parts of things would find everything organized. And if he could augment his penetration continually according to his need, he would always see in the same mass, new organs that were not perceptible at his preceding level of penetration. For it is impossible that one creature be capable of penetrating everything simultaneously in the smallest parcel of matter since the subdivision goes on infinitely. In this way, the apparent Chaos is but a sort of distance: as in a pond full of fish, or better, as in an army viewed from afar, from where it is impossible to distinguish the order observed there.[14]

Deus incondita non molitur, and he does not make chaos. Disorder and chaos are produced by human limitations, human subjectivity, and are relative to man. Because every existence is contingent, one can certainly assert that "the Earth and the Sun itself do not exist necessarily" and that the sun and the solar system may in the future "no longer exist." The universe has no true end, however, and its limitation, "at least in its present form," is decisive and essential.[15]

If we keep in mind Leibniz's anti-Cartesian position, discussed earlier, we can understand better an assertion that he offers in radical opposition to the theses of Lucretius, Descartes, and Hobbes:

> It is very true that what is not, has not been, and will not be is not possible, where possible is understood as co-possible. . . . You say further, Sir, that an infinite series contains all possible numbers. But I do not agree to this either. The series of square numbers is infinite, yet it does not contain all possible numbers. . . . The Universe is just the collection of a certain manner of co-possibles; and the current Universe is the collection all the existent possibles, that is, of all those that form the richest composite.[16]

To summarize Leibniz's view: (1) the real world is chosen by God from among all the possible worlds, and it represents the translation into reality of the best possible context from among copossibles; (2) matter assumes not *all* forms, but only those which are copossible with the archetype of world that God has chosen; (3) the problem of the transition from the original chaos to the organized world does not exist because *there has never been* chaos.

All of the terms of the problem became transformed: mechanism and determinism are not incompatible. It is possible to speak of the history of the physical world, of the formation of the solar system, of change, of time, of the history of the universe and the history of the earth, and still avoid the impiety of the Lucretian, atheist, and materialist traditions. Those positions are neutralized when chaos and disorder become relative, and a fairly large space opens up both for empirical research into the changes that have occurred and are occurring in the world and for the consequent recognition of a history of the earth and of the universe.

A design is present in the universe. The universe is the execution of a "program" and the explication of a primordial, perfect idea. Everything is *implicit* in that idea: does the explication or the unwinding of that idea add something to the perfection of the program? Does the temporal process, in which that idea unwinds, offer an increased perfection? Is it true, as A. O. Lovejoy would have it, that two systems of philosophy are thus somehow present in Leibniz? Or is Hintikka right to criticize this thesis extensively?[17] These are questions to leave to the specialists. What it is important to emphasize here is that the presence of that design makes the disorder only apparent. "The disorders took place within order," and even the earth's great initial convulsions gave rise to an "equilibrium." This view (which recalls many passages in Hutton) permits Leibniz to accept even the conclusions of Thomas Burnet's *Theoria Sacra* that had seemed the most revolutionary and perilous. It was true, as Burnet had declared, that we are living on ruins. Leibniz, citing Burnet, repeats this explicitly in the *Protogaea*, in the *Theodicy*, and in a letter to Thomas Burnet of Kemney.[18] There is a decisive difference, however: for Leibniz, those ruins are not, as for Burnet, the expression of a world in decay, and they do not express the least lack of "order and proportion." In particular, they are not evidence—as Burnet had thought—of the dawn of a new chaos, the emergence of disorder, or the presence of a new, unforeseen, and negative "intention" on the part of nature. Those ruins are not preparing a decline; they are not signs of a corruption or an end. The earth of Leibniz's history has never been a paradise, and that history is not, as for Burnet, a deterioration or a "degradation." Quite the contrary: it is changes and disorders that have led things to their present perfection:

> But after the fire, we must judge that the earth and the water caused no lesser devastations. Perhaps the crust, formed by the cooling, and which had great cavities under it, collapsed, so that we live on nothing but ruins, as Mr. Thomas Burnet, chaplain of the late king of Great Britain, among others, has so rightly noted. Many torrential rains and inundations have left sediments, traces and remains of which are found, and which show that the sea was once in places that are now very far from it. But these convulsions finally ceased, and the globe took the form that we see today. Moses hints at these great changes with few words: the separation of the light from the dark indicates the fusion caused by fire; the separation of the wet and the dry marks the effects of the inundations. But who does not see that these disorders

> served to bring things to the point at which they now can be found, that we owe to them our riches and our commodities, and that by means of them this globe has become proper for cultivation by our efforts? These disorders took place within order. Disorders, true or apparent, which we see from afar, are the sun spots and the comets: but we do not know what uses they may bring, nor what regularity they contain. There was a time when the planets passed for wandering stars; now their motion is judged to be regular. Perhaps the same will be true for the comets. Posterity will know.[19]

To believe that God made the world for us alone "is a great abuse," Leibniz declares in his *Discours de métaphysique* (1686). It also was true, however, that God "made all of it for us, and that there is nothing in the universe that does not regard us and that is not in keeping with the intentions that He has for us." Precisely because man's point of view is limited and partial, it should not be confused with that of God, for whom the universe is "like a whole that He penetrates with one glance." Man can try, laboriously, to reconstruct his own history and that of the physical world. He must be aware, though, that "we know only an insignificant part" of the eternity that extends infinitely; a part that corresponds to only the few millennia of which history has left us a record: he must realize that he is in a position similar to that of men born and grown to maturity in a dungeon or in the Polish salt mines in which one could live out his life believing that all light comes from lamps.[20]

What was true for human history was also true for the history of the earth. We live on the earth knowing only its surface, "and we do not penetrate its interior for more than a few hundred spans."[21] Leibniz has an extremely vivid sense of the historical and temporal dimension of nature. In his *Nova methodus descendae docendaeque jurisprudentiae* (1667), the history of the world is presented as both the history of an individual and, at the same time, the science of the species ("Cosmographia . . . est mundi tamquam alicuius individui historia, et simul tamquam speciei scientia").[22] It was not by chance that the *Protogaea* was written as the opening discourse or "preface" to a projected civil history of the house of Hanover. This preface was to serve as a description of the German state and of the entire earth before the beginning of all the possible human histories. Nor was it by chance that this text ended with the affirmation of a welding together of natural events and human history: "The nature of things fills in the lacunae of history for us; and in turn our history does nature the service of perpetuating to posterity knowledge of the remarkable works we have the privilege of contemplating."[23]

Leibniz's *Protogaea* had a curious destiny. It was written between 1691 and 1692, a little more than ten years after the publication of Burnet's *Theoria sacra* and before the appearance of the successful and much-discussed cosmological works of Woodward and of Whiston (1695–96). It was not published, however, until fifty-six years later, in 1749, the year of the publication of the first volume of Buffon's *Histoire naturelle*. In 1749 all that Buffon knew of this work of Leibniz's was the very brief, two-page extract that had been published, in January 1693, in the *Acta Eruditorum* of Leipzig.

Leibniz begins his *Protogaea* with a statement of principle: all that comes out of the hands of nature begins regular in form. Thus it happened with the earth. The mountains became folded and irregular at a later time. If the globe was liquid at its beginnings, Leibniz argues, it necessarily had a smooth surface. It is in conformity with the general laws of bodies that solid things originate in the hardening of liquid things. This is confirmed by the presence of solid bodies enclosed with a solid body, as, for example, one can encounter "the debris of the past—plants, animals, artifacts—now coated with a casing of stone." The casing that now appears solid was necessarily of more recent formation than the object it encloses "and it must have been first in a fluid state." This fluidity came from internal motion and high heat, which in turn came from fire, light, "that is, [from] the very subtle spirit that pervades all things." Thus we arrive at the moving cause, Leibniz concludes, that sacred history takes as the beginning of cosmogony.[24]

From the very first pages of the *Protogaea,* then, Leibniz starts from hypotheses of a "Cartesian" sort and from conclusions that Steno had reached. He sees the origin of the world—as confirmed both by Scripture and by reason—in the original separation of light from darkness (or active things from passive things) and in the consequent division of the passive things into liquid and solid bodies: "And we have an equal right to believe that bodies we now see dry and opaque were at first incandescent, that later they were submerged by the waters, and that they finally emerged, through a secret labor, in their present form."[25]

Globes, originally incandescent and luminous, similar to the fixed stars and to the sun, became planets or bodies that were opaque because of the scoria produced by the boiling, incandescent matter. Heat became concentrated at their interior and a crust hardened through cooling. These hypotheses seemed completely acceptable to Leibniz. Even Scripture hints that the earth contains a fire that will reemerge, and these conjectures are also reinforced "by the traces that still subsist of the primitive face of nature." The scoria produced by this fusion are turned into glass; if "the great bare bones of the Earth, those naked rocks, those imperishable flints, are almost completely vitrified, does this not prove that they sprang from the melting of bodies carried out by the powerful action of the fire of nature on still-tender matter?"[26]

Glass makes up the foundation of the earth and glass is present, as if hidden, in all other bodies and in their particles. The latter, corroded and fragmented by the waters, were subjected to repeated distillations and "sublimations" until they generated an ooze capable of nourishing plants and animals. At the earth's origins, when it was still incandescent, heat forced humidity into the air. As in distillation, water vapor appeared. This water vapor, when it came into contact with the cooled surface of the earth, produced the waters. They absorbed the acidic and alkaline salts present in the ashy residue of fusion, forming a sort of lye that flowed into the sea. The origin of the sea's saltiness can be assimilated to the processes of the chemicals that produce oils *per deliquium*, gathering them from bodies that have been heated and, in cooling down, attract humidity. The hardening of the earth's crust during the cooling process produced enormous blisters containing air or water. Because of differences in the matter involved and in the degree of heat, these masses cooled at different rates and the blisters collapsed to form valleys and mountains. Water from the ocean deeps joined the waters that flowed down off the mountains, causing inundations that produced sediments. When the same phenomena were repeated, these deposits were covered over by other superimposed sediments. Not all rock, but only the primitive masses, the "base" of the earth, resulted from the cooling that followed the first fusion. Other rocks, as strata serve to document, came from reconcretions that followed disintegrations, in their turn provoked by precipitations taking place at different times.[27]

Leibniz realized that conjectures on the "cradle of the world" contained the germs or seeds of a "new science that one could call Natural Geography." He presents his work not as a finished construction, but as an attempt to entrust to posterity the task of describing the strata found in all the various regions of the earth. He does claim, however, to have identified the "general causes" that lie behind "the skeleton and, so to speak, the visible bones of the Earth and its general structure." The mountain chains of the Himalayas, the Atlas, and the Alps made up this ossature and, together with the ocean basins, they gave rise to the *structure* of the earth. There were elements of stability in such a structure: it was the result of an initial formation process after which "the perturbing causes having been exhausted and brought into balance, a more stable state was finally produced" ("donec, quiescentibus causis atque aequilibratis, consistentior emergeret status rerum").[28] Once the globe had been consolidated into this structural equilibrium, which was in its broad outlines its present structure (Leibniz uses the terms "solidatus iam, ut nunc, globus"), subsequent changes could take place—changes such as cave-ins, inundations, the blocking of watercourses, the emergence or sinking of volcanos. These were due, however—the distinction is an important one—to "particular" and not to general causes.

Leibniz, as Giovanni Solinas has justly remarked, is much less "diluvian" than many of his contemporaries.[29] Indeed, he examines the Flood on the level of particular causes and he discusses, one after the other, the principal theses that had been advanced to explain the submersion of the entire surface of the earth. The causes that Leibniz discards in rapid succession are a shift of the earth's center of gravity, magnetic attraction, abundant rains, the rising of the waters of the ocean, the passage of a comet, or the moon's close approach to the earth. The hypothesis that Leibniz accepts as possible is the one that Thomas Burnet had formulated: the earth's crust broke apart where its support was weakest, enormous masses of water washed down from the mountain tops and, simultaneously, the waters contained in subterranean cavities and in the ocean depths flowed out from springs that had broken open and submerged even the mountains. Finally, when they found a new entrance to the "abysses," the waters abandoned those places on the earth which today appear as dry land.[30]

As we have seen, Leibniz repeats Steno's central theses, and he cites Steno's work both to explain the existence of mineral veins and the presence of fossil fish:

> It is probable that a great number of layers, which once were horizontal, as long as the outside crust of the Earth remained unchanged, were profoundly inclined as a result of the collapse of the ground or of large-scale cave-ins. Is it not natural, in fact, that during the passage from the liquid state to the solid state, either when the Earth first hardened, or later when the great inundations finally dropped their sediments, that every thing took its place according to its weight and that all, by the law of liquids, lay horizontal and level, a level that was subsequently destroyed by the superior force that shook the foundations of the Earth?[31]

We are living, Leibniz says (and he was referring specifically to Eisleben, in Saxony) on top of a sort of "double earth": one goes back to a time when the land was full of lakes populated with fish; another results from the accumulation of a great mass of soft soil. Because of the action of "fire, petrifying vapor ["gorgoneus vapor"], the effect of salts, or even time alone," this mass solidified into separate layers. Hard rocks came to be placed on top of these strata, then clay, and finally that common black earth that men now cultivate. Even if we were willing to admit that those lakes were subterranean before they were filled up, we must acknowledge that "the surface of the Earth as it was at that time has been turned upside down and completely transformed."[32]

The theory of strata led Leibniz to a resolute support of the thesis of the organic origin of fossils. His pages of demonstration of this thesis and polemics against those who rejected it are forceful, but they contain no notable theoretical innovations. It is often possible, Leibniz says, to see fish forms delineated in slate as exactly and neatly "as if an artist had inserted a sheet of engraved metal into the black stone."

> I have in my hands fragments that reproduce a mullet, a perch, or a minnow, sculpted into the rock. A little earlier, an enormous pike had been brought out of the quarry. His body was flexed and his mouth open, as if, surprised while alive, he had stiffened against the petrifying force. . . . When faced with these phenomena, most observers are content to say these are tricks of nature ["lusus naturae"], a term void of meaning, and they present the ichthyomorphic stones to us as indisputable examples of the whims of the "genius of things," hoping to eliminate all difficulties and prove to us that nature, that great artificer, imitates, as if in play, teeth, animal bones, shells, and serpents. . . . There is such a strong relationship between these supposed simulacra of fish and reality: their fins and their scales are reproduced with such precision and the number of these images in the

> same place is so great that we suppose a manifest and constant cause more willingly than a game of chance or I know not what generative ideas, vain terms to hide the prideful ignorance of philosophers.[33]

The thesis of the *lusus naturae* is refuted here, not only on the basis of a geological theory of the formation of earth strata, but also on the basis of a "metaphysical" premise: "chance never copies nature exactly" ("numquam perfecte veritatem casus imitetur").[34] In a world at the root of which there is a choice and which is articulated as a program, there is no room for chaos and the rule of chance is much reduced.

The fish-bearing slate, Leibniz continues, is "a hanging vein, to speak like our miners," extending for several miles in a horizontal layer. It is clear that fish from one and the same pond were crushed by a mass that rested on them. The fish imprints are therefore real *imprints* and they come from real fish: whatever explanation for this might be adopted, this much must be accepted as true. Anyone who does not accept this is operating not on the level of the tricks of nature, but on that of tricks "of men's imagination, which sees armies in the clouds and recognizes whatever modulations it wants in the tolling of bells or the rolling of drums."[35] Kircher and Becher, Leibniz points out, had insisted more than others on the admirable tricks of nature and on its miraculous formative powers. They "saw" mythological scenes in rocks; they identified Christ and Moses on the walls of the Baumann Grotto; they recognized Apollo and the Muses in the veinings of an agate, and the pope and Luther in the rock of Eisleben. For Leibniz this was like praising the virtues of the mandrake root, grown to resemble human form, or like the "superstitious women" who hope to predict the future by peering at an egg broken into a glass of water.[36]

One theme emerges from the *Protogaea* that merits particular emphasis: that of the chemical laboratory and the mechanical arts taken as models to explain the processes of the natural world. Leibniz's approach to the problem is typically Baconian:

> Many things familiar to mechanical artisans and to empiricists are familiar to the common people but are unknown to the learned, who consider such things miraculous if now and then they come to be outlined in books. From which it happens that while the mechanics are unaware of the possible uses of their observations, the learned, in their turn, are unaware that their desiderata might already have been satisfied by the work of the mechanics.[37]

Leibniz, as is known, expressed more than once his support of an idea that lies at the heart of Bacon's philosophy: that the work of technicians, artisans, and "mechanics" can cast light on scientific theories. He was also an impassioned supporter of the need to join together the wisdom of the "empiricists" and that of the "theorists" to arrive at a new culture. On the very pages of the *Initia et specimina scientiae novae generalis* on which he emphasized Galileo's and Harvey's debt to "mechanics" and invited the learned to write the great history of labor and technology, Leibniz also emphasized the significance and the importance of chemistry, which "is beginning to be torn from the hands of the charlatans and vagabonds and to be taken up not only by lucre-seekers [*lucripetae*] but also by light-seekers [*lucipetae*]." He insists on how much the life of man owes to this "most beautiful science"—the art of glassblowing, for example, and the art of metal foundry, which are just branches of chemistry. But chemistry was necessary not only to the enrichment of life but also to the understanding of life. What could be more useful, Leibniz exclaims, than the "knowledge of fermentations, of solutions, of precipitations, of the struggle of liquids to reduce these phenomena to determinate families, given that everything happens in the human body for similar reasons?"[38]

The theme of the practical workers and of the laboratory are both present in the *Protogaea:*

> Anyone who diligently compares the products of nature, wrested from the womb of the Earth, with the products of the laboratories (thus we call chemists' workshops) will accomplish an important task, in our opinion; for then the striking resemblances between the products of nature and those of art will shine before our eyes. Although the inexhaustible author of things has in his power different means to carry out his wishes, he nevertheless takes pleasure in constancy amid the variety of his works. It is already a great step forward toward an understanding of things to have found even one means of producing them [et magnum est ad res noscendas vel una producendi rationem]. Thus it is that geometers derive all the properties of a figure from just one mode of description.[39]

This comparison between the mechanical arts and geometry as forms of knowledge that *construct their object* demonstrates the presence in Leibniz of the thesis, highly influential in the culture of the seventeenth century, that knowing was like building. This thesis, as is known, was connected (in Bacon, in Descartes, in Mersenne, in Gassendi, and in Hobbes) with a new way of conceiving of the relation between nature and art. There is no essential difference, in this view, between the products of nature and those of art, and *ars* is not a simple *imitatio*, nor is it an activity that is *adulterina*. Obvious traces of this position can be seen in the *Protogaea* as well:

> Nature, in fact, is nothing but a greater art, and we cannot always clearly distinguish the artificial from the natural. It matters little, what is more, that an identical thing be forged by an alchemist or exhumed out of the bowels of the Earth by the miner. . . . Since most of the time things take disguised forms rather than changing their nature, it is less astonishing that there are so many things common to the laboratories and the mines.[40]

Leibniz has no intention of taking a stand on whether there is a practical possibility of generating metals through the art of chemistry; whether gold, silver, mercury, or salt can be produced *de novo* or destroyed. Just as art transforms clay into ceramics in the kiln, it is possible that nature, by means of a more powerful fire, may have transformed various mixtures or types of earth into slate, alabaster, and other types of stone. And it is possible that the metalic material that was dispersed in silt may have melted under heat and collected in the cavities once occupied by the fish bodies, since destroyed by time or by heat.

> The art of the goldsmith offers us something analagous, and it is with pleasure that I compare the hidden operations of nature to works that are manifestly man's. When they have wrapped a spider or some other insect in a material appropriate to this end, leaving a narrow opening, however, they harden the material in the fire. Then, with the aid of mercury that they pour in, they shake the ashes of the animal out through the opening and, to take its place, they pour molten silver through the same opening. Finally, breaking the hull, they find a little silver animal with his full apparatus of feet, antennae, and tiny hairs, astonishingly similar [to the original].[41]

In the passages on the problem of fossils, Leibniz touches on all of the traditional themes: the shells, the Horns of Ammon (ammonites), the *glossopetre*, the skulls and bones of animals. He takes pains to state his disagreement with those who claim that when the ocean covered the entire surface of the globe, "the animals that now people the Earth were aquatic, that they became amphibious as the waters receded, and that their posterity finally abandoned their former habitats."[42] This hypothesis ran counter to Scripture, and involved insurmountable difficulties. Still, Leibniz does not completely exclude the possibility of changes in animal species. There are some scholars, he says, who marvel at the presence of fossil species, "figured bodies whose analogue we would search for in vain in the known world."

> They say, for example, that the Horns of Ammon, which are classified among the nautilus, differ by their form and their size (for some are 33 centimeters in diameter) from all species that the sea now furnishes. But who has explored its utmost depths and its subterranean abysses? How many animals, formerly unknown, has the New World not offered us? And is it not presumable that, in the great upheavals that the Earth has undergone, a great number of animal forms have been transformed? [credible est per magnas illas conversiones etiam animalium species plurimum immutatas].[43]

Leibniz marveled at the many bones extracted intact from the Baumann Grotto and from Scharzfeld—bones "of marine monsters and of other animals of the unknown world [ignoti orbis animalia], which could not have been born in those places, but which had been transported there from the bosom of the ocean by the violence of the waters." Other bones found at Scharzfeld doubtless belonged to elephants, he notes, whether we believe that those animals "were once much more widely distributed over the surface of the globe than today, their nature or that of the terrain having changed, or that their bodies were carried by the violence of the waters far indeed from their birthplace." Also at Scharzfeld, teeth were found, often still in the jawbone, that were impossible to attribute to any known animal.[44]

The question of Leibniz's views on the mutation of animal species exceeds the limits of this study. It is appropriate to note, however, that his preformationism—which states that every being that has ever existed in nature has always existed there—does not at all exclude the possibility that an original embryo may have changed to become the "present animal":

> I am unable to speak in detail on the generation of animals. All that I believe I can assert is that the soul of every animal has preexisted and has been in an organic body, that finally, through a vast series of changes, evolutions, and involutions, it has become the present animal.[45]

Contemporary science, Leibniz continues, shows that organic bodies are never produced by chance or by putrefaction, "but always by seeds, in which there was undoubtedly some pre-formation." Before conception "there exists not only the organic body, but also the soul of the body: in a word, the whole animal." Since our determinations of species are

> provisory and proportional to our knowledge, it is possible that in some time and in some place in the universe the animal species are, have been, or will be more subject to change than they are among us at the present and several animals that have something of the cat, like the lion, the tiger, and the lynx, may have been of one single race and may now be like new subdivisions of the former species of cats.[46]

From this point of view it was also not to be excluded that terrestrial animals once present in the region had disappeared "either because of changes in climate or for other causes that it would be difficult to guess *in ea rerum remotarum caligine.*"[47]

Leibniz was to return more than once to the problems he addressed in the *Protogaea*. In 1698 and in 1699, in two letters addressed to Thomas Burnet of Kemney, he expressed a vaguely positive opinion of Woodward and Whiston, adding that he had only read a review of the latter's book. In both of these letters he presents his theses in a detached, occasionally almost joking way, reducing his impressive labors to his hypothesis on the formation of sea salt:

> I have not yet seen the New sacred Theories of the earth. You know, Sir, that I have one *à ma mode* and that I think that the sea is an *oleum per deliquium*, in the manner of Chemists. . . . You know that I too have my chimeras on this subject and that I believe that the sea is a sort of *oleum per deliquium*.[48]

What is most striking in the first of these two letters is Leibniz's concern to keep his distance from any sort of "sacred history" and from any attempt to "explain in detail the History of Genesis" or to determine "if the days of the hexamaeron are years or much longer periods." Leibniz claims that he never wanted to rival the various sacred histories of the earth: it was better in those matters "to let the theologians decide" and, "in the meantime, I will be satisfied with explaining things according to reason, in such a way that they not contradict Holy Scripture, for *verum vero non dissonat*."[49] But the problem was a delicate one and the terrain treacherous. As late as 1749, and only on the basis of the extract published in 1693 in the *Acta Eruditorum*, Buffon, for example, felt impelled to set himself up as censor to Leibniz. When one discusses the remote past of the earth, Buffon writes, the seeming verisimilitude is in direct proportion to "force of imagination." To give as certain, like Whiston, Buffon continues, that the earth was once a comet, or to claim, like Leibniz, that it had been a sun "is to assert what is equally possible or impossible." To go on to say that in other ages the sea covered the entire earth was to ignore a point as important as "the instantaneous creation of the world."

> For, if that had been the case, it must of necessity be allowed, that shells, and other productions of the ocean, which are still found in the bowels of the earth, were created long prior to man, and other land-animals. Now, independent of Scripture-authority, is it not reasonable to think that the origin of all kinds of animals and vegetables is equally ancient?[50]

As we have seen, Leibniz was by no means in agreement on this point. In a letter to Maximilian Spener (1710), he states, in fact, that "all of the terrestrial globe was once covered by the sea before the appearance of man [ante ortum hominis]." In that letter he had also accepted the "Lucretian" hypothesis that had been set aside as dangerous in the *Protogaea:* "It is possible to believe that some terrestrial animals have descended from marine or amphibious animals that were abondoned by the sea and had changed over a long period of time to the point where they no longer tolerated water."[51]

In the letter to Bourguet of 22 March 1714, Leibniz briefly summarizes his thoughts on the origin and formation of the earth. Our globe was once similar to a burning mountain and it was at that time that the minerals were formed that we discover today and can reproduce in our foundaries. Rocks are scoria or vitrifications produced by that ancient melting process; sand is the pulverization of that glass; the sea is an *oleum per deliquium* derived from the cooling that followed calcination.

> There are three materials that are distributed widely over the surface of our globe (that is, the sea, the rocks, and sand), explained quite naturally by means of fire, which it would not be at all easy to account

> for by another hypothesis. This water once covered the entire globe and caused many changes in it even before Noah's Flood. I am therefore quite favorable either to the sentiment of M. Descartes, who judges that our earth was once a fixed star, or to one of my own, that it could have come out of a fixed star, since it could have been a molten piece or a great blotch thrown out of the Sun, toward which it tries to fall.[52]

There was no reference in the text of the *Theodicy*, published four years earlier, to the Flood, or to Descartes, or to the origin of the earth from either a star or the sun.[53] Perhaps inserting the "Cartesian" hypotheses and Burnet's hypothesis into a metaphysical and theological framework based in the recognition of the order and harmony of the world did not seem to Leibniz to offer sufficient guarantees. His theses, when one looks beyond the many reservations expressed in the *Protogaea*, implied, or could imply, not easily acceptable consequences. His attempt to tell the story of those changes that "Moses suggests in few words" in the Bible was not, after all, an easy undertaking. Perhaps behind Leibniz's decision to leave the text of the *Protogaea* unpublished there lies something other than the little care he showed for many of his writings or the failure to carry through the monumental project of a great history of the duchy of Brunswick and of the house of Hanover. In the letter to Thomas Burnet of Kemney, cited on several occasions, immediately after declaring that he did not intend to deal with the Genesis story in detail, that such decisions are the theologians' affair, and that *verum vero non dissonat*, Leibniz adds a few lines:

> I am happy if the authors whose sentiments are dangerous are refuted, but I do not know whether it is opportune to establish against them a sort of inquisition when their false opinions have no influence on customs, and, although I am very far from the sentiments of the Socinians, I do not believe it just to treat them as criminals.[54]

Socinian, as Bayle clarified in the entry under "Spinoza" in his *Dictionnaire*, is a "generic term" by which are designated— "even without having read either Socinus or his disciples"—all who have no faith in the mysteries of Scripture. Burnet was certainly an author *theologis suspectus*, and Brucker so considered him in his *Historia critica philosophiae*. The qualification of Socianian had already been attributed to Burnet very early on and was to be reconfirmed, many years later, in the article "Unitaires" of the great *Encyclopédie* of the philosophes. And Leibniz knew well not only Burnet's *Sacred History of the Earth* but also his *Archeologia philosophica*.[55]

10·Burnet's Heritage

WILLIAM Whiston's *New Theory of the Earth* was published in London in 1696. It was dedicated to Newton and it presented three theses concerning the cosmos: (1) the earth had been formed as a result of the cooling of a nebulous comet of the same mass as the earth but of incomparably greater volume; (2) we call the Flood the high tides and outpouring of internal waters provoked by the earth's passage through the tail of a comet (later identified as Halley's) six times bigger than the earth and twenty-four times nearer to the earth than the moon; (3) the final conflagration will be set off by the approach of a new comet—or by the same one—which will make the waters disappear and the earth be reconsolidated as it was in the beginning.[1]

The great cosmological hypothesis that Whiston formulated (and which Buffon was to summarize accurately) was an attempt to examine Burnet's description and to enlarge on it. Whiston's attitude toward the Bible was very different from Burnet's, however. Whiston's intention was to avoid with equal care the "literal" interpretation, which ended up discrediting Scripture by making it seem far removed both from common sense and from scientific fact, and the "allegorical" interpretation, which regarded the words of Moses as fables not intended as description and with not the slightest relation to truth, and which catered (as Spinoza and the libertines claimed) to the limited "Capacities of the Jews" whom Moses addressed. Moses did not give "a nice and philosophical account of the Origin of all Things," but rather "a historical and true representation of the formation of our single earth out of a confus'd chaos and of successive and visible changes thereof."[2] Whiston declares himself firmly convinced that the opposition of "the literal to the true" in Scripture, of "the obvious and natural to the rational and philosophical interpretations of the Holy Scriptures," was wholly contingent. He looked forward to "that happy time" when doubts would be removed and reason and revelation would "reciprocally bear witness to, and embrace each other." In the meantime, however, he felt constrained to acknowledge some nearly insurmountable difficulties: the "Vulgar Scheme of the Mosaick Creation," as well as being full of temporal distortions, "represents all things from first to last so disorderly, confusedly, and unphilosophically, that 'tis intirely disagreeable to the Wisdom and Perfection of God."[3]

A new theory was thus needed to substitute for the "vulgar scheme" of the Mosaic account. Whiston points out that Genesis was usually taken to refer to the universe, whereas, he suggests, "the *Mosaick* Creation extends no farther than this Earth and its Appendages, because the Deluge and Conflagration, whose Boundaries are the same with that of the *Mosaick* Creation extend no farther." Furthermore, it rejected the anthropomorphic vision of the universe. How indeed could one think that an innumerable series of suns and planets could have been created only for the convenience of our little earth? And what are we to say of the fact that for six thousand years no one had even dreamed of the existence of the satellites of Jupiter and Saturn? Must we not ask

> if 'tis an instance of, or consistent with the Divine Wisdom, to make thousands of glorious Bodies for the sole use of a few fallen and rebellious Creatures, which were to live for a little while upon one of the most inconsiderable of them! . . . That so vast and noble a system, consisting of so many, so remote, so different, and so glorious Bodies, should be made only for the use of Man, is so wild a Fancy, that it deserves any other treatment sooner than a serious confutation.[4]

Whiston was a disciple and admirer of Newton. He intended, as we have seen, to keep his distances from both a literal and an allegorical and euhemeristic interpretation of the Bible. A letter that Newton wrote to Burnet on 13 January 1681 helps to clarify Whiston's position, both in relation to Newton and in relation to Burnet. Burnet had raised difficulties and had attacked theologians who saw a physical reality in the hypothesis of the six days. In his response to Burnet, Newton insists on a fundamental thesis: Moses spoke of "realities." His description, deliberately adapted to the capacities of understanding of the people, was not ideal or poetical, but was a *true* description:

> As to Moses I do not think his description of ye creation either philosophical or feigned, but that he described realities in a language artificially adapted to ye sense of ye vulgar. Consider therefore whether any one who had understood the process of ye creation & designed to accommodate to ye vulgar not an Ideal or poetical but a true description of ye creation as succinctly and theologically as Moses has done, without omitting any thing material wch ye vulgar have a notion of or describing any being further than ye vulgar have a notion of it, could mend that description wch Moses has given us.[5]

The Mosaic account, as Newton saw it, is therefore a description of true things carried out in language adapted *by art* to the limited capabilities of his hearers: "For Moses accommodating his words to ye gross conceptions of ye vulgar, describes things much after ye manner as one of ye vulgar would have been inclined to do had he lived & seen ye whole series of wt Moses describes." Thus Moses knew the real structure of the physical world and the truths of astronomy; he knew the true story of the creation of the world. But he limited himself, by a series of stylistic expedients, to presenting the truth in a comprehensible manner. He simplified; he did not falsify. He was a commentator and a popularizer, not an inventor of fables or allegories.

Whiston, in all probability, knew of Newton's letter to Burnet. Certainly he was aware of Newton's ideas. He declares, as we have seen, that his line of thinking is not unlike Newton's. In reality, however, his position was rather different from his master's and his attitude toward Burnet was quite ambiguous. The author of the *Telluris theoria,* Whiston declares, went so far as to suppose that the holy writers "only secur'd the fundamental and general verities involving the rest under, and explaining the whole by a way of speaking, which was mystical and mythological," which was "fitted more to the needs of men than to the reality of things." If Burnet considered this censure too severe, Whiston continues, "the learned Author . . . will, I hope, see reason to excuse, and not be displeas'd with me, when I have own'd, as I must ingenuously do, that in accusing him, I condemn myself, for I myself, in great measure, have thought the same things."

The difficulties involved in the biblical text were so great, Whiston continues, it appeared full of "such absurd incongruities," the history it narrates seemed so impossible—and Burnet's account, on the contrary, seemed so ingenious and probable, so in harmony with the most ancient traditions—that "'twas not easy for me to deny all assent to that very conclusion, which yet on further enquiries and discoveries, I think not unworthy of the foregoing censure."[6]

Whiston's retraction was little more than a formality, and he was not in the least disposed to follow in the direction indicated by Newton. Whiston persisted in his view of the "morality" of Moses' teachings to his people, and he repeated the libertine and Spinozist image of a lawgiver Moses who was using the creation story in the greater aim of ensuring the obedience of his subjects:

> The designs of *Moses* . . . were not the gratifying the Curiosity, or satisfying the Philosophical Enquiries of a few elevated Minds, but of a more general and useful Nature; namely, to inform the *Jews* and the rest of the world, that all the visible frame of Heaven and Earth was neither existent from all Eternity, nor the result of blind Chance, fatal Necessity, nor unaccountable Accidents, but the Workmanship of God Almighty. . . . To affect their minds, by this means, with the awfullest Veneration for the God of *Israel* . . . to shew them the unreasonableness of all sorts of Idolatry. . . . In short, the main design was to secure Obedience to those Laws he was about to deliver from God to them.[7]

Whiston emphasized that the Jews of Moses' time were to be seen as a simple, "primitive" people: "We do not find in our Learned and Inquisitive Age, such a ready Comprehension and Reception of Truths in Philosophy among Men," so we "may easily imagine how any natural Notions relating to the Constitution and Original of all the Bodies in the Universe must have been entertain'd among the rude and illiterate *Jews*."[8] As with many other "cosmologists" of his time, Whiston was well aware of the discussions of the history of peoples and the origins of civilization. He conjectures that the inhabitants of China "are the offspring of Noah himself after the Flood and not derived from any of his other posterity, Shem, Ham, or Japhet." There were, he says, good reasons to believe that the first monarch of China, Fohi, was Noah himself. He notes the characteristics of the Chinese language and writing system, and does not exclude the possibility that Egyptian civilization preceded the Flood.[9] He constructs a chronological table "containing the Hebrew, Phoenician, Egyptian and Chaldean antiquities compared together before and after the deluge."[10] Whiston's ideas on the last of these were decidedly heterodox theses. Whiston has too often been identified as one of the many who contributed to the "edifying version" of Newton's celestial mechanics. His position cannot be reduced to generic categories quite so easily, however. His "fantastic" hypothesis of the earth's encounter with a comet was to meet with large success. His reading of the Pentateuch and his interpretation of the figure of Moses had been strongly influenced by Burnet, by Spinoza's thought, and by Richard Simon.

11·Blind Chance and Admirable Design

AS Domenico d'Aulisio was to see clearly, the "blasphemies" of Burnet were drawn from the physics of Descartes. The theses outlined in the *Telluris theoria*, in any event, stood in total antithesis to the attitudes and doctrines most widespread in England and in European culture at the end of the seventeenth century. In his will, dated 22 July 1691, Robert Boyle stipulated a bequest of fifty pounds a year as remuneration for a "learned theologian" who would give "eight sermons a year for proving the *Christian Religion* against Infidels, *viz*, Atheists, Theists, Pagans, Jews, and Mohametans." Newtonian theories on the structure of the universe and on the structure of matter became in the Boyle Lectures, initiated by Richard Bentley in 1691–92, powerful arms against the Epicureans and the freethinkers, and against the "enthusiasts," or those who preached a popular millenarianism in connection with the revolution of 1688.

Newton's natural philosophy, then, could be *used as an ideology*: it could serve as proof of the existence of an omnipotent God and, at the same time, as a weapon to combat freethinkers, atheists, and "enthusiasts." The providential intervention of God, as Margaret C. Jacob has noted, appeared to be "essential to the maintenance of an ordered universe—and indeed to the maintenance of a post-revolution settlement as it was conceived by the Low Churchmen."[1] Without God, the universe would dissolve into chaos. In a sermon entitled "A confutation of Atheism from the Origin and Frame of the World" given on 7 November 1692, Richard Bentley attempted to prove that the system of the heavens and the earth had not existed from eternity, that matter was not eternal, and that the world could not have been caused by the fortuitous or mechanical motion of particles in space or in chaos. In his sermon, "The Atheistical Hypothesis of the World's Production," he asserts that the "Origin and Frame of the production of the World . . . must be the selfsame thing," since this hypothesis states that to arrive at "the present frame of things" one needs only the "mechanical affections" of matter, with no intervention whatsoever on God's part. "Mutual Gravitation or Attraction," he goes on to say, is "an operation or vertue or influence of distant Bodies upon each other through a empty Interval." There are no "Effluvia" or "Exhalations" to convey or transmit gravity, and this power "cannot be innate or essential to Matter: And if it be not essential, it is consequently most manifest . . . that it could never *supervene* to it unless impress'd and infused into it by an immaterial and divine power." Without universal gravitation, the entire universe would be neither stable nor permanent; it would be like "a confused chaos, without order or beauty." Gravitation is not itself mechanical; it is "the finger of God, and the execution of the divine law."[2] John Ray was also to praise the perfections of the world as an expression of the infinite wisdom and unattainable perfections of God and to insist on the "admirable Contrivance of all," on the "adapting of all the Parts of Animals to their several Uses," and on the order and harmony of the universe in which God has placed man.

It would be all too easy to dwell on the image of the "harmonious universe" and to multiply examples of its use, pointing out its importance also in the history of physics and biology. It would be just as easy to document the many attempts to interpret Genesis literally and realistically that were characteristic of Newtonian and Neoplatonist circles. The texts that relate to the Newtonian ideology and to the universe/machine of the Newtonians and the more specifically cosmological texts have been widely studied in recent years, both in broadly synthetic presentations (Collier, Nicolson, Westfall, Casini, Jacob) and in studies that contrast the materialistic Hermeticism of John Toland to the harmonious picture drawn by the Newtonians (Jacob, Giuntini). Many texts in the controversy on Burnet have been examined by Nicolson. Rather than returning to pages that have already been analyzed, it seems more appropriate to pause to consider in some detail a position that was in many ways exemplary—that of John Keill, first professor of Newtonian physics at Oxford, in his *Examination of Dr. Burnet's Theory of the Earth* (1698).

Several points should be stressed first, however. Unlike Boyle and Newton, Bentley or John Ray, Thomas Burnet had conceived of the world, not as the manifestation of a harmonious design, but as a "Great Ruin" and a renewed chaos. He had seen in the earth a sort of refusal or a *recrementum* on the part of a nature without order or plans. He had rejected the idea that divine revelation coincides with natural understanding. Finally, he had rejected the possibility that the Book of Nature coincides with the Bible, denying that the latter gives an accurate description of what the former contained. Newtonian circles had fully displayed their disagreements and their deep-rooted opposition on these points. At the beginning of his *Essay upon the Ancient and Modern Learning*, William Temple compared Burnet's work on the antediluvian age with Fontenelle's on the plurality of worlds. The comparison, as far as certain underlying themes were concerned—

the acceptance of the doctrine of an infinite universe and the rejection of a nature constructed in function of man—was completely justified. Temple's essay dates from 1690. That same year Erasmus Warren published his *Geology or a Discourse concerning the Earth before the Deluge*, in which he punctiliously opposes scriptural passages to every one of Burnet's affirmations. The problem, Warren declares, did not lie in establishing "meerly whether the World we live in, be the same *now*, as it was of *old* before the Flood: or whether there be not as much difference betwixt its *primaeval*, and its *present* State." The problem was rather to know whether "some sacred and *revealed Truths;* or gay, but groundless *Philosophic Phancies;* shall be preferred." In 1694 Jean Graverol explicitly accused Burnet of Socinianism and Spinozism in his *Moses vindicatus*. Two years later, in 1696, Archibald Lovell called Burnet a freethinker and declared himself worried about the pernicious effects that Burnet's book might have on the younger generation.[3]

When John Keill's book came out in 1698, the controversy was nearly a decade old and the "cosmological" works of Whiston and Woodward had already appeared. Keill vehemently attacks the "world-makers," who build imaginary worlds, and the "flood-makers," who loose imaginary deluges.[4] When Keill accuses Burnet of paying insufficient attention to experiment and observation and reproaches him for his excessive "fantasy," he does so not only in defense of a rigorous definition of a science based in patient observation and contrasted to over-audacious speculation. His attack also arose out of an orthodox interpretation of Newtonianism, and it was directed at adversaries who are easy to identify.

Keill launches his study with a full-scale attack on the "moderns." They claim to "perceive the *intimate essence* of all things, and to have discovered Nature in all her works, and can tell you the true cause of every effect, from the sole principles of *matter and motion*." Who, he asks, are these moderns "who can inform you exactly how God made the World" and who are as "extravagant and presumptuous" as the philosophers and poets of antiquity? Keill takes on all of his many adversaries at once. He attacks Spinoza for affirming that there is one substance in the universe and that all particular beings are modifications of that substance; he attacks Hobbes for holding that an incorporeal substance is a contradiction in terms and therefore denies its existence; he attacks More for speaking of a fourth dimention of an "essential spissitude" of souls; he attacks Malebranche for holding that we perceive not things, but only the idea of things and he declares Malebranche's philosophy "a solid piece of non-sense." The extraordinary presumption of these new philosophers, according to Keill, had been greatly encouraged by Descartes, "the first world-maker this Century produced," who declared that at the beginning God created only a certain quantity of matter and motion, and who claimed to derive the world "by the necessary laws of *Mechanisme*, without any extraordinary concurrence of the Divine Power" and to explain animal life itself by means of combinations of determined quantities of matter and motion.[5]

Keill did not agree with William Wotton's positive judgment of Cartesianism when Wotton said admiringly of Descartes: "By Marrying Geometry and Physics together, he put the World in hopes of Masculine ofspring." Such praise was not due Descartes, Keill insists, whose theory of vortices, as Newton demonstrated, was decidedly false. It was Galileo and Kepler who deserved praise since they, with the help of geometry, discovered truths of physics much more valuable than all of the volumes of Cartesian philosophy. The cosmology-builders and flood-makers, Keill continues, took a mechanistic philosophy as their base. They then denied the presence of any divine power in the world and interpreted all of the great changes in the history of the earth—which the Bible presents to us as miracles—as "the necessary consequences of natural causes, which they pretend to account for." An interpretation of these changes and of the Flood based on mechanical causes, Keill points out, ends up favoring the positions of the atheists and reinforcing the view of a world that exists from all eternity. When it swept away mankind, the Deluge also destroyed the records, traditions, and "memorials" of previous ages. Only one couple of "ignorant country people" saved themselves from the universal catastrophe, and their offspring repopulated the world and reinvented the arts and sciences.[6]

Keill's aim was exactly the opposite of that of many of his predecessors. He was not interested in making the biblical account agree with the truths of natural philosophy, nor in constructing hypotheses in natural philosophy that avoided contradiction with the Bible. *After Burnet* what was needed was the inverse operation: the defense of religion. If the Scriptures report a great number of miracles made by the hand of God,

> why ought we then to deny this universal destruction of the earth to be miraculous? . . . These flood-makers have given the Atheists an Argument to uphold their cause; which I think can only be truely answer'd by proving an universal Deluge from Mechanical causes altogether impossible. . . . This I intend to do by shewing that their Theories are neither consonant to the established laws of motion, nor to acknowledged principles of natural Philosophy.[7]

The method of the "world-makers," on the one hand, could be linked to the teaching of Descartes. On the other, it did not obtain results that could in any way be connected to what we learn in the Mosaic account. Furthermore, Keill states, he has shown that this method "is repugnant also to the laws of nature and gravitation." Only arguments drawn from reason and philosophy could be used against Burnet. Every reference to the Mosaic text was useless because Burnet "denyes the truth of his narration, which he imagines to be invented by that excellent Lawgiver to please and amuse the Jews: I have therefore in this treatise only made use of arguments which are drawn from philosophy."[8]

Keill thus downgrades the "Cartesian" cosmologies of the last decade of the century to works of science fiction and opposes them by citing the worth of Newtonian science, the certainty of its laws, and the rigor of its definitions. Behind his attack on the romantic cosmogonic hypotheses of the "world-makers" and his appeal to the positive qualities of Newtonian physics lie, in reality, three operating premises: (1) the histories of the earth and the cosmos cannot be completely explained on scientific bases, but have also been shaped by some miraculous events; (2) the truth of the biblical account is absolutely beyond discussion; (3) the presence of final causes in nature must be acknowledged, and the anthropomorphic point of view is perfectly legitimate, even in a context of natural philosophy.

Acceptance of this third point was the only way to avoid impious atheistic solutions to these problems and to return to God. Robert Boyle and John Ray, Keill insists, had already amply demonstrated the utility and necessity of final causes both for science and for religion. The earth, with all its mountains, is not a product of chance: it serves for the life of man and it reveals the wisdom of God. Not even the position of the earth's axis can be explained by mechanical changes of place: God, from the moment of creation, placed the earth in the position most advantageous to man. The cosmologies of the "world-makers" were nothing but unsupported hypotheses, and such hypotheses should be banished from science. Buffon's harsh judgment of Burnet's work ("an elegant romance, a book which may be read for amusement, but can not convey any instruction" can already be found in Keill's pages: Burnet's book, he says, is written in a pleasant style and with clever rhetoric, and it can be read as "a philosophical romance." Keill, on the other hand, was writing "only to those who might perhaps expect to find a true philosophy in it. They who read it as an Ingenious Romance will still be pleased with their entertainment." The accusation of writing romances instead of science was, during those years, quite widespread: even the *Chronologia* of the great Newton had seemed to Whiston little more than a "sagacious romance."[9]

The polemic concerning Burnet was widespread, as we have seen. Aside from Keill, it was to involve John Woodward, William Whiston, John Ray, Jonathan Swift, and Alexander Pope. We can even find echoes of the reputation for impiety that followed the author of the *Telluris theoria sacra* for more than a century in a popular ballad, which credits Burnet with daring to assert

> That all books of Moses
> Were nothing but supposes. . . .
> That as for Father Adam
> And Mistress Eve, his Madam
> And what the Devil spoke, Sir
> Twas nothing but a joke, Sir
> And well invented flam.[10]

A very complex cultural world was in fact hiding behind Burnet's "ingenious romance," a world steeped in Cartesian mechanics, in Platonism, and in Democritan philosophy, open to the theories of deistic and libertine currents of thought and prone to absorbing Providence into nature. "Blinder mechanism and blinder chance" was how Richard Bentley defined Burnet's system in a bitter attack on the folly and unreasonableness of atheism (1692). In Naples during those same years Domenico d'Aulisio, in his *Scuole Sacre*, compared Burnet's "blasphemies" with Spinoza's, declaring the works of both men to be founded in the atheistic materialism of Cartesian physics:

> It is well known that Benedict Spinoza and Thomas Burnet, a worthy doctor, drew their blasphemies from no other source than the physics of Descartes, in which space, which has always been and will be always, is none other than matter—which however is eternal—and where the laws of motion have mechanically produced the world out of matter, and what is in the world has thus arisen without architect.[11]

To understand the reasons for such a bitter attack and the significance of the thesis (as in Keill) of a "miraculist" explanation of the Flood, we should keep in mind that only a few years earlier Burnet's work had been used precisely to subject the concept of miracle to an all-out attack. In his work *Miracles No Violation of the Laws of Nature* (1683), Charles Blount brought together a series of texts taken from Burnet's work, without warning his readers that they were citations and without revealing the composite and anthological character of his work. The sense of these texts was that the order of nature reflects the divine order and the Holy Scripture considers miraculous facts that instead arise from the order of nature and that can be completely explained by natural means. Blount passes with no transition and almost without explanation from these statements of Burnet's to passages from Hobbes (*Leviathan* 1:37) and from Spinoza (*Tractatus* 6) which declare that only persons of limited culture and little intelligence take as miracles those facts that learned men can instead explain without even holding them to be worthy of particular admiration.[12]

Eternal matter and its motion, in Burnet's view, engender the world. The machine of the world needs neither watchmaker nor architect: if matter is eternal, all the forms that it assumes are produced by time. Matter *plus* motion plus time give rise to the reality of the world. Blinder mechanism and blinder chance oppose the vision of the world the Newtonians proposed in their image of nature as the sanctuary of God. For William Derham, for example, the way in which the continents and the waters, the rock strata and their alterations are distributed over the earth constitutes an evident proof of the presence of a "divine hand" and the existence of an infinitely wise Providence. Even the regular distribution of males and females—according to John Arbuthnot, who championed the Mosaic text to oppose Woodward's cosmology—is a sign that Providence operates in the world.[13]

All that exists is beneficial for the earth and for man, and "the whole chorus of nature raises one hymn to the praises of its Creator." This last expression, which he puts in the mouth of a deist, is David Hume's. Hume's *Dialogues on Natural Religion* (1749, published 1779) offer a crystal clear picture of the fundamental choice between the Newtonians' emphasis on teleological finality—which insisted on the perfection of the eye and the hand, on the correspondence between the sexes, on the adaptation of animals—and the image of a world which may have been produced and destroyed an infinite number of times and in which "the stronger prey upon the weaker, and keep them in perpetual terror and anxiety." When Cleanthes, the "deist," argues against the eternity of the world and offers "convincing proofs" of "the youth, or rather infancy, of the world," rejecting the idea of "a total convulsion of the elements," the skeptical Philo responds that there are no valid arguments against the thesis that such upheavals took place in the earth's past and that, to the contrary, over all the earth one could find "strong and almost incontestable proofs . . . that every part of this globe has continued for many ages entirely covered with water."[14]

The "incontestable proofs" taken from the "shells" had much reinforced the picture of a world in mutation, which appeared to many as radically different from the image that insisted on finalism and on the perfection of nature. In the later seventeenth century the debate about these shells had gradually broadened until it was transformed into a cosmology and a cosmogony. In this way the debate was both resolved and absorbed. To some extent, it was swept along. Total systems had been constructed—often on extremely fragile foundations—on the basis of the ambiguous documents of the earth's past and daring hypotheses and global constructions had had free rein. The relation between theory and reality had been lost, or had threatened to become lost. Hume also said that, unlike Galileo when he began with the moon and "proved its similarity in every particular to the earth" and unlike the "cautious proceedings of the astronomers," analogies had been too easily drawn in cosmogony. How could one claim to prove "any such similarity between the fabric of a house, and the generation of a universe?" "Have you ever seen nature in any situation as resembles the first arrangement of the elements?" Have worlds ever "been formed under your eye?"[15]

The skeptical Voltaire was to ask very similar questions when he encountered the unbridled "analogies" of Boulanger.

12·Crustaceans and Volcanos

ANTONIO Vallisnieri (1715 and 1721)—like Anton Lazzaro Moro somewhat later (1740)—was far from giving free rein to analogies in his writings and very circumspect when it came to cosmogonic hypotheses. The works of both men appear to be firmly anchored in the specific periods of geology, in the discussions of the strata of the earth's surface, in the "marine bodies," in the formation of mountains, and in the "natural" characteristics of the Flood.

Vallisnieri was well aware of how difficult it is to shake old notions out of people's heads:

> Ha gran forza una vecchia opinione
> E grand' arti vi vuole e gran fatica
> A levarla dal capo alle persone.[1]

The "old opinion" he speaks of was of course that the "marine bodies" found in the hills came from "the same virtue that generated them in the sea." How could one still believe, Vallisnieri exclaims, in a virtue that "without the egg, whips up organic bodies and, like a magician's little figures, turns out real fish, real shells, real oysters, real worms . . . to be brief, real, very real, archi-real parts of animals?" In addition to the works of his contemporaries, Vallisnieri knew and made use of the writings of Kircher, of Burnet, and of Woodward. In general he keeps his distance from Burnet and greets some of Woodward's ideas with caution. On the question of the Flood, he declares, it is easy to indulge in ingenious rhetoric and "amplify with hyperbolic exaggerations such a striking and so extravagant an effect," and it is easy to confuse the laws of nature and turn them upside down. We should look to see whether such descriptions, "fit for frightening the weaker sex and the ignorant common people, correspond to the solid reflections, the powerful reasons, and the carefully pondered and just observations of the experimental philosophers."[2]

When it came to Burnet's "hyperbolical exaggerations," the wiser course was to imitate Woodward, fall back on "the sacrosanct, occult mysteries of the omnipotent arm of God," and recognize that natural agents were not sufficient to explain the Flood. Nevertheless, Vallisnieri adds, it should be possible

> to search for some other more natural and simpler cause, one more in conformity with the inviolable laws of the Great Mother who, even in her oddities and her errors, knows her own limits and neither knows how to nor is able to escape out of those limits, except on the explicit command of the Most High.[3]

How, Vallisnieri asks, can one accept Burnet's hypothesis of a disintegration and liquefaction of the earth? If it is true that the "immense, jagged crests that seem to abut with eternity and with the sky" were ground to dust, why did the same thing not happen to the trees, to the fragile shells of the crustaceans, and to the many "very tender plants now found, with their leaves intact, between one stone and another"? The rock strata in the hills reveal the action of many, many inundations, not just of one, since the materials that make them up are not deposited in order of weight.

> By the laws of gravity it seems indisputable that the very first materials to sink to the bottom should have been the metallic ones, then the marblelike materials, the rocklike, the soillike, continuing thus, by degrees, so that the lightest were the last, until the waters remained limpid. Wherefore the surface of the earth finally lay veiled, covered, and as if plastered over with an ultra-fine silt, as we often see happen after an inundation. . . . This being so, the hills and the plains would

> have been formed of few but very regular layers. . . . Nevertheless, anyone who is not blind can see that the structure of the mountains, where they are worn to the rock or laid bare, is very different from that of the plains, where deep wells are dug. He will see a layer of round or blunted stones . . . and above this another layer of smaller stones, and in a third a layer of sand, and finally earth; and he can see this order repeated again in other places, and again above these, in other places up to the top of the mountain, which shows clearly that the mountain was made in several different times and by several inundations, not just by one.[4]

Stratification appears to Vallisnieri to be "the elegant, organic, and natural structure of the machine of the Earth." The "holy laws of nature" are immutable and uniform because what derives from an immutable wisdom is in its essence immutable. Do we not also find layer on layer in "plants many years old, the cortex or shells of crustaceans, and in the roots and bulbs of many flowers?" Vallisnieri's description of this "stratified nature" echoes the idea of the admirable order and perfection of the physical world. The laying down and the positioning of the strata "could not have been otherwise" if we think of the purpose and the placement of the mines, springs, rivers, plants, and animals. Everything was destined "to be in such a place and not in another" to avoid confusion and observe "an order and a rule of distributive justice for all."

> God therefore formed the world in layers, and he also set laws for the elements, for motion, and for nature, so that his fundamental and essential layers would always be conserved, and so that the alterations, which we see occurring through subterranean fires or other causes, would all be directed to his purposes, even if these changes were unimportant details, considered in relation to that great whole, and were of little importance to the massy bulk, so to speak, of the most admirable structure ordained by Him. Indeed, God chose to establish an order in such a way that if some things were ruined, others would continually be regenerated, but that the first bases would for ever be solid, and the primal, immeasurable framework of the mountains would never be unhinged or removed.[5]

The world, for Vallisnieri, is an organized body and contains what Hutton was later to call "reproductive operations" or reparatory mechanisms. The stratification of the world is the work of God and reflects the wisdom of the creator of the universe. The world's structure is solid and immutable: all changes take place only within this immutability, and in no way touch its essence. This is one reason why Vallisnieri saw Burnet as "a visionary, although famous." Vallisnieri knew well that when we approach problems connected to "things that happened so many centuries ago, we all play at guessing games." He would have liked to remain faithful to his image of the experimental scientist who, if he must "guess," tries to do so "modestly," being satisfied with "what is simplest and most probable," trying not to go too far astray and working to produce "histories, and not romances." He warns against "fabricating" the world's past to suit ourselves, against claiming that marvels took place that never did so, and inventing inexistent miracles.[6]

It was difficult, however, to avoid romantic hypotheses if one tried to explain—as Vallisnieri did in the second of his *Lettere critiche* (Critical Letters)—why antediluvian men lived for so very long. After the Deluge, he suggests, the air was for some years less healthy, full of "corrupt, very heterogeneous and tumultuous particles." The earth long remained "drenched, filthy, muddy, and in many places, because of the imponded and stagnant waters, even fetid." This caused changes in men's blood and a weakening of both the "fecundating spirit" of the males and the "little machines of the children-to-come enclosed in the females' ovaries."[7]

Vallisnieri's works undoubtedly inherit much from Galileo. Even without explicit references, a passage like the following should document his influence clearly:

> We must in all things reflect that the senses often are fooled if they operate without judgment, and judgment is just as fooled if it operates without the senses. We must set them in harmony, and both of them with nature, which does not function with such clever machinery as some think, but operates with simple, clear laws; laws at least as different from our own as nature's knowledge is far from our own.[8]

Even more than in statements of methodology, Galileo's influence is evident in Vallisnieri's work on the origin of springs *(L'origine delle fontane)*, in which he opposes the theses of the "moderns" to the doctrine of the circulation of water within the earth. He also refers to Descartes and to the necessity of questioning all things and cites Bacon on the need to descend from the high tower of generality to look at particular details in nature with "scrupulous exactitude," abandoning "vain speculations." Vallisnieri draws on different traditions, and he uses the "moderns" in various ways in his quarrel with traditions. Beyond the cautiousness of the experimental philosopher, he holds firm, as we have seen, to a faith in an "admirable chain" maintaining all things and tying together all the parts of the "great machine" of the world. The image of the universe as a "watch worked with divine art" leads directly to an admiration for the Sovereign Wisdom that formed it, and Vallisnieri cites Petrarch to echo that all the beautiful things that adorn the world are good when they leave the hands of God:

> Tutte le cose di che 'l mondo è adorno
> Uscir buone di man del Mastro eterno.[9]

Vallisnieri contrasts the "philosophers of nature" to "philosophers on paper" and declares that a philosopher must first look and then reflect. He believed in the simplicity and immutability of nature, and speaks of a "chain of nature." Liebniz and Malebranche are present in his pages, as is Bacon, not only in rules of methodology, but also in the idea of a great collection or natural encyclopedia that would take the form of a true "history" as the Lord Chancellor interpreted the term. Many themes interweave in Vallisnieri's work, not always coherently welded together into a unified vision.[10] Beyond his belief in particularized experience, beyond his optimistic picture of a natural theology that guarantees solidity to the structures of the world, guaranteed by the Creator's chosen purposes, we see reemerging in Vallisnieri the theme of the Deluge as a decisive catastrophe—a sort of second fall that carries within it the origins of the sufferings, the imperfections, and the ills that men endure.

> The terror from still seeing themselves in the midst of so much water, the imprisonment, so to speak, for a year, the terrible [fact of] the life of a whole world submerged, the anguish of [seeing] kin killed, that natural repulsion and suffering of the soul at [the sight of] a spectacle so deadly, so dismal, so horrifying, so disgusting [as] innumerable corpses of both men and cattle; the horror, even after the Deluge, of seeing that they were alone and as if abandoned in a desolate desert—[all this] without any doubt contributed much to destroying the perfect harmony of [men's] blood and to disturbing the regular movements of their spirits and their humors. Thus the very principles of generation were impaired, and thus those defects and those unhealthy dispositions were passed on to their unborn children, and penetrated them so very deeply that this hereditary misfortune lasts even to now and will last to the end of all time, since we all come from that infected lineage, though it is now enormously multiplied and divided throughout the world.[11]

The work of the Venetian cleric Anton Lazzaro Moro, unlike that of Antonio Vallisnieri, followed the form and procedures of a treatise. The two volumes of his *De' crostacei e degli altri corpi marini che si truovano su' monti* ("On crustaceans and the other marine bodies that are found on the mountains") were published in Venice in 1740. A German translation was published in 1751 and reprinted in 1775.[12] The first volume (approximately half the work) is devoted to a systematic and detailed refutation of Burnet's and Woodward's systems. The second volume, in addition to reflections on the scientific method and on the characteristics of nature, contains the demonstration and defense of a "new proposition":

> The marine animals and plants the corpses or relics of which are found today either on or within certain hills, born, nourished, and grown in marine waters before those mountains were raised up over the surface of the sea, were thrust where they now exist, petrified, for the most part, at the time those mountains, coming out of the bosom of the water-covered land, were raised to the height at which we now see them.[13]

Moro accepted the invitation offered in Vallisnieri's *Critical Letters* and he refers to Vallisnieri as to a master: the question of the fish, crustaceans, "and other petrified sea products" would be "sent back into the fray as if no one had yet written of such matters." The problem was necessarily connected to that of the structure of the earth and of the origin of that structure. It was not so much a matter of investigating which particular species of animals and marine plants were present in hillside rock formations; it was important above all to explain why they were there. The problem of the fossils could not be isolated; it was intimately connected with the investigation of "what was the origin of the islands, the mountains, and all the layers close to the surface of the Earth."[14]

Moro uses a series of arguments to refute the explanation based on the Flood: the waters of the Deluge could not have been the same as the waters that initially covered the earth; the rain that fell for forty days and nights would not have been sufficient to cover the highest mountains, etc. He concludes that the Deluge "is not the fruit of natural causes"; it was not explicable by "natural principles." It must be understood as "one of the most solemn prodigies of Omnipotence," which can produce events "outside of the ordinary laws of nature." The Flood was comparable with "God's reversal of the Sun at Joshua's request, with the darkening of that same Sun at the death of Christ, with the multiplication of the loaves and the fishes worked by Christ through the hands of the Apostles."[15]

A prominent place among those who had tried to "explain" the Flood obviously belonged to Thomas Burnet, who "with prodigious extravagance" had invoked the laws of hydrostatics to explain the transition from original chaos to the organized world. But how, Moro wondered, could those laws "be put into practice" when "everything was muddled and confused and matter had not taken on any form; when nothing was more dense, nothing more rarefied and, consequently, no one thing was lighter than any other thing?" When Burnet placed his "crust" over the waters, he not only had upset and thrown into confusion "all of the laws of hydrostatics; he had also neglected the fact that stones and marbles are often found located *above* soils, sands, and clays, which are composed of lighter bodies.[16]

Moro discusses and analyses each one of Burnet's theses: the initial oval shape of the earth, the quantity of water contained in the "abyss," the calculation (which Moro found erroneous) of the amount of water in the Flood. Burnet's system was "fantastic and vacillating in its very principles." It was "in all and for all purposes null, fallacious, and groundless." It did not agree with the present state of the world, did not explain where the waters of the Flood came from, did not clarify why they subsided, did not say where the primigenial earth ended up "after its disintegration," and, finally, was in disagreement with the Genesis account.[17]

One of the principles of Burnet's system—that the world came out of an original chaos—seemed to Moro most particularly unacceptable. Moro's image of nature was in the Galileian and Newtonian tradition: the physical world, which is made of "an admirable and highly ordered disposition of parts"—parts not only regularly disposed but "distributed with wisest prudence"—could not have originated in disorder. Every animal, every plant is "organized right down to its seed." Nature, which is "the art of God," reveals "an infinite wisdom that rules, disposes, and moves all." As Newton had affirmed and Vallisnieri had repeated, nature is *sibi semper consona:* it is "most simple in its working" and "works nothing in vain." Its laws are "uniform, general, and constant," and it is "exact and invariable in their execution." Nature is a positive force: it is more interested in constructing than in destroying and for this reason the "sinkings of the Earth" that snuff out life are less frequent than the liftings up that permit life to be generated and affirmed. From this image of a simple and always unified nature Moro drew a methodological precept that he applied consistently, down it its ultimate and most paradoxical consequences: the explanation of *one natural fact or event* "puts us in full security that the others like it were produced by the same cause."[18] The unexpected emergence of an island in the Cyclades was to become for Moro that one natural fact that by itself explained all other similar facts.

Woodward's system seemed to Moro less fantastic than Burnet's. Even if it claimed to be founded in fact, it too, however, mixed together the true and the false and it engendered propositions "that seem true and are false, seem real and are imaginary, seem certain and are uncertain, seem clear and are obscure, seem coherent and are repugnant." Woodward's system, Moro continues, was founded on three hypotheses: (1) the earth is stratified, not only on the surface, but in depth; (2) the earth was originally in a liquid state; (3) there exists within the earth a great abyss containing an enormous quantity of water which, through a network of cracks, is in communication with the waters of the oceans. Not one of these three hypotheses seemed to Moro well founded. In particular, it could not be true that unexpected flood crests in rivers were caused by a suddenly increased flow in springs when pressure built up from the waters gushing out of the great abyss. Woodward, Moro says, would certainly have found in the flood that occurred in Friuli in 1692 "a most felicitous proof of the existence of his Abyss and of the need to blame it for these events." For Moro, however, the causes of the flood waters of the Tagliamento, "which flowed over its banks (4 October 1692) with such an extraordinary flood crest that memory could furnish none to equal it," lay in the collapse of a part of Mount Uda (or Monte di Resto) that fell "right across the channel of the Tagliamento."[19]

Burnet and Woodward offered neither satisfactory explanations nor acceptable proofs of their assertions. The existence of the crustaceans and the shells in question (the "marine-mountainous bodies") could not be explained either through recourse to the Flood or by supposing that the sea had originally covered the mountain tops. If we should suppose the sea to have reached that height, in fact, we would lack "a sufficient cause to make it subside and a place to lodge the water." The theses of Aristotle, Eratosthenes, and Strabo among the ancients, of Cesalpino, Fracastoro, Scheuchzer, and Leibniz among the moderns, all encounter this insuperable difficulty. Vallisnieri, Moro notes, had already put an important question to Leibniz: "When, then, and where did this horrible chasm open up to absorb and engulf within the Earth half a globe's worth of water?" Moro adds an objection to Vallisnieri's: Why are the marine-mountainous bodies not found on *all* the hillsides?[20]

> May Sig. Leibniz's partisans explain with satisfactory arguments this diversity between one hill and another, indeed, between one part

> and another of the same hill; may they explain to where, when, and how all that water—which according to them occupied such a large part of the world—could possibily have retired. May they explain how the fish, as the water subsided, did not follow its course, but allowed themselves to be imprisoned among the marbles and in the dark depths of the metal mines. May they explain also how the crustaceans and the marine plants came to be fixed in those impenetrable depths, and how the marbles could so very exactly receive the impression of crustaceans into their very substance. Then let them brag that their sententious opinions are true.[21]

Moro had full confidence in the possibility of finding one unified explanation: the "lithified or calcinated" marine products had been transported or buried where they are now found "all in the same manner, by the same or a similar cause." Moro's "new proposition" was founded, as we have seen, on the observation of an island that emerged without warning out of the Aegean Sea in 1707.

> Monday 23 March 1707, at daybreak, there was observed in the bay of the island of Santorini, between the two Bracian islands, commonly called the Greater and Lesser Cameni, something that resembled a swimming rock, and that was at first believed to be a wrecked ship. Several sailors went in haste to identify the supposed ship, but they soon discovered to their immense surprise that it was a reef beginning to come up out of the depths of the sea. The next day several other persons . . . attempted to land on the same reef, which was moving and moreover growing visibly, and they carried away several comestible things, among others some oysters of an extraordinary size and of exquisite flavor. . . . Two days before the birth of this reef the entire island of Santorini was shaken by an earthquake soon after midday, which can only be attributed to the movement and detachment of that great mass of rock which the Author of nature had hidden from our eyes for so many centuries. . . . This reef, without other disturbances, continued to grow until June 4, at which time it already occupied a space nearly a half mile long and stood twenty-five feet over sea level.[22]

Moro was convinced that he had found in this sea-born island and in the life forms among its rocks the "Ariadne's thread" to resolve the problems of fossils, the origin of mountains, and the origin and structure of the earth. The uniformity of nature guaranteed this example as sufficient to found his thesis, so Moro sought a series of confirming examples. In 1538 Monte Nuovo had emerged out of the countryside near Pozzuolo. Pliny and Strabo testify to many islands (Delos among them) that emerged from the sea, as well as to many islands that had become peninsulas when new strips of land emerged.[23]

Moro was a plutonist: *all* islands have emerged out of the sea, irrespective of their size. What had happened at the Santorini reef had happened in Sicily,

England, Madagascar, Java, and Borneo. Continents were nothing but large islands. The force that led to the emergence of land bodies, to their "expulsion," was the subterranean fire present within the earth: "All islands and all mountains, even the continents, indeed, all lands destined by Providence for the use of man have been expelled from the bosom of the terrestrial globe into the open air by means of subterranean fires, and have been raised above the water that formerly covered them."[24]

Asia, Africa, Europe, and the Americas "are none other than great islands or peninsulas." This generalization, which lumps together batholites and tuff, the island of Santorini and Great Britain, seemed to Moro perfectly legitimate. In this work, however, Moro introduces the important distinction—generally attributed to the later works of Giovanni Arduino (1759) and Johann Gottlob Lehmann (1775)[25]—between *primary mountains* (composed of "great rock masses" of unstratified rock) and *secondary mountains* (composed of "layer on layer, of one or various sorts of materials"). On one important point he disagreed with Vallisnieri: the different strata or "crusts" found in the hills were not the deposits of a series of successive inundations. For Moro they were lava or material deposited, layer on layer, by the eruptions of the primitive mountains:

> It is not to be understood that those inundations were of water, but only of those materials (although Vallisnieri thought otherwise) of which each of those crusts is composed. Since the primitive mountains were born at the beginning of all things, when they were thrust out of the bosom of the earth by subterranean fires and raised above the surface of the water that formerly covered all, they vomited various sorts of materials from their open mouths and caverns and these materials, either as running rivers or as rain falling from above, ran down and spread, one after the other, over the slopes of those mountains, just as we sometimes see it happen in Vesuvius, Etna, and in other similar flaming mountains. Thus many table lands and sediment deposits came to be formed in those low lands, some composed of one sort of material, some of another, and some of various sorts of materials.[26]

Moro offers an objection to his own theory: If the mountains and the strata of the earth are all composed of "materials expelled from inside," must we admit the existence of a great empty space inside the globe? Epicureans and Gassendists "would not find themselves incommoded" by this idea, but the Peripatetics would see in it proof that the entire system was false. Moro preferred not to take a stand on the "philosophical vacuum": whether the void existed or not, *de facto*, islands were formed and *de facto*, Etna and Vesuvius emit materials. But he admitted that there were other answers on the theoretical level to satisfy those who denied the void, who were numerous and revered (and reverend—*reverendi*) for the habit they wore. Two models of what happened on the earth could be constructed:

> First, whatever space might have opened up by means of the blazing subterrnaean fire among those deep-lying materials . . . remains filled by the fiery igneous material until some opening or hole is made in the outer surface. When this happens and the fire has diminished, the space remains filled with air or with other volatile fluids. In this manner every danger of vacuum is avoided. Second, if Empedocles' opinions should prove true, who imagined that within the rather thick hull of the Earth all of the internal cavity is full of fire, it is possible to say that the vacuum is avoided when each mountain or island is born, because when the portion of the earth on which that mountain or that island in formation is raised and moved away from the center, another portion of earth in another part of the terraqueous globe sinks toward the center.[27]

Moro claimed to have invented a system that was "natural, true, certain, and incontestable." In it "all the effects of one sort are attributed to one and the same cause," apparently different effects "are recalled to one same principle," and nature appears always "constant and uniform in its operation." By the end of the work, his system, outlined "in a synthetic manner," takes on the lines of a veritable history of the formation of the earth.[28]

As Moro outlines it, the earth at the beginning had neither "prominences nor mountainous features." Its surface, totally covered with fresh water, was "completely flat and rounded . . . and consisted wholly of stone." The action of internal fires made the primary mountains emerge out of the rocky surface. Many of them split apart, erupting from their apertures "earth-like materials in great abundance: sand, clay, stones, some hard, some liquid, metals, sulphurs, salts, pitch, and every sort of mineral." Like gigantic rivers, these materials descended into the water below, while some also fell from the air. Salts and pitch remained in the water, which became salty, and formed the sea bottom. The deeper the water became, the more "the extent of its surface shrank because the land was either rising from the deeps or earth, vomited from the mountains, was descending in great quantity into the sea." The secondary mountains were then born, some of them "from under the sea." In the sea and out of the sea the first plants were born, and "when the vegetation was grown, the animals were born of the fertile earth," marine animals first, and then, "when the dry land became wooded and covered with green," the terrestrial animals, and finally man, who "was an inhabitant of that first and most ancient earthly surface." New mountains sprang out of the recesses of the earth, some from the dry zones, bearing no "marine productions," and others from the sea bottom, bearing "marine productions . . . either in their internal or external parts or in both together." The ancient fertile lands, before they were covered with new layers, remained in the open air. For this reason anyone who excavates can find "those trees, those grasses, and the bones of those animals that lived there and flourished in those obscure times." The soils that were deposited subsequently were often different from the soil that lay beneath. For this reason "the later soils produced plants and animals different from those that the ancient earth previously produced; and thus it happens that excavating to those deep strata we find certain races of trees and the bones of certain races of animals no example of which now lives and thrives on the inhabited upper surface."[29]

The new strata appeared "not all, and not everywhere at one time or in one season." This explains why, when wells or mines are dug, "not only one, but several levels can be found with clear signs of having been inhabited."

Like all who constructed systems, Anton Lazzaro Moro too substituted another account for the Genesis account. His appeals to God's will in the lighting of the subterranean fires and his citation of the Bible notwithstanding, his pages have a Lucretian tone and flavor. Moro wrote before Benoît de Maillet, before Buffon, before Guettard, before Boulanger. His work is dense, full of problems, hypotheses, and important observations. The problem of the comparison between periods of biblical chronology is absent from his pages. Nevertheless, Moro drew from his investigation of crustaceans and of those world-building volcanoes the image of a nature in mutation. Man operated from time to time and during the course of time on *one* of the layers of the earth, and what he has constructed and is now constructing will one day be only a "relic" buried by new layers.

> Future time will perhaps show that Nature is not tired of making this sort of changes: it will be the lot of the men of future centuries, living on new layers that will probably occur over the present surface of the earth, when they excavate some deep hole in the ground, to find fragments and remnants of the natural or made things that we use now on this sunlit suface; just as we now, digging in some deep place, find the relics of those things that the ancient inhabitants of those plains, now buried beneath us, once used.[30]

13·Systems, Romances, Theories

IN the years during which Hume was writing his *Dialogues on Natural Religion* (1749), the great geological and cosmogonic discussion that Buffon had set off in the 1680s had not completely died down. In 1748 the *Telliamed* of Benoît de Maillet was published; the year after saw the publication of Leibniz's *Protogaea* and of the first volume, almost completely devoted to the history and the theory of the earth, of Buffon's *Histoire naturelle générale et particulière*, and in 1752 Jean-Etienne Guettard published his notes entitled *Quelques montagnes de la France qui ont été des volcans*. Between 1750 and 1752 Nicolas Antoine Boulanger wrote the manuscript of his *Anecdotes physiques de l'histoire de la nature*, in which he considered, as the original title indicates, "the origins of the valleys, the mountains, and the other external and internal irregularities of the terrestrial globe, with physical observations on all the vicissitudes that seem to have befallen it."[1] In 1755 Boulanger wrote the article "Déluge" for the *Encyclopédie* and, in 1766, his *Antiquité dévoilée par ses usages* was published. These problems are also reflected in the articles "Fossiles" (1757) and "Terre" (1765) by d'Holbach and in various of Diderot's and Voltaire's pages and comments.

All of the ingredients of a typically "Burnetian" vision of the history of the earth are present in the *Telliamed* of Benoît de Maillet: Cartesian mechanics, the rejection of final causes and anthropomorphism, the thesis of the eternity of matter, the affirmation that the entire surface of the globe was for immemorial ages covered with water, the doctrine of the formation of the continents and the mountains after the water receded, and, finally, the "Lucretian" thesis of the origin of animal and human life from "seeds" and from marine creatures. For Maillet, the formless mass of the earth enclosed endless miracles, unknown facts, and a nearly incredible variety of phenomena. Only the idea that "God by a single word, in an instant, produced the world out of nothing" had led men to imagine foolishly that "this inhabited globe came from His hands exactly in the state in which we see it." Matter is eternal, and there is a perpetual motion in the very substance of the universe. Like a great living organism, the universe forms within it the causes of its own annihilation. Globes that once were luminous become opaque and vice versa: "This continual circle of revolutions is formed and renewed perpetually in the vast immensity of matter."

The idea that matter and motion may have had a beginning seemed to Maillet not only repellent to reason, but without foundation in the sacred texts. The phrase, "In the beginning God created the Heavens and the Earth" was only an incorrect translation of the Hebrew:

> The words used in that language signify only 'made or formed the Heavens and the Earth,' and that the correct translation of the Hebrew phrase, is 'When God made the Heaven and the Earth, Matter was without Form'; in a word, the Septuagint have interpreted the Hebrew word *Barach* by the Greek *epoiēse*, which simply means no more than 'made' or 'formed.' According to the remark of the learned Burnet, the word 'create' is a new term, invented only a few centuries ago to express a new idea which, therefore, has no equivalent expression in all the ancient languages such as Hebrew, Greek, or Latin; therefore, your Bible assumed the preexistence of matter, which God put in action from all eternity, and of which he formed the heavens and the earth.
>
> But, set aside the religious opinions and concentrate on the truth of things and on what reason discovers in them, I should tell you that, not being able to understand how matter and motion had a beginning, I must consequently believe them to be eternal.[2]

In addition to Thomas Burnet, de Maillet used Scilla, Steno, Huygens, and many other authors, but his fundamental views are Cartesian, particularly as drawn from Fontenelle. In his view, a continual transmigration of matter—dust particles and water particles—took place among the vortices that make up the universe. Pushed to the edge of a vortex by its light weight, an extinguished sun received deposits of matter and became enriched by the debris of other globes. In this way new strata formed on all such worlds—strata of sand, clay, and mud, and then of different sorts of rock, marble, and slate, and the various types of minerals. Finally, after a series of vicissitudes, the mountains emerged out of the sea and, last of all, life appeared: first marine, then animal, then human life. Perhaps the earth entered into the vortex of the sun when the moon was already present there. Perhaps worlds that were once inhabited are buried within the earth, one on top of the other: "entire cities, durable monuments, and all that we now observe on the surface of our globe." The epochs of the world are incalculable: We should be content not to "fix a beginning to that which perhaps never had one. Let us not measure the past duration of the world by the standard of our own years." The empirical proofs of this grandiose cosmology were drawn, once again, from the fossil shells and from the disposition of rock strata. "Who will be so obstinate as to refuse the truth arising so clearly from this discovery?"

In the lengthy and detailed attack on the *Telliamed* contained in two of the eighteen volumes that make up *La religion vengée ou réfutation des auteurs impies* (1757–62) the usual themes crop up: a plan, an architect, and a purpose to oppose this image of the world, plus an appeal to man as the aim of all creation:

> This person supposes that the origin of our globe lies in the necessary course of a preestablished mechanism, and that the same will happen concerning its end. . . . The author of the *Telliamed* pretends in vain to avoid the charge of impiety. His system openly opposes the history narrated in the book of Genesis. . . . It is vain to combat the Holy Scriptures with all one's force and then claim to conform to them. . . . The system of *Telliamed* is perhaps what is most opposed to the whole of Sacred History.[3]

During those same years Laurent François, preaching against the "Spinozists," insisted that the earth is miraculously divided into seas and continents. Life on the earth "awakes in us the idea of a greater Wisdom," and he asks rhetorically who dug the basins to contain the waters and who fixed their rate of evaporation. In the distribution of the parts of the earth, he answers, a design is present, and the end of the universe is man: "The Naturist will easily concede that man knows how to turn everything to his own profit, but he cannot concede, within his system, that all has been made for man. And this is a new argument for the falsity of his system."[4]

The idea that the sea could have once covered the earth seemed to François Nonotte as well (1772) a totally extravagant notion: Cyrano de Bergerac "in his voyage to the Moon tells us nothing more marvelous." In its substance, Nonotte adds, the world "is such today as it was two or three thousand years ago," and some little heap of gravel formed on some beach is certainly not enough to document the earth's changes. Once it was the fashion to "write romances of gallantry; today they are written about physics," and such was Buffon's *Theory of the Earth*. How, Nonotte asks, can anyone believe he is speaking seriously when he asserts that the planets were formed out of the incandescent body of the sun? How could we believe in his fantastic account of the earth's meeting with a comet? How could anyone accept the idea that the Atlantic Ocean could have been "formed" or that the Alps, the Caucasus, and the Pyrenees "could have been raised gradually out of the earth"?[5]

Baron d'Holbach knew very well that the defense of plain good sense, the accusation of writing romances, and the quarrel with audacious hypotheses often hid intentions much less noble than an appeal to experience and realism. There are hypotheses, he wrote in his *Système de la nature* (1770), "that will undoubtedly seem daring to those who have not meditated sufficiently on nature." In particular, there are those that regard the revolutions that took place on the earth "in times so distant that history has not been able to transmit details of it to us." Perhaps a meeting with a comet has more than once produced convulsions on our globe. Perhaps the human species has more than once been nearly totally ravaged and the men who escaped destruction were in no condition to preserve their wisdom for posterity, nor "the opinions, the systems and the arts anterior to the revolutions of the earth." Perhaps the survivers formed a new race of men, who "by dint of time, experience, and work" gradually revived the inventions of the primitive races.[6] Perhaps we ought to see in these periodic renewals

> the true source of the imperfection of our learning, of the vices of our political and religious institutions, over which terror has always presided—and of that inexperience and the puerile prejudices that keep man everywhere in a state of childhood, in a word, so little disposed to consult his reason and to listen to the truth. To judge by the weakness and the slowness of his progress in many respects, one could say that the human race is no more than coming out of its cradle, or that it was destined never to reach the age of reason and virility.[7]

It is possible that in 1757 d'Holbach was already in a position to draw on the work of Johann Gottlob Lehmann, *Versuch einer Geschichte der Flötzgebürgen* (Investigation into the History of Stratified Mountains; 1756), which was translated into French in 1759 under the title *Essai d'une histoire naturelle des couches de la Terre* in the third volume of the *Traités de physique, d'histoire naturelle, de minérologie et de métallurgie*. But even if the "scientific" theses that d'Holbach sketched in the geological articles of the *Encyclopédie* were based on widely known theories, they are certainly not to be separated—not even concerning their technical aspects—from the the general vision of the world to which he gave full expression in his *Système de la nature*. The article "Fossiles" distinguishes between *fossiles natifs*, formed in the ground, and nonland fossils, *étrangers à la terre*. He cites, among the ancients, Xenephon, Herodotus, Strabo, Eratosthenes, and Ovid. Among the moderns d'Holbach refers to Burnet and Woodward, noting that they had restated the thesis of the lands emerged from the sea which had been abandoned (except in the case of Avicenna) during the "centuries of ignorance." Experience proves, d'Holbach says, that the marine bodies found underground were not thrown there by chance, since we find grouped together there bodies that we still constantly find together. We find these bodies, furthermore, "not in the least disposed as if having fallen according to their specific weights": often heavier bodies are found in the upper strata, and heavy and lighter bodies are often found mixed together. Even without doubting the reality of the Deluge, d'Holbach continues, there is nothing to prevent us from conjecturing that the earth has undergone other revolutions. To explain the size and the quantity of fossils, we need to "think of a very long sojourn of sea waters, of many centuries, and not of a passing inundation of several months, as was that of the Flood, according to Genesis."

The earth's layering, d'Holbach wrote in the article "Terre (couches de la)," has always stimulated the imagination of the physicists. Many have theorized that substances were deposited at the first in an enormous fluid mass; others have turned to the Flood, and this hypothesis, "more ingenious than true, has a great many supporters." But also in this case, a brief period of flooding cannot be taken as the cause of such a complex phenomenon. We must think in terms of seas that for many centuries occupied now inhabited lands. In another article, "Terre (révolutions de la)," d'Holbach speaks of changes and revolutions and advances the hypothesis of a series of revolutions during the "night of ages," repeating Burnet's image of a world in ruins and placing the testimony of sacred history and pagan fables on the same plane:

> History has transmitted to us the memory of a great number of these revolutions; but there is an even greater number of them that remain in the night of ages and of which we have evidence only through the debris and the ravages, traces of which we see in nearly all parts of the globe we inhabit. Thus Moses transmitted to us in the Genesis the memory of the universal deluge; and profane history has spoken to us of the deluges of Deucalion and Ogygus, but no historical monuments have informed us of other signal revolutions, which considerably altered the surface of the *Earth*.

D'Holbach refers explicitly to the many revolutions anterior to the ones the Mosaic account narrates, citing the flattening of the earth at the poles, the displacement of the earth's axis, and earthquakes. His position resembles catastrophism, but he does not completely exclude the idea of the continuing action of various causes. It is in theorizing the ongoing process of radical change that he reiterates Burnet's image of the rubble and the ruins of the world. He rectifies this image immediately, however, appealing to an eternal nature that recreates what it destroys: "We see all of these causes, often joined together, acting perpetually on our globe, and it is therefore not surprising that the earth offers us at almost every step a vast collection of rubble and ruins. Nature is busy destroying on one side, in order to act to produce new bodies on the other."

D'Holbach's discussion of the earth, its changes, and its earthquakes begins to broaden out. It is not only a technical argument, but touches on the problems of time and the meaning of history, on the images of nature, on man, and on the relationship between man and nature. When we see the prodigious effects of earthquakes, d'Holbach says,

> it is natural to regard them as the principal cause of the continual changes that occur on our globe. History has recorded some revolutions . . . but most of them and the more considerable of them are buried in the night of earliest antiquity. Thus we can speak of them only in conjectures, which seem, however, well founded. Thus there is every reason to presume that Great Britain was torn from the European continent and Sicily similarly from the rest of Italy. . . . Is it surprising, after that, that the astonished traveler can no longer find the seas, the lakes, the rivers, and the famous cities described by ancient geographers, of which today no trace remains? Why should the fury of the elements have respected the ever-fragile works of the hand of men while it shakes and destroys the solid base that serves as their foundation?[8]

Man is a work of nature, exists in nature, and is subject to natural laws, from which he cannot escape: "Man is a production made within time, specific to the globe we inhabit" and the result of the laws that govern it. It is true that existence is essential to the universe, but the combinations and the forms that matter can assume are not essential to it. Hence even if the material elements that make up the world have always existed, "the earth has by no means always had its current form and properties." We can imagine that the earth was once a mass that detached itself from some other celestial body, or an extinct comet that once occupied other regions of space where other forms of life were present. In any event, whichever of these hypotheses we adopt, we must conclude that "plants, animals, and men can be regarded as productions specifically inherent to and proper to our globe, in the position and in the circumstances in which it is currently found; these productions would change if this globe, by some revolution, were to change its place."[9]

Animal and human life is tied to *this* earth and it is neither absurd nor inconsistent to imagine that man and the animals might disappear from existence. Why should these precise animals be indispensable to nature, d'Holbach argues. And why could nature not continue its eternal course without them? Does not everything change around us? And do we not ourselves change, and the universe with us?

Suns and planets are extinguished, other suns and planets come into existence and man,

> an infinitely small portion of a globe that itself is but an imperceptible speck in all immensity, thinks that it is for him that the universe is made, imagines that he must be nature's confidant, flatters himself that he is eternal, and calls himself King of the universe.
> O man! Will you never understand that you are an ephemeral thing?[10]

Some say that nature must be "herself intelligent, or must be governed by an intelligent cause," d'Holbach continues. But intelligence is a faculty "characteristic of organized beings, that is, constituted and combined in a determinate manner," whose behaviors come to be so designated. Nature cannot be called "intelligent" in the same sense as some of the beings that nature produces are called intelligent. To have intelligence, to have designs and projects, you have to have ideas, and to have ideas, you need organs and senses.[11] Nature has been called industrious, but understanding how it has produced a stone or a metal is just as difficult as understanding how it could have produced an organized brain like Newton's. We call a man industrious when he can do things that we are unable to do. But nature "can do everything, and if a thing exists, that is the proof that nature could make it." In reality, when we speak of the industriousness of nature, we draw a parallel between nature and ourselves and we use a strange sort of reasoning:

> The works of nature that astonish us the most do not belong to nature at all; but are due to a workman intelligent like ourselves, but whose intelligence [we judge] in proportion to the astonishment that his works produce in us, that is, [proportionally] to our weakness and to our own ignorance.[12]

Wisdom and folly are for d'Holbach qualities founded in our judgment. Is not the world full of a thousand things that seem mad to us? And are those creatures for whom we like to think the world was created not often foolish and unreasonable? And the marvelous animals, who are praised as the work of an immutable God, are they not perhaps changing continually, and do they not end up destroying one another?[13]

Attacks on the overly audacious hypotheses of the "naturalists" and on the cosmologies and the theories of the earth had not been launched only by reactionary writers like François and Nonotte and did not come only from the champions of tradition. It was in direct polemics with the Lucretian vision of Thomas Burnet and in defense of Newtonian views that Voltaire popularized in France the image—fairly close to Derham's or Bentley's—of a universe of order and harmony. The introduction to Voltaire's *Essai sur les moeurs* (1756, but begun twenty years earlier) opens with a paragraph devoted to the "changes in the globe." It is yet to be established, Voltaire argues, whether the earth on which we now live was once different from its present state. Voltaire admits this possibility: perhaps the world "has undergone as many changes as States have suffered revolutions." It seemed to him certain that the sea once covered immense streches of land that are today under cultivation or dotted with cities, and there is no shore that time has not advanced into or retired from the sea. If many stars of the Milky Way have vanished before our eyes, why should we be surprised if the earth has undergone continual changes?[14]

Voltaire did not have in mind great and decisive changes, and he doubted that the mountains emerged from the oceans. If Atlantis ever existed, he says, its disappearance "could be the greatest of these revolutions." The shells found on the mountains are probably the lodging place *(le logement)* of "the little crustaceans" that lived in fresh water lakes. The *glossopetre* and the ammonites were not shark's teeth or organic residue: "they seemed to me terrestrial fossils." Some, he exclaims, have dared to say that the mountains have been born of the sea and that the entire globe has been burned and became a glass ball: "these imaginings dishonor physics; such charlatanry is unworthy of history."

Voltaire was a sworn enemy of the new geology and the new cosmology. He had philosophical reasons for this—as a Newtonian and a deist, advocate of final causes and apologist for an admirable design in reality; as an adversary of materialism, advocate of science as common sense and opponent of general hypotheses; as an anti-Cartesian and foe to the pretension of "constructing worlds" with pen and word. As the leading Voltaire scholar has written, he was even motivated by the conviction "that all problems could be resolved by enlightened common sense,"[15] which is a singularly weak epistemological premise, even in the context of the science of his times.

The idea that one could construct a cosmology in order to explain the existence of a few shells seemed to Voltaire a symptom of authentic madness. In his *Eléments de la philosophie de Newton* (1737, rewritten in 1741) he wrote that if the only possible way to reason about objects is by analysis, if "one must never make hypotheses" nor ever "invent principles," how can one ever imagine the internal structure of something like the earth, of which we know only the exterior? Given that roses do not grow on oak trees and soles do not swim out of beehives, how could anyone seriously reject the idea that "all species are invariably the same" and that "they have been determined by the Master of the world"?[16]

In the *Dissertation sur les changements arrivés dans notre globe et sur les pétrifications qu'on prétend en être encore les témoignages* that he submitted in 1746 to the Academy of Science of Bologna, Voltaire states three theses that he would take up again in 1768 in three other works, *L'homme aux quarante écus*, *Les calimaçons (Dissertation du physicien de Saint-Fleur)*, and *Des singularités de la nature.*[17] Voltaire's thesis—which is already contained in the title of the *Dissertation*—is of a disconcerting simplicity: fossils are not evidence of or vestiges of organisms that had formerly lived on the earth: they are simply stones or objects. Whoever interprets them as "evidence" is in error, although the error is typically philosophical, since it is founded in the pleasures of the imagination, the pleasure of constructing systems, and the taste for hypotheses:

> There are errors that are unique to the people; others that are unique to philosophers. One error of the latter type perhaps is the idea of many physicists all over the earth who see evidence of a general upheaval. . . . The extraordinary, the vast, [and] great transmutations are objects that sometimes please the imagination of the wisest of men. Philosophers want to see great changes on the stage of the world, just as the people want them in [theatrical] spectacles.[18]

To put a stop to these excessive pleasures of the imagination, Voltaire works out an explanation based on the dump heap. A stone is found that bears the trace of a turbot, a petrified pike is found, shells of a species native to the waters of Syria are found in France and in Italy and some draw the conclusion that the sea once covered the mountains:

> It was more natural to suspect that those fish, carried by some voyager, went bad, were thrown away, and became petrified in the course of time. But this idea was too simple and too little systematic. . . . Is it a completely fantastic idea to recall the innumerable flocks of pilgrims who left on foot from St. James in Galicia and from all provinces to go to Rome by the Mont Cenis [pass] bearing shells on their hats? They came from Syria, from Egypt, from Greece, as from Poland and from Austria.[19]

There was an even simpler explanation for the ammonites: they were not shells of vanished or unknown species, but rolled-up snakes, or stones formed in that strange manner by nature. Reptiles "almost always form a spiral when they are not in movement; and it is not surprising that, when they become petrified, the stone takes the formless figure of a volute." We are almost back to the medieval doctrine of the *lusus naturae*, which Voltaire reinterprets, in 1768, in updated form:

> What! Nature, who forms stones [in the shape] of stars, volutes, pyramids, globes, and cubes, could it not have produced stones that very roughly resemble fish tongues! I have walked over a hundred ammonites of a hundred different sizes, and it always astonishes me that some refuse to allow that the Earth produces these stones, [the Earth] which produces grains and fruits surely more marvelous than volute-shaped stones.[20]

Voltaire's adversaries, whom he names explicitly, were Palissy, Descartes, Burnet, Woodward, Buffon, and de Maillet ("descendant of Thales," or *monsieur le créateur*). They are, in general, all those philosophers who saw in the Pyrenees, the Alps, and the Appennines "the ruins of a world that has changed form several times," who constructed "beautiful systems," who "believe in creating a universe with words," who "destroy and reconstruct the earth as they please just as Descartes gave it form," and who "put themselves nonchalantly in God's place and who have created the universe with the pen as God once created it with his word."[21]

Voltaire contrasts his view of the world as the revelation of a purpose to the "cracked and broken" world of Burnet's *Sacred Theory of the Earth:*

> An author who has made himself more famous than useful by his theory of the Earth, has claimed that the deluge turned all of our globe upside down, formed the rocks and the mountains out of the rubble, and threw everything into an irreparable confusion; he sees nothing but ruins in the universe. The author of another and no less famous theory [de Maillet] sees in it nought but orderly arrangement and assures us that without the deluge this harmony would not subsist. Both of them see mountains only as the effects of a universal inundation. . . . When I first examine the mountains, which Dr. Burnet and so many others regard as the ruins of an ancient world dispersed here and there without order, without design, like the wreckage of a city destroyed by cannon fire, I see them instead arranged in an infinite order from one end of the universe to the other.[22]

It made no sense to Voltaire even to pose the problem of the formation of the world because there was no hope of "discovering the means that God used to form the world." One could only investigate whether this globe was or must one day be "absolutely different from what it is." Here, "all you need is to have eyes," but Voltaire saw mountains with eyes that were very different from Burnet's. Mountains seemed to him "a piece essential to the machine of the world," and essential to animals and plants alike.[23] In the first place, he argues, one must distinguish between isolated mountains and the great mountain chains that are to the world "what bones are for bipeds and quadrupeds" and that "serve to reinforce the earth and to water it." The mountain chains that crown the two hemispheres, the more than six hundred rivers that descend from them, the affluents that flow in to swell those rivers after having fertilized the surrounding countryside, the thousands of springs that furnish water to plants and animals—all that "does not seem to be the effect of chance or of the declination of atoms, any more than do the retina that receives light rays, the crystalline lens that refracts them, the anvil, the hammer, the eardrum, the blood courses in our veins, the systole and diastole of the heart, that equalizer of the machine that makes life." Voltaire accompanies this reiteration of the organic image of the world as a great animal with a corresponding glorification of final causes—although he warns against overreliance on them—and with related invective aimed at Epicurus and Lucretius:

> All of the pieces of the machine of this world seem to be made for each other. Some philosophers affect to scoff at final causes, rejected by Epicurus and Lucretius. It is rather Epicurus and Lucretius, it

> seems to me, at whom we should scoff. . . . According to them, the mouth is not made for speaking and for eating, the stomach for digestion, the heart for receiving the blood of the veins and sending it to the arteries, feet for walking or ears for listening. Those people, however, permitted tailors to make suits to clothe them and masons houses to lodge them; and then they dared to deny to nature, to the great Being, to the universal Intelligence, what they granted their least significant workmen.[24]

Nature, as Newton had asserted, is *sibi consona* and likes uniformity and constancy; it is only our imagination that loves great transformations. The idea of a change in the earth is false. In the same manner and for the same reasons, the theses that proclaim the immersion of lands by the oceans and the mutation of living species are fallacious:

> There is, then, no system which can give the least support to this idea so widely prevalent, that our globe has changed its face, that the ocean has occupied the earth for a very long time, and that men have formerly lived where porpoises and whales are today. Nothing which vegetates or which is animated changes; all the species have remained invariably the same; it would be very strange indeed that the grain of the millet had eternally conserved its nature, and the entire globe varied its.[25]

"Systems" constructed by philosophers are worthless. There is only one system: that of the Great Being "who has given to each element, to each species, to each genus its form, its place, and its eternal functions." The widespread passion for genealogies notwithstanding, Voltaire announces, "there are few men who believe they are descended from a turbot or a cod." Men "have not been fish" and "all that is, probably is by immutable laws."[26]

In 1768, two years before the publication of d'Holbach's major work, in the same year in which Voltaire was writing his most polemical works attacking systems *bâtis sur des coquilles* and Robinet was adding a fifth volume to his treatise, *De la nature,* Maupertuis published the *Avant-propos* to his *Essai de cosmologie* (first published in 1750), which was to open the first volume of a new edition of his works. He drew a clear line of demarcation and placed all philosophers and scientists of his time on one side or the other:

> All of today's philosophers form two sects. Some would like to submit nature to a purely material order and exclude from it any intelligent principle—or at least would like to avoid all recourse to that principle in the explanation of phenomena—and to banish *final causes* entirely. The others, to the contrary, make continual use of such causes, discover the views of the Creator throughout nature, and penetrate his designs in the most minute phenomena. According to the first, the Universe could do without God; at least, the greatest marvels we observe in it are no proof of His necessity. According to the second, the very

> smallest parts of the Universe are as many proofs of God: his power, his wisdom, and his goodness are painted on butterflies' wings and spiders' webs.[27]

Between the end of the seventeenth century and the middle of the eighteenth, shells (as well as eels) had—as Voltaire said—"made new systems blossom." D'Holbach and Voltaire regarded both shells and "systems" with very different eyes. D'Holbach made revolutions and transformations of the earth one of the essential points of his vision of the world. The construction of "systems" was in Voltaire's eyes either an expression of philosophic charlatanry or, at best, a deceptive and useless effort.[28]

To contrast the "mechanists" and the "Newtonians," however, does not exhaust all the possible ways of taking a stand on the problems of fossils or the history of the earth. Although the distinction between the opponents and the proponents of finality touches a central point, the picture, in the 1760s, was more complicated—sufficiently complicated to leave a good deal of room for theses that came to clearly negative conclusions regarding the "modern" doctrines in geology and the organic origin of fossils.

Themes drawn from Leibniz and Diderot, reflections on geology, embryology, psychology, metaphysics, and eschatology are all present, in a truly singular mixture, in the five volumes of Jean Baptiste Robinet's *De la nature* (1761–68), which Arthur O. Lovejoy rightly called a "draft for a philosophy of *évolution créatrice*," and in the *Palingénésie philosophique* (1770) of Charles Bonnet. Robinet applies the principles of the fullness of nature, the perennial activity of the infinite Cause, and the great chain of being to their limit. All that is *possible* has the highest probability of being *real*: "I have formed so vast an idea of the work of the Creator that from the fact that a thing can exist, I infer readily enough that it does exist." Robinet's credulity, the oddness of so many of his convictions, and the illustrations of stones in the shape of human organs and of plants trying to assume human form that grace his pages come of a specific metaphysical premise. As if reinvigorated by what Leibniz had brought to them (or a particular interpretation of Leibniz), some of the themes most characteristic of the "Hermetic tradition" and the "magical" vision of the cosmos spring back to life: vitalism, universal sympathy, the analogy between beings, the ladder of nature, the presence of the All in even the most minute and apparently insignificant aspect of reality. But it should be noted that the themes and views found in Robinet do not represent just a brief parenthesis, nor have they an ephemeral life: many of his pages anticipate themes that would have singular success in the romantic philosophies of nature from Goethe to Schelling and beyond.[29]

In Robinet's universe, perfectibility does not concern man alone; it is a cosmic and universal law, and the progress of the universe is strictly connected (as it was in Leibniz) with the mathematical principle of the infinite divisibility of the continuum. Nature is always the same but is, at the same time, nonpermanence, process, and change. On many fundamental points Robinet's thought was close to Charles Bonnet's, but on one point there is a gulf between them: Bonnet accepted the idea that the ladder of being could be subdivided into reigns or classes. This negated the principle of continuity, and Robinet opposed every attempt at division or separation within the fundamental law of continuity: the quality of any thing was, in some degree or in some measure, possessed by all things.

Starting from these premises, Robinet (1) worked out an explanation of the formation of geological strata and of mountains based on the doctrine of the "generation" of minerals and metals; (2) denied the validity of the thesis that "fossils" were the petrifications or imprints of animals that had lived in the past, and identified the fossils instead as "forms" assumed by the mineral world as a "preparation" for man. One of his principal targets was—once again—Thomas Burnet, who had spoken of the earth as a confused mass of broken and shattered bodies, had seen discontinuity and disorder in a reality that Robinet saw instead as continuous and orderly, and who had denied the existence of ends and of a "design" in nature.[30]

For Robinet only an age-old prejudice had led people to consider minerals as "raw bodies" and had induced the learned to abandon the law of the uniformity of nature, which was instead the first element of the idea of the whole, without which philosophy could not exist. He opposed a thesis of "fossil germination" to the explanations that the various systems had elaborated: "Stones engender stones, metals produce metals, just as animals engender their like, as plants engender plants, by seeds, grains, or eggs, for these words are synonymous."[31]

Thus, just as with shells, Robinet explains, fish scales, bones, teeth, horns, nails, and even minerals "have an organic structure." Quarries and mountains are masses "produced by a very great number of stones that have vegetated one on top of the other and one beside the other in every direction." On the plains the rock beds are generally horizontal because atmospheric pressure, which is equal on all points of the level surface, compresses the embryonic minerals uniformly, both as they begin to fecundate and as they develop. Once the first stratum is formed, a second process of generation will produce a second stratum like the first. If, however, there is a superabundance of "germs" in the lower strata, they will produce conic forms, which will develop into pyramids. Then atmospheric pressure will no longer be sufficient to assure the uniform distribution of the "germs" and successive strata will be formed at an angle:

> A chain of mountains is formed from several rocks which were initially at a considerable distance from one another; little by little as they teem a great number of generations made them increase to the point where they meet and attach. . . . Each mountain contains rocks of the same race with a similar structure; and often those of two neighboring mountains are of different species. This is what Mr. Scheuchzer has observed, although he has ideas totally opposed concerning the origin of mountains.[32]

The fact that the fossils—and Robinet uses the term indifferently for both true fossils and minerals—had no visible reproductive organs constituted no problem: many insects had none either. And who could exclude the possibility that male and female mineral seeds existed? What was true of fossils was naturally true for the planets and for the celestial globes, which were "animated by a life of their own, with the power to produce other similar [bodies]." Although it was "obscure and hidden," the animal nature of fossils and planets was nonetheless evident. The answer to Buffon, who insisted that matter is insensible and inactive, lay primarily with the phenomena of magnetic attraction.[33]

It was not true in the least, Robinet insists, that fossils are petrified plants, animals, or animal parts. With what proofs can one claim, for example, that these objects are "the petrified tooth of an unknown animal!" Undeniably, the fossils "represent" such things, but why must we conclude from this that they really are petrified plants, animals, or animal parts? Most of the hypotheses that had been advanced to explain fossils, Robinet declares, refer to the revolutions that the earth has undergone and to *empreintes* of the outlines of plants and animals. These were completely absurd theses:

> We would laugh at the simplicity of a savage who knew nothing of the art of painting, and who would say, at the sight of a blackened portrait, "there was once a man there of bones and flesh like me, but time, with the help of some cause which I suppose without being able to identify it, has destroyed the substance of that man and only his delicate lineaments now remain." This is exactly the reasoning of the naturalists. Still, this portrait, even if we suppose it to be blackened by time, resembles a man more than many figured stones resemble the plants and the fish of which they are believed to be the impression.[34]

Nature, not art, Robinet argues, has engraved human faces on agates, and no one who examines those objects has ever imagined them to be heads petrified by time. Why should not "an ordinary stone [as] naturally bear the image of a fish as that of a man"? For Robinet, prejudices and presuppositions had been at work in the matter. What was really astonishing was that men had not yet come to recognize either the real organization of the earth or its animal nature.[35]

Actually, Robinet continues, all beings were conceived and formed on the basis of one *unique design* and within this continuous whole they represent *infinitely graduated variations* of this design. Man is the point of arrival of the cosmic process; he is the "perfect form" for the realization of which nature has run through the whole gamut of variations of intermediate forms. Fossils and minerals, like plants and animals, are documents of the *apprenticeship* that nature goes through to form man. This is why Robinet devotes pages and pages to the illustration of stones, shells, and fossils shaped like brains, skulls, feet, kidneys, ears, eyes, breasts, hearts, and male and female sex organs. Working in the medium of stone,

"Nature in working these stones, really models the different forms of the human body."[36]

Buffon had every reason to consider Burnet's, Whiston's, and Woodward's ideas singular. He judged their works to be "hypotheses made by chance" that rested on unstable foundations and that mixed physics with fable. When faced with uncertain and unknown fact, he declares, "it is easier to imagine a system than to construct a theory." Whiston's theses seemed to him both to contradict faith and to be "influenced by the enthusiasm of system." Whiston "strangely jumbled divinity with science." Burnet, Buffon declares, had in reality written "an elegant romance, a book which may be read for amusement, but cannot convey any instruction." Woodward, finally, had "wanted to build an immense edifice upon a foundation less firm that sand." All three of their "hypotheses" had one point in common: they held that at the time of the Deluge, the earth changed its form internally and externally. But these "speculators" were unaware of a fundamental truth: if the earth, before the Deluge, was inhabited by "the same plants and the same animals [that] continue still to exist," it must have been "nearly the same as now." These philosophers' greatest error lay in having blended "bad philosophy with the purity of divine Truth" and in having equated the problem of the earth's past to that of the Deluge.[37]

Buffon's thought moved quite evidently on a very different plane. He contrasted the "thousand physical romances" founded on superstitions and that "give scope to the imagination" to a method capable of taking the world as it is, observing all of its parts, and proceeding by induction from the present to the past. Nature is constant, he declares, and its laws are immutable. There is no need to appeal to miracles, unusual means, unexpected catastrophes, the appearance of new planets, or "convulsions of nature." Events do not necessarily occur at one precise moment; they can take place slowly and "naturally" during the course of time. If we keep in mind what happens slowly and day by day, we can find "plausible reasons." In one of the most inspired passages of his *Theory of the Earth*, Buffon outlines this new manner of considering the events of the earth's history with exceptional clarity:

> We ought not to be affected by causes which seldom happen, and whose effects are always sudden and violent; they do not occur in the common course of nature; but effects which are daily repeated, motions which succeed each other without interruption, and operations that are constant, ought alone to be the ground of our reasoning.[38]

Everyday phenomena and constant operation: in comparison to the system builders, the paradigms had changed radically. In Burnet's pages the history of the earth was a series of catastrophes and revolutions. Buffon does not exclude earthquakes, upheavals, and inundations, nor does he exclude the idea that at one time "the Earth's materials were less solid." But his "history" takes a tranquil course:

> The flux and the reflux of the ocean have produced all the mountains, valleys, and other inequalities on the surface of the

> earth; . . . currents of the sea have scooped out the valleys, elevated the hills, and bestowed on them their corresponding directions; . . . the same waters of the ocean, by transporting and depositing the earth, etc. have given rise to the parallel strata; . . . the waters from the heavens gradually destroy the effects of the sea by continually diminishing the height of the mountains, filling up the valleys, and choaking the mouths of the rivers; and, by reducing everything to its former level, they will in time, restore the earth to the sea, which, by its natural operations, will again create new continents, interspersed with mountains and valleys, every way similar to those which we now inhabit.[39]

It seems as if nothing is happening, but instead, in the long ages of nature and by the gradual and uniform action of many agents, all is destined to change. We can express only imperfect and approximate judgments on the revolutions and changes in nature:

> The defect of historic records deprives us of the knowledge of particular facts. We desiderate both time and experience. We never consider, that, though our existence here be extremely limited, Nature proceeds in her course. We are ambitious of condensing into our momentary duration both the past and the future, without reflecting that human life is only a point of time, a single fact in the history of the operations of God.[40]

Even Buffon, however, attentive investigator of everyday processes that he was, affected certain highly imaginative hypotheses—for which Voltaire was to reproach him bitterly. To explain the Flood and the shift of the earth's orbit William Whiston made use of a comet. Buffon set himself to solving the problem of the origins of the universe. Newton had removed this problem from the purview of science, seeking instead a scientific explanation for the "primitive and regular position of the planets," for the "most beautiful system of the sun, planets, and comets"—a system that, as Newton had said, was "not to be conceived" as due to "mere mechanical causes" but could only "proceed from the counsel and dominion of an intelligent and powerful Being." To resolve this problem Buffon restates Whiston's hypothesis and supposes that a comet imposed force on the material of the planets:

> This force of impulsion was certainly communicated to the planets by the hand of the Almighty when he gave motion to the universe; but we ought, as much as possible, to abstain in [natural philosophy] from having recourse to supernatural causes; and it appears that a probable reason may be given for this impulsive force, accordant with the laws of mechanics, and not be any means more astonishing than the changes and revolutions which may and must happen in the universe.[41]

A cause the effect of which would be in harmony with the laws of mechanics: this was the quest for a way to make the universe intelligible within time. (As we know, Kant would take a stand in clear opposition to Newton in just this search.) By choosing this path, as Marcella Renzoni has justly pointed out, by "reintroducing the cosmogonic problem through mechanical causes, Buffon managed to graft into Newton's perfect system of the universe Descartes's insistence on going back to the very origins of the universe."[42]

Whiston's fantastic hypothesis acquired a new consistency in Buffon's pages. Moreover, even if he criticized them bitterly (and deservedly so), Buffon devoted many pages to Burnet, Whiston, and Woodward and was right to consider them to some extent privileged interlocutors. In taking up Descartes's endeavor, they attempted to construct precisely those hypotheses on the origin of the universe (drawn from Newton's science) that Newton himself had refused to construct.

In the article "Mosaïque et chrétienne philosophie" of the *Encyclopédie*, Diderot groups Burnet with Dickinson and Comenius and sees in his work one of the most characteristic examples of that monstrous mixture of "theology" and "systems" that taught men to "reason when they should believe and believe when they should reason." Out of this mixture soon came "a host of bad Christians and bad philosophers." Religion, Diderot wrote later, in 1753, "spares us many a digression and much effort." If religion had not illuminated us on the origin of the world, "how many different hypotheses we would have been tempted to mistake for the secret of nature!"[43] Some of these hypotheses—even if they were what we would call today science fiction—had indeed been worked out, and with notable success. They had also provoked some notable digressions. Burnet, Whiston, and Woodward—in agreement on this question—had declared that the "origin of things" *must be debated* and that this discussion should in some way be connected to discussions in physics and biology on the "subsequent course of nature."

The Humes or the Voltaires could raise all the questions they wanted, but they were powerless to empty the discussions about cosmology of their meaning, nor could they stop their spread. Who can say that all "romances" among hypotheses have had and have no effect whatever on the growth of scientific knowledge?

14·Boulanger and Vico

WHEN he read Boulanger's cosmological generalizations, his pages on the fires, devastations, earthquakes, and darkness that had reigned on the earth, and his description of an age of terror for men, who "wandered through the ruins of the world at the mercy of all the torments that seemed to persecute them" and who considered themselves "objects of the hatred and the vengeance of an infuriated nature," Voltaire noted in the margins: "qui te l'a dit?" "as tu des memoires de tout cela?" and "peut on etablir tant de sottises sur de si legers fondements?"[1]

The cosmology builders, however, then as now, were not very sensitive to appeals to everyday experience. The enterprise that Voltaire had deemed *insensé* when he was discussing the "systems constructed on shells" was destined to have a long life in spite of the questions he raised and Hume had raised, and was to affect culture deeply and provide the background of many a discussion of natural history. This was to become evident not only in Maillet, in d'Holbach, and in Buffon, but also in the works of Boulanger in which he attempted to make a connection between the natural mutations of the earth and the myths, customs, and fables of the ancients:

> The infinite multitude of shells scattered over the whole of the Earth shows us the most common and best known natural monuments of the submersion of our continents. Later, we found the historical monuments of this in the earliest of the events narrated in the history of the Hebrews, and even in the traditions altered or embellished by fables, which identified the Earth as the daughter of the Ocean.[2]

Boulanger's great intellectual adventure started from shells: from marine *tuyeaux*, from corals, and from sponges. Jean Jacques Rousseau, who met Boulanger at the house of a friend, thought him a man whose "exalted imagination no longer saw anything in nature but shells, and he ended up really believing that the universe was nothing but shells."[3] Boulanger makes frequent mention of Voltaire's *Eléments de la philosophie de Newton* in his attempt to combat deistic Newtonianism. The examination of fossils and the history of the earth seemed to Boulanger to involve pushing the origin of the world back in time and interpreting Genesis not as a description of the creation, but only as a means used to teach men that the world had a divine origin. The pyramids of Egypt, he wrote in the article "Déluge" in the *Encyclopédie*,

> as monuments go back almost to the birth of the world; yet already decomposed shells are present in the formation of the stones that were used to build them. What an enormous succession of centuries this formation supposes! And how can we explain this phenomenon without admitting the eternity of the world? Can one explain the presence of the marine bodies in the stones of the pyramids by one cause, and the presence of the same bodies in our rocks by another cause?

Even Boulanger's major work shows traces of the "geological" origin of his great investigation of the history of civilizations, of myths, and of the ancient terrors that lie at the root of human behaviors:

> The physical scientist [*physicien*] has seen and called attention to the authentic monuments of those ancient revolutions; he has seen them everywhere engraved in indelible characters; when he has excavated the earth he has found accumulated debris, moved [from its original location]; he has found immense heaps of shells on the tops of mountains today as far distant from the sea as possible; he has found unmistakable

> remains of fish in the depths of the earth. . . . These facts, unknown to the vulgar, but now well known to all who observe nature, force the physical scientist to recognize that all the surface of our globe has changed; that it has had other seas, other continents, another geography, and that the solid ground that today we occupy has formerly been the sojourn of the ocean. To doubt of the reality of these facts would be to belie nature, who has herself raised everywhere the monuments that attest to them.[4]

Boulanger's work has often been compared to that of Vico, either to fantasize about "plagiarism" (on the basis of Galiani's comments) or to point out differences and similarities between them. We can perhaps add one or two touches, in line with the interests of the present study, to the picture of the "differences" that Franco Venturi sketched more than thirty years ago.[5] The best starting point is perhaps in Boulanger's constant concern to *fondre et confondre* study of the earth's history and that of the history of mankind. The very first lines of the *Avant-propos* of his major work confirm this: in reading the history of ancient peoples and the history of the peoples that modern geographical discoveries have acquainted us with, Boulanger says, we note "that almost all the nations of the earth have had and still have traditions that have been transmitted to them of the changes that formerly took place in nature."[6]

The great crises of the earliest human history reflect the earth's great catastrophes, or *malheurs*. The uncertain documents of the history of those dark ages refer to geological or natural events that men hid from themselves and still hide—but also recall—in myths and fables. Struggle with a hostile environment and defense against a terrible nature make up a great part of man's past. Religion was born as diluvian symbolism, as an attempt to commemorate a natural event from which man was saved; morality was born as man underwent the fatal experience of the fragility of his sojourn. Society is, above all, a means of defense against an inimical nature. Diderot clearly saw this close connection between *nature* and *culture* in the *Antiquité dévoilée*. Boulanger, he says,

> saw the multitude of different substances that the earth hides in its bosom and that attest to its antiquity and to the incalculable series of its revolutions under the star that illuminates it. He saw the changed climates and the lands, now burned by a perpendicular Sun, now scarcely grazed by its oblique and swiftly passing rays and heaped with eternal ice. . . . He saw the nourishment of a present world growing on the surface of a hundred past worlds. . . . After having considered everywhere the traces of the earth's misfortune, he sought its influence on [the earth's] ancient inhabitants: hence his conjectures on societies, governments, and religions. But these conjectures had to be verified by comparing them with tradition and with histories. . . . He was inclined to a belief that the savages descended from wandering families, confined to the forests by their terror in the first great events.[7]

In spite of Vico's famous appeals to lightning and the heavens, his allusions to wildlife, "which, thanks to the recent Deluge, must have been plentiful," to "nitrous salts" and "exhalations" of the earth, he found this sort of problem alien to his interests. The sections on the Deluge in the *Scienza Nuova Prima* (96–99) and in the definitive edition of the *Scienza Nuova* (369–73) speak only of the behavior of men after the Deluge, in the age in which women, "wild, indocile, and shy," behaved like female animals and let their young "wallow naked in their own filth," abandoning them as soon as they were weaned. When Vico speaks of "fragments of antiquity" or "various marble remains," he is referring to stones sculpted by the hand of man that bear traces of the magical characters of the Chaldeans and the hieroglyphics of the Chinese and the Egyptians. When he speaks of fossils, they are not shells or imprints of prehistoric fish; they are not *natural objects*, but the *traces of a human presence:* remains of skeletons and of ancient armor. There are innumerable seventeenth- and eighteenth-century texts that begin with a sentence like: Particularly on the hills do we find . . . followed by a reference to fossils shells. In Vico, what follows speaks of human relics and the skeletons of giants: "great skulls and bones of an unnatural size," the immense arms of the old heroes, "preserved along with the bones and the skulls of the ancient giants."[8]

Man alone has a history in Vico. Nature has no history of its *own*. Man does not emerge *out of* nature (as, for example, in Herder); he is placed from all time *within* nature. In the *Scienza Nuova* the theme of the relation between investigation of the natural world and investigation of the world of human history emerges many times. In his attempt to make a science of history, a science equal to the science of nature and superior to it in certitude, Vico undeniably appealed to both the great empirical tradition and to the tradition of rationalism. His appeal to Bacon's method did not exclude a recourse to definitions, to axioms, to accepted aphorisms. The new science "proceeds exactly as does geometry" and in order to understand it we have to "have formed the habit of reasoning geometrically" and reduce our mind "to a state of pure intelligence free from any particular form." As he constructs his science, Vico leaves room not only for generalizations, but also for pure theoretical models, for "the wholly metaphysical task, abstract in its idea," and he distinguishes between the "hypotheses" and a "truth meditated as an idea, which then with authority will become fact." Vico's attitude toward historical knowledge, from this point of view, is not too unlike the attitude of modern thought toward the natural world, once modern thought had come to emphasize the value of experience as more than a simple, passive registration of fact, and that of theory as an instrument capable of verifying experience.[9] All this is true, but it is also true that the relation between investigation of the natural world and investigation of the historical world often takes the form of an opposition in Vico.

Vico seems to make a choice between clear alternatives: the investigation of nature had already been accomplished, and few opportunities remain for further achievements; the new world to explore was human history. It was here that possible successes lay. Here one could truly *know*. In the *Dipintura preposta al frontespizio* (frontispiece) of the *Scienza Nuova* "the globe, or the physical, natural world" appears supported only on one side. It is as if it were balanced precariously over the altar, shown so because "until now, the philosophers, contemplating divine providence only through natural order, have shown only a part of it," without considering that globe "in respect of that part of it which is most proper to men, whose nature has this principal property: that of being social."[10] Vico's New Science seemed *new* to him because it put an end to an attitude toward nature and to natural order that had characterized *infin ad ora* the investigations and the efforts of scholars.

Those investigations, in Vico's eyes, had little chance of success. Natural science was in reality a dead end. The terms *meraviglia* and *seriosamente*, which Vico uses in this context, should be taken at full value: "Whoever relects on this cannot but marvel that the philosophers should have bent all their energies [*seriosamente si studiarono*] to the study of the world of nature, which, since God made it, He alone knows."[11]

The old inquiry, which offered no further hope, must be replaced by a *new* inquiry: the meditation "on the world of nations or civil world," the world that men, since they had made it, "could hope to know."

"The nature of things is nothing but their coming into being," Vico declares, but his aphorism pertains to man, *not* to nature. However we may want to evaluate the modernity of Vico's philosophy (and on this point I disagree completely with Pietro Piovani), it is undeniable that Piovani rightly spoke of it as a "philosophy without nature." He says: "What is important is that the natural and the human, for this new way of philosophizing, remain separate. . . . The nature that is the protagonist of the *Scienza Nuova* is human nature. . . . The philosophy of culture is proposed [in Vico] as the heir of a defunct philosophy of nature." As H. P. Adams noted many years ago, myths, for Vico, are not imagistic reconstructions of natural events, but of happenings in social life: they are "mythologized politics."[12]

Once again Burnet enters into the picture. Nicola Badaloni, in his *Introduzione a Vico*, has called scholarly attention to Burnet's work, remarking justly that Vico's commentators and other historians of Italian culture of those years have shown no interest whatever in this personage. Badaloni not only has important things to say regarding two of Burnet's fundamental doctrines (on the origin of the earth and on the state of the dead); he also points out clearly the multiple relations between Burnet's themes and those characteristic of the Academy of the Investiganti, and he realizes the importance of Vico's reference to Burnet. Badaloni says:

> In a passage of the *Scienza Nuova Prima* Vico refers to Burnet. The difference between the two thinkers lies in the fact that in the cyclical movement of human affairs, Vico stops at some beginnings and refuses to go beyond them. This is why Vico does not take up the problem of origins, which Burnet had taken up, connecting the history of the formation of nature with that of the formation of man.[13]

It seems almost as if this difference were more a diversity of tasks than a diversity of attitudes or visions of the world. According to Badaloni, Vico and Burnet share the thesis of a cyclical course of events in human history, the former limiting himself to man's formation and the latter connecting the problem of the formation of civilization with that of the formation of natural reality. Searching for basic differences or opposing ideas in the two thinkers would be fruitless since all we would find is more or less ample treatment of themes and differences in their approach to specific problems.

In reality, Vico does not limit himself, in sections 97 and 98 of the *Scienza Nuova Prima,* to just *not taking up* the problems Burnet had treated; he takes an explicit stand *against* Burnet's doctrines. The context within which he sets his rejection is very significant. The "poverty of conventional languages" of the first peoples, their recourse to a mute language that made use of "objects, which at first must have been natural, and later carved or painted"—a use that subsequently developed into hieroglyphics—proves, according to Vico, that the Chinese "boast in vain of enormously ancient origins," that their history counts no more than four thousand years, and that the Deluge preceded the formation of the first nations by a relatively short time. "This poverty of articulate languages among the first nations," Vico says, "which was common throughout the universe, proves that the Flood occurred not much time before."

Precisely this "proof" that the Deluge occurred in relatively recent times—which also showed the vanity of all attempts to prolong the history of the world beyond the four thousand years allowed by the Vulgate—seemed to Vico to provide a "demonstration of this truth of the Christian religion: that Noah and his family were preserved from the Flood and that their antediluvian literature was preserved by the people of God, even during the period of slavery in Egypt." It was in the name of this proof of the recentness of the Deluge and the continuity of antediluvian culture through the Hebrew people alone that Vico opposed "Thomas Burnet's fantasy of a capricious revolution of the earth." Vico declares his demonstration a "true resolution" of that theory, which seemed to him to be built on ideas drawn from van Helmont and from the physics of Descartes:

> This poverty of articulate languages among the first nations, which was common throughout the universe, proves that the Flood occurred not much time before. Moreover, in demonstrating this point we provide a true resolution of Thomas Burnet's fantasy of a capricious revolution of the earth, the grounds of which he derived first from Van Helmont, then from Descartes' *Physics*. According to Burnet, because the Flood dissolved the Southern part of the earth more than the Northern part, the latter retained more air in its bowels and was therefore more buoyant and on a higher plane than the former. The Southern part therefore sank into the ocean, thus causing the earth to incline away from a plane parallel to that of the sun.

The summary that Vico gives here of one of the theses advanced in the *Telluris theoria* is fairly accurate. The inclination of the earth's axis, according to Burnet, does not date back to origins; it does not go back to the creation. It is the result of purely mechanical causes, of a great natural catastrophe. The earth's equator originally coincided with the ecliptic. The catastrophe of the Flood and the consequent irregular distribution of the air, water, and land masses displaced the earth's axis. What Vico summarizes was certainly not the central point of Burnet's doctrine; nevertheless, when he joined those "fantasies" and the tenets of Cartesian mechanics, Vico accurately grasped the rigidly mechanistic nature of Burnet's doctrine and, at the same time, refused the image of a Flood that had occurred in very remote times.

In the third edition of the *Scienza Nuova* (1744), Vico mentions, in one way or another, more than a hundred authors who wrote between the beginning of the sixteenth century and the 1740s. Only thirty-seven of those authors had been cited in the pages of the *Scienza Nuova Prima*. During the years between 1725 and 1744, the years of the various drafts and the many rewritings of his masterpiece, Vico also enriched it with a host of citations and references (in a few cases even to books he had not read). Only three authors mentioned in the *Scienza Nuova Prima* fail to appear in the definitive edition: they are Burnet, van Helmont, and Leclerc. These can really be reduced to two, as van Helmont is mentioned only incidentally in the passage regarding Burnet. These are significant absences.

One thing, however, is certain: Burnet's "system" and his vision of the world had repeatedly been compared to those of Lucretius, Hobbes, and Spinoza, and he had been accused of impiety and atheism. Vico, who perhaps only knew the *Telluris theoria* indirectly, but who certainly knew of its reputation and its basic orientation, and who certainly had had word of both the controversies it had given rise to and of d'Aulisio's condemnation of it, intended to stay as far away from it as possible.

15·*Le sombre abîme du temps*

THE chronological tables of the history of the earth that Buffon adopted and rejected, one after the other, between 1749 and 1778 have been patiently reconstructed by Jacques Roger in his preface to the critical edition (1962) of Buffon's *Les époques de la nature* (1778). It took the earth about 3,000 years to solidify and 35,000 years to cool; it reached its current temperature in the earth year 74,382; about 45,000 years from now (at 168,000 years from its formation) everything will again be frozen and life will have disappeared from the earth. These figures, published in 1775, are repeated in the *Epoques de la nature*. A study of the manuscript drafts, however, documents Buffon's hesitations and incertitude. He thought at one point that he had underestimated the action of the *causae latentes,* so he abandoned the "short" chronology for a "long" chronology, speaking of 117,000 years (instead of 3,000) for the solidification of the earth, of another period that lasted 700,000 to a million years instead of 25,000 to 30,000: thus the age of the earth became nearly three million years instead of about 75,000. But the new figures remained in manuscript form, and the new chronology, so extended that it might have seemed inconceivable to Buffon's contemporaries, was withheld. Buffon declares that he first presented

> a shortened outline of time spans. This abbreviated version, or rather this small scale was necessary to me to preserve the order and the clarity of the ideas, which would have been lost in darkness [*des espaces obscurs*] if I had abruptly presented the outline of time spans on the scale that I use today and which is forty times longer than that of my first table. . . . I have even fashioned my last table on a scale forty times greater in accordance with an estimate that is but unpropitious and overly weak, for I am quite persuaded that in reality the *causae latentes* of Newton [were underestimated] . . . and that alone would have augmented our scale tenfold more, and would have given me 10 million years instead of 600 thousand for the duration of our age. But, once again, although it is quite true that the more we stretch time, the closer we get to the truth and the reality of the use that nature has made of it, still, we must shorten it as much as possible in order to conform to the limited power of our intelligence."[1]

Buffon was certainly not moved by religious motives when he lowered his estimates from three million years to 75,000. Even though he took the trouble, in the published text, to declare his system purely "hypothetical" and thus not in contradiction with "the immutable axioms of revealed truth," he had still lengthened by about 69,000 years the time scale that biblical tradition admitted. Falling from 12,000 meters is not much different from falling 1,000 as far as risking your life is concerned. The reasons for Buffon's caution were of another sort: he felt, and Boulanger agreed with him, that his contemporaries of the end of the 1760s were not yet capable of imagining the *sombre abîme* of so interminable an antiquity for the world.

The parallel between civil history and natural history—which, as we have seen, was not new—appears on the very first page of the *Epoques de la nature,* expressed with singular power and striking literary qualities:

> As in civil History written documents are consulted, medals sought, ancient inscriptions deciphered . . . so in Natural History it is necessary to rummage through the archives of the world, to draw from the bowels of the earth old monuments [and] collect their debris. . . . It is the only means of fixing a few points in the immensity of space and of placing a few milestones on the eternal path of time. . . . Civil

> History, bounded on one side by the shadows of a time fairly close to our own, extends to the other only on the small portions of the earth occupied successively by peoples mindful of their tradition. While Natural History embraces all spaces equally, all times, and has no limits other than those of the universe.[2]

In the seventh and last of the epochs of nature—and here Buffon directly reflects the influence of Boulanger—there appeared men who, "trembling on an earth that trembled under their feet, naked in mind and body, exposed to the injuries of all the elements, victims of the fury of ferocious animals," were all "penetrated with the common sentiment of a bodeful terror." After they learned to sharpen flints, light fires, clear the land by means of stone axes, and make clubs and bows and arrows, rafts and canoes, men gathered in small "nations" composed of a few families, which in turn gave rise, in the northern regions of Asia, to the first larger societies. Nature's hostility marked human history permanently: the memory of the earth's misfortunes—devastations, fires, earthquakes—became "almost eternal." Fear and superstition "from that time on took over the heart and mind of man for ever." Man is to this day subject to ancient terrors; he is still "barely reassured by the experience of the ages, by the calm that followed those centuries of storm, in sum, by his acquaintance with the effects and operations of nature."[3]

The archives of the physical world, for Buffon, were enormously richer than those that preserve the documents of civil history. The history of the earth seemed vastly greater than man's history. Beyond human history stretched a nearly boundless territory, in which fossils were time's uncertain and difficult documents and thus must be deciphered and understood.

The idea of time had come to be part of the idea of nature. Diderot claimed, in 1753, to have "inserted" the idea of succession into his definition of nature. Could what happened for the individual, who grew and died, not perhaps also happen for the species? If faith did not teach us "that the animals came from the hands of the Creator as we see them now," could we not conjecture that the elements of animality have existed from all eternity, "scattered and mixed within the mass of matter"? Again, Diderot continues, could we not conjecture

> that the embryo formed of these elements, passed through an infinity of organizations and developments . . . had successively motion, sensation, ideas, thought, reflection, consciousness, sentiments, passions, signs, gestures, sounds, articulate sounds, a language, laws, sciences, and arts, [and] that millions of years went by between each of these developments?[4]

These millions of years tended to become a commonplace during the 1750s, albeit within very limited circles. Three years before the publication of Diderot's *Interprétation de la nature*, Thomas Wright, clockmaker and professor, had also "introduced time" into his concept of nature. In Wright's *Original Theory or New Hypothesis of the Universe* (1750), the sun is not at the center of the universe. The sun and all the stars of our galaxy are in motion, more or less on the same plane, placed within a great wheel, "disk," or ring. This explains our view of the Milky Way, which we see looking toward the margins of the disk, and of a sky that is more or less "thick" with stars. In Wright's work, as in the *Allgemeine Naturgeschichte und Theorie des Himmels* of Kant (1755) and in the *Cosmologische Briefe über die Einrichtung des Weltbauses* of Lambert (1761), the very ancient image of the sky as a "vault" of stars came to an end and a completely new chapter opened in the history of cosmology.[5] Wright thought, with Kant, that there were other habitable worlds, and at the beginning of his work he refers to Bruno's theses (which he knew of through Toland's work) and brackets Bruno's name with those of Huygens and Newton.

Many familiar themes return in the *Original Theory:* the nobility and beauty of the Mosaic account, its elegance, its skill in adapting a "sublime" subject to the understanding of the populace, the Jews' unfamiliarity with an Egyptian culture that Moses, to the contrary, knew well, and the backwardness of the Hebrew people, "who had not yet learnt to make use of their reason." We can also see cropping up again the attack on the vision of the physical world based on chance—a "fatal Rock upon which all weak Heads and narrow Minds are lost and split upon" and that should be avoided "not only as the Nurse of Atheism, but as the dreadful Father of Despair."[6]

Wright also explains that he was trying in his work to enlarge his readers' ideas on creation and on space and to illustrate his notion of time as coexistent with motion and connected to our "sensibility" as terrestrial creatures:

> As Distance is the Measure of Magnitude and of all Extent, and helps our Imagination to the Ideas of Space, so are progressive Moments the Measure of Velocity, and makes us sensible of Duration: And as Space may be extended through all Infinity, so Time may be continued as to Eternity. This Succession of temporal Ideas impressed, or excited in the Mind, as an Effect of Matter in Motion, producing a perpetual Change, both of Objects earthly and celestial, enables us not only to reflect upon past Vicissitudes of Nature, but from their regular Courses, known Order and Returns, predict Phaenomena to come, and prove the periodical Effects of Nature's constant Laws so just and certain, that Time may be said with Truth to co-exist with Motion.[7]

"Our general Judgement of Duration" is in some manner tied to our "Sensibility" of the "Solar, or rather a natural Year," and its "constituent Parts, both horary and diurnal." The time periods of the universe were quite another affair: Wright notes that Saturn's year corresponds to about 29 terrestrial years; the great comet of 1680 makes its periodic return every 575 of Saturn's years (corresponding to 16,675 solar years); the general motion of the stars (or the Great Saturnian Year), tied to the precession of the equinoxes, completes its revolution every 25,920 years; and the *Vortex Magnus* of the entire universe "cannot be made in less than a Million of Years."[8]

The fixed stars are not a scattered swarm, without physical order, but constitute a *system* that is analogous to the solar system. Wright does not reach the point, as Kant does later, of applying this notion of system (Kant was to call it a "systematic constitution") to the nebula and interpret them as galactic "systems" external to our galaxy. He holds that the rotation of the universe takes place around a "focus" or "center of creation . . . a primitive Fountain, perpetually overflowing with divine Grace, from whence all the Laws of Nature have their Origin." Wright does not presume to say "what this central Body really is." He is certain, however, that it takes the universe about a million years to rotate around it.

In this universe, enormously vaster than the traditional one and which extends in time to the confines of eternity, many of the traditional categories lose their meaning: "Even the total Dissolution of a System of Worlds, may possibly be no more to the great Author of Nature, than the most common Accident in Life with us." It is probable, in the universe at large, that "such final and general Doom-days may be as frequent there, as even Birth-Days or Mortality with us upon the Earth."[9]

Wright's ideas on the Milky Way and on the structure of the cosmos were summarized, as is known, in the January 1751 issue of the *Freye Urtheile* of Hamburg, and that review was read by Kant. Kant accepted the thesis of the galaxy as an enormous system governed by the same laws that regulate planetary motion, and he took over the idea of inhabited planets although, in contrast to Wright, he also extended the analogy to the other galaxies and attempted a "Cartesian" explanation of the origin of the universe. His explanation tried to account for the passage out of chaos "into a regular order and into an arranged system," enjoying the pleasure, "without having recourse to arbitrary hypotheses, of seeing a well ordered whole produced under the regulation of the established [by Newton] laws of motion." Attraction and repulsion, the original *plenum*, and vortices were sufficient to explain the *formation* and the *history in time* of the solar system. In this sense, even remaining within a decidedly theistic position, Kant could affirm that "the theory of Lucretius or his predecessors Epicurus, Leucippus, and Democritus, has much resemblance with mine." The "hand of God" that, as Buffon put it, "donna le branle à l'univers," was eliminated from cosmology, and the principal basis of Newtonian apologetics was struck at its very roots. Beyond the Milky Way other star systems, other Milky Ways and other universes opened up. Newton's static cosmology gave way to an evolutionist cosmology that attempted to account for the origin of the universe and for its stages of development, as well as for the distribution in space of the various astronomical systems and for the physical principles that operate within their systematic organization.[10] The spatial and temporal structure of the universe was taken on as an "object" of science. God's act of creation retreated into a sort of "antefact" of nature, and the created world became infinite in time as well as in space:

> There had mayhap flown past a series of millions of years and centuries, before the sphere of the formed nature in which we find ourselves, attained to the perfection which is now embodied in it; and perhaps as long a period will pass before Nature will take another step as far in chaos. But the sphere of developed nature is incessantly engaged in extending itself. Creation is not the work of a moment. When it has once made a beginning, with the production of an infinity of substances and matter, it continues in operation through the whole succession of eternity with ever increasing degrees of fruitfulness. Millions and whole myriads of millions of centuries will flow on, during which always new worlds and systems of worlds will be formed after each other in the distant regions away from the centre of nature and will attain to perfection. . . . And if we could embrace the whole of eternity with a bold grasp, so to speak, in one conception, we would also be able to see the whole of infinite space filled with systems of worlds and the creation all complete. But as, in fact, the remaining part of the succession of eternity is always infinite, and that which has flowed is finite, the sphere of developed nature is always but an infinitely small part of that totality which has the seed of future worlds in itself, and which strives to evolve itself out of the crude state of chaos through longer or shorter periods. The creation is never finished or complete. It has indeed once begun but it will never cease. It is always busy producing new scenes of nature [*Auftritte*], new objects, and new worlds. The work which it brings about has a relationship to the time which it expends upon it. It needs nothing less than an eternity to animate the whole boundless range of the infinite extension of space with worlds, without number and without end.[11]

Why, Buffon had asked himself, does the human spirit seem to *lose its way* in time more than in space or in the consideration of measure, weight, and number? More than a century earlier, Kepler had manifested his "secret and hidden horror" of Bruno's infinite space, where "we feel ourselves lost," as in a prison. Pascal had given expression to the feeling of annihilation and abandonment in the same infinity: "The whole visible world is but an imperceptible speck in the ample bosom of Nature." When he compares himself to all that exists, man feels himself "lost in this remote corner of Nature" and, closed into the "tiny cell where he lodges," he must learn to "weigh at their true worth earth, kingdoms, towns, himself." What is man in the infinite?[12] In the second half of the eighteenth century, not only the history of the earth, but also the history of men was on its way to becoming a "tiny cell" in an ongoing time that flowed from a semi-infinite past toward dimensions unmeasurable.

16·Hutton: A Succession of Worlds

THE debate over fossils that took place in the late seventeenth century and during the course of the eighteenth was to remain for a long time tightly interwoven with discussion of the history of the earth, the formation of the universe, and the history of man on the earth. The starry heavens, William Herschel wrote in 1784, must no longer be thought of as the concave surface of a sphere seen from its center. The regions of the heavens, into which we can penetrate with the aid of powerful telescopes, must be considered just as "a naturalist regards a rich extent of ground or chain of mountains, containing strata variously inclined and directed, as well as consisting of different materials." Pierre Simon de Laplace, in a note in the fifth edition of his *Exposition du système du monde* (1824, first edition 1796), cited the fossil bones described by Cuvier to sustain that the "tendency to change" concerned the solar system as well.[1]

During the same year in which Herschel published his *Observations Tending to Investigate the Construction of the Heavens* (1784), the first volume of Johann Gottfried Herder's *Ideen zur Philosophie der Geschichte der Menscheit* appeared. The first chapter bears the title: "Our Earth Is a Star among Stars"; the third: "Our Earth Has Undergone Many Revolutions ere It Became What It Is Now"; the sixth: "The Planet We Inhabit Is an Earth of Mountains, Rising above the Surface of the Waters." The second volume of this work discusses the earth as an "organic laboratory" for the formation of living creatures, the vegetable and animal kingdoms, and man as an intermediate and pivotal creature among the animals. The third volume compares the structure of plants and animals with that of man and investigates the physiological structure and the instinct of animals. To reach a discussion of the society of his time, Herder *passes through* cosmology, biology, anthropology, and prehistory. He *reaches* man through nature. Nature and society come to be seen as elements of one single—and providential—process of development.

The reliance on providence, finality, and Mosaic history would for many years to come continue to condition the discussion of fossils, the formation of the universe, and living species. It is interesting to consider, in this context, the *Theory of the Earth* that James Hutton presented as a paper before the Royal Society of Edinburgh in 1788 and later published, in complete form, in 1795. Although it is true that this work is "the earliest comprehensive treatise which can properly be considered a geological synthesis rather than an imaginative exercise,"[2] it is appropriate to end a discussion that began with the fanciful writings of Athanasius Kircher with a few words on Hutton. There is a passage of Hutton's cited as a matter of ritual and reproduced not only in all studies of the history of geology but also in the introductions to many current treatises on geology itself:

> For having, in the natural history of this earth, seen a succession of worlds, we may from this conclude that there is a system in nature; in like manner as, from seeing revolutions of the planets, it is concluded, that there is a system by which they are intended to continue those revolutions. But if the succession of worlds is established in the system of nature, it is in vain to look for any thing higher in the origin of the earth. The result, therefore, of this physical enquiry is, that we find no vestige of a beginning,—no prospect of an end.[3]

As a good Newtonian, Hutton conceives of the earth not as a collection of material objects moved by physical forces, but as a "system." The "succession of worlds" that has changed and will continue to change the aspect of the terrestrial surface is the result of an equilibrium between the action of water and that of the internal heat of the earth. In the same manner, the system of the planets results from an equilibrium between gravitational and centrifugal forces.

Again as a good Newtonian, Hutton holds that research into the *origin of the entire system* of nature makes no sense and that the question should be resolved—just as Newton had declared regarding the entire system of the planets—by recourse to divine intervention. In Hutton's investigation, the physical world is seen as a machine that gradually becomes organized for human ends—that is, in view of becoming "an inhabitable world."

> When we trace the parts of which this terrestrial system is composed, and when we view the general connection of those several parts, the whole presents a machine of a peculiar construction by which it is adapted to a certain end. We perceive a fabric, erected in wisdom, to obtain a purpose worthy of the power that is apparent in the production of it.[4]

The world, for Hutton, can be considered as an "organized body." We must look for the presence in it of a mechanism that Hutton refers to indifferently as a "reproductive operation," a "reproductive power," or a "reforming operation." This power from time to time takes on the task of repairing "a ruined constitution" of this world, thus assuring—beyond the destruction and the disappearance of worlds that succeed one another in time—duration, stability, and equilibrium to the entire system of the succession of worlds. Although the present earth is the result of the destruction of a more ancient earth, there is a process of repair implicit in that destruction. What is presumed is an equilibrium:

> If no such reproductive power, or reforming operation, after due inquiry, is to be found in the constitution of this world, we should have reason to conclude, that the system of this earth has either been intentionally made imperfect, or has not been the work of infinite power and wisdom.[5]

Both of these possibilities, in Hutton's eyes, were obviously unacceptable. One of the fundamental tasks of science, he insists, lies in indicating the ends for which the system has been constructed. It is legitimate to consider final causes:

> We live in a world where order every where prevails. . . . In a theory which considers this earth as placed in a system of things where ends are at least attained, if not contrived in wisdom, final causes must appear to be an object of consideration, as well as those which are efficient. . . . Chaos and confusion are not to be introduced into the order of nature, because certain things appear to our partial views as being in some disorder.[6]

The overly benevolent efforts of some contemporary geologists notwithstanding, Hutton's earth cannot be represented as a "system" capable of self-regulation. The physical world, for him, has been "peculiarly adapted to the purposes of man" and the earth, following an age-old tradition, is compared to the body of an animal in which healing follows partial destruction.[7] The many pages—passages of fundamental importance—that Hutton devotes to rocks and strata, to fossils, to "actualism" and to time must be read in view of their relation to these more general views and to this image of the universe and of science.

"We are not to look for nature," Hutton says, "in a quiescent state; matter itself must be in motion, and the scenes of life as a continued or repeated series of agitations and events." Even if Hutton believes that inactivity "is commonly, but erroneously attributed to material things" and that every material being "exists in power and energy,"[8] however, Hutton the Newtonian is far from Diderot, from d'Holbach, and from all those who had theorized on the superiority of process over quiescence, of flow and vital becoming over stasis and the immobility of structures. Diderot himself realized full well that if the state of beings is in "a perpetual vicissitude," if nature "is still at work," if the state of each of the interlocked phenomena is "without permanence," then "there is no philosophy" and "all our natural science becomes as transitory as words."[9] Hutton, on the contrary, imagined an indefinite series (indefinite in the sense that in the current state of science it was impossible to establish limits either *a quo* or *ad quem*) of "scenes of life" or of "great events happening in succession," but he conceives of such scenes or events as *situations that follow one another*. Science does not study the flow itself, but the series of situations. As his impassioned "propagandist," John Playfair, said in his *Illustrations of Huttonian Theory of the Earth* (1802):

> Amid all the revolutions of the globe, the economy of nature has been uniform, and her laws are the only things that have resisted the general movement. The rivers and the rocks, the seas and the continents, have been changed in all their parts, but the laws which direct these changes, along with the rules to which they are subject, have remained invariably the same.[10]

This firm conviction and the premise that the operations of nature are "invariably the same" is connected with Hutton's so-called uniformism or, better, actualism[11]—that is, to the thesis that we can infer the nature of processes that occurred in the past by analogy with the processes observable and at work in the present. Thus, all of the geological forces active today were active in the past as well, and their action is slow, continuous, and uniform. Sedimentation on the ocean floors, the consolidation and raising of strata, the origin of basalts and granites from the hot nucleus of the earth, the changes in strata by the operation of pressure and subterranean heat, the continuing processes of erosion and disintegration by wind and water: all of these processes have changed the aspect of the earth, they constitute its past, and they permit us to foresee the continuity of future changes. To understand "the natural operations of time past" we must then "examine the construction of the present Earth." We can thus predict the future course of events and throw light on the operations through which "a world so wisely ordered goes into decay" and how "the waste of a habitable land upon the globe" can come to be "repaired."[12]

But the scale of nature's time is very different from that of human history: the earth today is more or less similar to the earth of the ancients, but the formation and destruction of continents requires a quite different scale:

> Our fertile plains are formed from the ruins of mountains; and those travelling materials are still pursued by the moving water, and propelled along the inclined surfaces of the earth. . . . The immense time necessarily required for this total destruction of the land must not be opposed to that view of future events, which is indicated by the surest facts, and the most approved principles. That Time, which measures every thing in our idea, and is often deficient to our schemes, is to nature endless and as nothing; it cannot limit that by which alone it had existence; and, as the natural course of time, which to us seems infinite, cannot be bounded by any operation that may have an end, the progress of things upon this globe, that is, the course of nature, cannot be limited by time, which must proceed in a continual succession.[13]

Fossils become the clocks that can measure the long periods of natural history. The "relics of sea-animals of every kind in the solid body of our earth" permit us to trace "a natural history of those animals" that includes "a certain portion of time." Those solid bits were formed in an epoch after the one in which those animals inhabited the seas. If we can describe the operations through which those solid parts were formed, we will have a way "for computing the time through which those species of animals have continued to live."[14]

Human history moves in shorter ages. As far as man's history is concerned, the traditional measures of time can be accepted. From Hutton's point of view, which was that of the slow and uniform changes of nature, the secular controversies on human history did not have much importance, and the "Mosaic" thesis of the limited antiquity of man could be accepted in all tranquillity:

> If we are to take the written history of man for the rule by which we should judge of the time when the species first began, that period would be but little removed from the present state of things. The Mosaic history places this beginning of man at no great distance; and there has not been found, in natural history, any document by which a high antiquity might be attributed to the human race. But this is not the case with regard to the inferior species of animals, particularly those which inhabit the ocean and its shores. We find, in natural history, monuments which prove that those animals had long existed; and we thus procure a measure for the computation of a period of time extremely remote, though far from being precisely ascertained.[15]

Hutton was a patient man, capable of accumulating an enormous number of facts and of interpreting them rigorously. He was a complete stranger to fanciful hypotheses and was faithful to his massively Newtonian premises. His "foundation" of the science of geology undeniably signals a point of no return. He was also a devout Christian who did not deny the creation of the world, nor final causes, nor the work of Providence, and who still offered proofs of a natural time vastly greater than that of human history. Furthermore, he had a solid and calm faith, which shines through in many of his pages, in the patient operation of a reason subject to error but able to learn, even from its errors, to push farther:

> Our object is to know the time which had elapsed since the foundation of the present continent had been laid at the bottom of the ocean, to the present moment in which we speculate on these operations. The space is long; the data for the calculations are, perhaps, deficient: no matter, so far as we know our error, or the deficiency in our operation, we proceed in science, and shall conclude in reason.[16]

In contrast to Buffon, Hutton was not in the least interested in the formation of the world. He did not preface his account of the history of the earth with a history of the formation of the solar system, and he kept clear of all discussion of comets. He simply stated that the data and the knowledge currently available did not permit the finding of either the vestiges of a beginning in nature or the indications of an end. The creation and the apocalypse described by Saint John could not be the object of science. But Hutton did not set this affirmation in a revolutionary context. He insisted, as we have seen, on ends and on Providence, on the presence of man in the universe and on the adaptability of the earth to human ends. When he declared that we find in nature "no deficiency in respect of time, nor any limitation with regard to power," he quickly added, "but time is not made to flow in vain; nor does there ever appear the exertion of superfluous power, or the manifestation of design, not calculated in wisdom to effect some general end."[17] But this was not enough: in his most philosophical work, *An Investigation of the Principles of Knowledge* (1794), Hutton addressed detailed criticisms to Hume's negation of the principle of cause. Furthermore, he went out of his way to distinguish his position from any other position that aimed at dissolving God into nature, he took a stand against the equating of matter with extension and impenetrability, and he criticized Rousseau for his pedagogical spontaneity and attacked atheistic philosophies with considerable force.[18]

All of this, however, was not sufficient to render Hutton's theses acceptable to a culture that tended, after the great trauma of the French Revolution, to judge scientific theories primarily by their threat to morality or the likelihood that they would subvert religious tradition.[19] In the *Geological Essays* of Richard Kirwan (1799), in the writings of Robert Jameson, in the *Letters to Dr. Hutton* (1790) and in the *Lettres sur le christianisme* of Jean Deluc, Hutton was subjected to ferocious attacks and was charged with destroying religion and diffusing the impieties of the baron d'Holbach.

In reality, in addition to purely theoretical dichotomies within the discipline of geology (neptunism vs. volcanism, uniform vs. catastrophic change), the great dichotomy between deism and materialism that had characterized the period between Burnet's work and d'Holbach's (1681–1777) reopened in a different form during the years between the beginning of Hutton's activity and the publication of the *Origin of Species* (1788–1859). Once again the image of a nature as the "temple of God" (as Herder wrote in 1772) or the "temple of life" (as William Paley wrote in 1802); once again "intelligent design" and a harmonious and benevolent nature that assures man his "just place" within reality were challenged by chance and change, by hard necessity, and by an emphasis on struggle, waste, and discord. An edifying science found itself challenged by the image of a science which, as Darwin was to write, could not offer the slightest consoling notion concerning the dignity of man. The texts analyzed by Charles C. Gillispie in *Genesis and Geology* offer an impressive documentation of the existence of this dichotomy and of the dramatic terms in which it was expressed between the end of the eighteenth century and at least the middle of the nineteenth.

The history of this controversy is really quite different from the history we are summarizing. But even in this great new controversy, which opposed two images of nature, the dispute over time was to occupy a central position, as has often been noted. When Sir William Thomson (later Lord Kelvin) revived Buffon's theory of the progressive cooling of the earth, he recast Jean Baptiste Fourier's calculations on the thermic history of the earth in light of the then recently formulated second law of thermodynamics. Kelvin saw in potential energy the only possible antecedent to the present arrangement of the solar system. In this view, which also put in doubt Kant's and Laplace's hypothesis of the original nebula, the sun could not be older than 500 million years, the solidification of the earth's crust occurred from 100 to 200 million years ago, and the appearance of life, following a reduction of the high temperatures incompatible with it, went back several million years. Kelvin says: "The necessity for *more time* to account for geological phenomena than was generally supposed to be necessary, became apparent to all who studied with candour and with accuracy the phenomena presented by the surface of the earth."[20]

James Hutton and Charles Lyell seemed to have been defeated. The long time periods of the champions of uniform change were in crisis. Darwin was faced with a serious and apparently insuperable difficulty: he had declared that "the number of intermediate and transitional links, between all living and extinct species, must have been inconceivably great." He had calculated at three hundred years (an inch per century) the erosion of the cliffs at Weald, and he had made broad use of Lyell's time scale. Before the "odious spectre" raised by Kelvin, Thomas Huxley had made several concessions to the catastrophists. Darwin, too, eliminated many passages devoted to geological time periods in the sixth edition of the *Origin of Species* (1872) and conceded that the evolutionary process during the first ages could have taken place on a shorter schedule of periods.

> With respect to the lapse of time not having been sufficient since our planet was consolidated for the assumed amount of organic change, and this objection, as urged by Sir William Thomson, is probably one of the gravest as yet advanced, I can only say, firstly, that we do not know at what rate species change as measured by years, and secondly, that many philosophers are not as yet willing to admit that we know enough of the constitution of the universe and of the interior of our globe to speculate with safety on its past duration.[21]

Kelvin tied his physics to an image of science and to a philosophical point of view that were decidedly antimaterialistic. "Overpoweringly strong proofs of intelligent and benevolent design lie all around us," he wrote. "All living beings depend on one ever-acting Creator and Ruler." From this Newtonian point of view, the work of Darwin, which left very little room for an external Ruler and his benevolence, could not help but seem to him inadequate as a philosophy. Kelvin saw in the methodologies of astronomy and cosmic physics the "essence of science," and he consequently held that biology met with "prodigious" difficulties in "successfully acting up to this ideal." Although he rejected Darwin's "hypotheses" as too audacious, Kelvin had clear convictions concerning the origin of life: "Our own bodies, as well as all living plants and animals, and all fossil organic remains, are organized forms of matter to which science can point no antecedent except the Will of a Creator."

In his *Lectures on Some Recent Advances in Physical Science* (1876), Peter G. Tait forcfully repeated these same conclusions:

> We are led to a limit of something like ten million years . . . or (say at most) fifteen millions of years . . . as the utmost we can give to

> geologists for their speculations as to the history even of the lowest orders of fossils.
>
> But I daresay many of you are acquainted with the speculations of Lyell and others, especially of Darwin, who tell us that even for a comparatively brief portion of geological history three hundred millions of years will not suffice!
>
> We say—So much the worse for geology as at present understood by its chief authorities.[22]

For nearly forty years, as Stephen Toulmin and June Goodfield have written, the limitation Kelvin imposed on the antiquity of life "had the same cramping effect on evolutionary thought that the Biblical time-scale had had on Hooke's and Steno's geology." After the discovery of radioactivity (in 1903) "Kelvin's limits on the age of the Sun [as a sort of accumulation of carbon which is being consumed] have been swept away by our understanding of the thermo-nuclear process." Thus Kelvin's rigorous calculations lost all significance. By the beginning of the 1930s, Kevin's figure for the age of the earth could be multiplied a hundredfold.[23]

In our post-Darwinian world, the history of the universe, the history of the solar system, the history of the earth, and the history of the human species are entities constructed on vastly differing chronological scales: five billion years for the history of the earth, from two to three million years for the appearance of man, about a million and a half years for the use of the first tools, from twenty to fifty thousand years for the appearance of *homo sapiens*.[24] That nature by far precedes man, that man emerged out of nature, that much of "nature" is still present in his behavior—all of this is now part (or should be part) of common sense, even of the common sense of people who have only the vaguest ideas on stratigraphic paleontology or population genetics. For this reason it is worthwhile to recall that for many, many centuries the history of man was conceived as coextensive with the history of the earth. An earth not populated by men long seemed meaningless, like a reality that was somehow "incomplete." This is why the "world makers" and the constructors of fantastic world systems, while they struggled to bring together their investigation of the *origin of things* and the investigation of the *course of nature*, contributed to laying the foundations not only of new knowledge, but also of a new image of man and of nature.

2
THE AGES OF HUMAN HISTORY

17·The Egyptian Culture of Moses

IN the impressions that extinct plant and animal species had left in rocks, Athanasius Kircher traced the letters of mysterious alphabets and pictures of celestial bodies and looked for analogies, similarities, and correspondences. His extravagant interpretations—and the writings of many other followers of "reactionary Hermeticism" in the seventeenth century—hid a clearly defined cultural aim: the revival of the magical and Hermetic vision of the world and its projection as a still valid alternative to the image of the natural world shared by "mechanistic philosophy" and modern science. An organic philosophical and theological system and its attendant cosmology, which was in radical antithesis to the new physics, should, on the one hand, take the place of a traditional Christian scholasticism that by then seemed insufficient, and, on the other, it should absorb *some* of the more important theoretical discoveries—and if possible *all* of the technical acquisitions and mechanical inventions—produced by the new science. During the very years that saw the consolidation and the triumph of the new astronomy and the new physics, Kircher was trying (and his efforts were also "political") to revive all of the characteristic themes of Renaissance alchemy, magic, symbolism, and naturalism. His universe was pervaded by an elementary dynamic force, the emanation of divine truth, that operates within all phenomena and is revealed through a universal symbolism. For Kircher, knowledge of the universe demands that we pass from common intellectual knowledge, founded in the limited variety of phenomena, to an intuitive and total knowledge, secret and esoteric, revealing ever-deeper coherences, opening the way to unity and to knowledge of the All. The interest in the Egyptians and the "hieroglyphic method" was intimately tied to this vision of the physical world: divine truth, in Kircher's thought, finds its highest and most complete manifestation in Christianity, but it is immanent in all religions and all philosophies, and it was first revealed to men within the ancient wisdom of the Egyptians.

For Kircher, history is not growth, development, or progress. Progress lay in a return: truth and the way to truth required the retrieval of a truth that had long lain buried, the return to a remote age, the rebirth of a faraway, long-lost knowledge, of an extremely ancient, pre-Christian wisdom born on the banks of the Nile that had inspired Pythagoras and Plato. The verities of the Hermetic texts, written in the mysterious symbols of the obelisk of Heliopolis, hint at truths of Christianity and the concept of the Trinity; the Egyptian magic cross is a symbol of Christ's cross.[1]

Kircher's grandiose and confused efforts seemed to fulfill the hopes that Francesco Patrizi had expressed at the end of the preceding century, when the work of the Jesuits seemed to have made it possible to diffuse Hermetic philosophy, and Patrizi had turned to Gregory XIV to ask him to reject Aristotle and the Scholastics at last and to adopt Hermetic philosophy as the one authentic Christian philosophy. But the idea of the return to and the revival of this extremely ancient, lost wisdom, carried to its logical extremes, might prove to be fraught with dangerous implications. Anyone who held that Hermes Trismegistus had preceded Moses might also believe—as Patrizi had sustained—that the Thrice Greatest had spoken of the Trinity more clearly than Moses himself: "The *Pimander* contains the creation of the world and of man, which is nearly the same as in Moses. And it narrates the mystery of the Trinity in a much clearer fashion than Moses himself."[2]

If the truth of the Christian religion was in some way hidden in Egyptian hieroglyphics, if the creation account in the *Pimander* could be seen as equivalent to the Mosaic account, if behind the symbols of pagan religions (as Hermetic writers had claimed) a universal revelation lay hidden, and if this revelation was *already present* at the roots of the life of humankind even before it became completely manifest in the pages of Scripture, then the unique, inimitable value of the holy texts might be questioned and the way be opened to the impious theses of the libertines and the Spinozists and to the collapse of all distinctions between sacred and profane history. Might not the Hermetic texts, or texts close to Hermeticism, appear perilously close to the deistic thesis of a Christian religion that, as Matthew Tindal wrote in 1730, "has existed from the beginning" and that God "both then and ever since" made accessible to men, so that Christianity, "tho' the Name is of a later date, must be as old, and as extensive as humane nature"?[3] That people as prominent as Tindal and Toland showed evident sympathy for pro-Egyptian ideas is undeniably significant. This way led to a view of Moses as one of the many philosophers of antiquity who drew on an even older tradition. This is what invariably occurred in all appeals to an original store of truth as the source of partial truths embodied in varying historical forms:

> In spite of the fact that Moses, concerning the origin of things, did not speak explicitly of atoms, I hope, nevertheless, by means of arguments that I have collected, to have demonstrated that corpuscular philosophy is the most ancient of all, and that it flourished not only a thousand years before Democritus, but even before Moses himself, probably in Egypt, Phoenicia, Arabia, and Idumaea.[4]

The harsh judgment of many an author, Vico among them, of books as seemingly innocuous as those of John Marsham (1671) and John Spencer (1670 and 1686) can only be explained in this cultural context. Furthermore—and this is important to emphasize—the thesis of a contact between Hebraic civilization and Egyptian civilization and that of a transfer of elements of Egyptian culture to the Jews was fatally linked to the assertion that the Hebrew people did not develop historically in isolation.

John Marsham, in his *Canon chronicus aegyptiacus, hebraicus, graecus* (1671), sketched the broad outlines of a comparative history of the Egyptians, the Chaldeans, the Greeks, and the Jews. The importance of the similarities he found in customs and legislation concerning homicide, abortion, incest, blasphemy, matrimony, etc. led him to the thesis of the priority and the superiority of Egyptian civilization: "Even from the Holy Scriptures we know that the Hebrews were long inhabitants of Egypt, and not unjustly can we conjecture that they had not completely abandoned Egyptian customs and had conserved some remnants of Egyptian culture."[5]

Marsham is speaking here of a true acculturation. He contrasts the wisdom of Egypt to the rough and uncertain character of the Hebrew *gens*, a nomad people without homeland, with no laws and no juridical structures, transformed into a "republic" or a "nation" by Moses, who had been brought up in Egypt. Marsham also credits the Egyptians with the practice of circumcision, unknown to Abraham when he entered into Egypt. Furthermore, Marsham ties the doctrine of the immortality of the soul to Egyptian civilization, presenting it as a *nobilissimum inventum* of the Egyptian priests, given that the Jews would have been *minus solliciti* by what concerned men's destiny after death.[6]

John Spencer's *De legibus hebraeorum* (1686) begins from a similar point of view. Spencer analyzes the Mosaic juridical system, considering it as one of the many organizational structures of the ancient world, and emphasizing the similarities between the Hebrew laws and those of other peoples. He considers it obvious (his expression is *clare patet*) that the Egyptians were already famous for their religion and their science *diu ante Mosis tempora*. The Egyptians had both sacred and civil ceremonies, and it was unthinkable to Spencer that the Jews could have influenced them: "No one who is not supinely credulous could hold that the Egyptians, through a pure desire to imitate, suddenly rejected their own institutions to conform to the customs of the Hebrew people, who came of servile blood."[7]

The Egyptians gave the Jews not only idolatrous beliefs, but the better part of their regulations, their customs, and their cults. Spencer presents the Mosaic laws as operating on two levels, the ethical and the social and practical. He sees Moses the legislator in libertine terms, as an astute "politician" who, for political purposes, invented religious practices and restored the ancient customs, which sprang from human needs and requirements and not from divine inspiration:

> I will try to prove that Abel, Noah, and the others who preceded Moses offered sacrifices spontaneously, and that therefore the rite of sacrifice did not draw its origin from some divine precept, but from convention and human choice [*instituto et arbitrio humano*]. Since, furthermore, a profound darkness covers the events anterior to Moses, I do not presume to be able to demonstrate that men initially offered sacrifices with a pious soul and led to do so by natural religion.[8]

Since some leading Vico scholars have never opened Marsham's and Spencer's works, they state that these books in no way contain the "pro-Egyptian tendencies" that Vico ascribed to them. Antonio Corsano, starting from this judgment of Nicolini's, went so far as to state that Vico, in his treatment of the earliest civilizations, in reality transferred his own "audacious comparisons" to those "naïve precursors" and those "by-then distant English scholars."[9] Naïveté aside and without questioning the existence of that audacity, it should be emphasized that these theses seemed particularly dangerous to contemporary readers, even many years before Toland carried them to their ultimate consequences in his *Letters to Serena*.

In a review that appeared in the *Acta Eruditorum* of 1684, Marsham's book was classified as "profane" and to be used with extreme caution. There were two reasons for this: the author traced the origin of Hebrew rites to the Egyptians, and he was inclined to question traditional chronology.[10] As the anonymous *Avertissement* that prefaces Richard Simon's *Histoire critique du Vieux Testament* (1685) puts the problem, Marsham's work "seems to have no other end but that of insinuating into the mind of the reader that all of the religion of Moses and of the Hebrews was taken from that of the Egyptians." But it was not only this "insinuation" that was suspect. Marsham went farther than religion permitted and went beyond the limits of orthodoxy when he sustained that the Egyptians were an older people than the Jews:

> In his most scholarly *Canon chronicus*, Marsham toils mightily over the Egyptians, whose antiquity he sustains not only before the Greeks, but before the Hebrews as well, and beyond what religion can admit [*ultra quam per religionem permitti poterat*]. . . . Let us now pass from the Hebrews to the Egyptians, once a most civil people. It is in fact likely that the laws of the Hebrews came down to them, in spite of the fact that John Marsham, an English lord, holds that many laws and many rites came down from the Egyptians to the Hebrews, an affirmation that has something of the sacrilegious [*quae sententia nescio quid profani habet*].[11]

In 1704 Melchior Leydekker compared Bochart's and Horn's theses to those of Marsham and Spencer—"who had abused their erudition." Could we not state, Richard Simon asks, four years after the publication of Leydekker's work, that Moses, who was trained in the knowledge of the Egyptians, "took from them the best of what he found in their books"? Can we not find in the *Zohar*, one of the most ancient Hebrew books, statements taken from the Platonists, from the Chaldeans, and from the Egyptians? Among the many works projected by John Woodward we find listed (we are in 1726) a book on the wisdom, the arts, and the religion of the ancient Egyptians. The aim of this projected book was to defend the "Mosaic institution" and to refute Marsham's and Spencer's theses, according to which "some part of this institution was taken from the Egyptians."[12]

Even a cursory reading of the works of John Toland makes clear the grave consequences of reading such texts and the extent to which these statements (of Hermetic and Neoplatonic origin) concerning the superiority of the ancient and mysterious wisdom of the Egyptians could become "revolutionary." Styling his work after Fontenelle's *Entretiens sur la pluralité des mondes,* Toland inserts these theses into a broader "materialistic" context in which we can see legacies from Bruno and Spinoza and references to the *Archeologia philosophica* of the "great Dr. Burnet."

In Toland's *Letters to Serena* (1704), Christian beliefs seem to be integrated (often totally dissolved) into the patrimony of thought and the institutions typical of other early civilizations. The prevalent thesis according to which the pagans arrived at the discovery of the soul's immortality "from the antient books of the Jews" was, in Toland's eyes, "altogether groundless." Those books do not discuss immortality, he argues, and it makes no sense to talk of a chronological priority of Hebrew wisdom because it is manifest from the Pentateuch itself "that many Nations had their several Religions and Governments long before the Law was deliver'd to the Israelites."[13] Spinoza, in his *Tractatus,* and La Mothe le Vayer in his *Dialogues,* had already advanced the thesis that the origin of the idea of immortality should be sought in a pagan culture. In Toland's work this thesis is joined to a double rejection: of the priority in time of the Hebrew religion and of the idea that religions originated in one primal religion.

Toland was a firm advocate of the primacy of the Egyptians, who were, in his eyes, "the Fountain of Learning to all the East." Their religion and sciences flourished "long before the Law was deliver'd to Moses, which is an indisputable Testimony of their Antiquity before any Nation in the World."[14] The Jews, on the contrary, were "of all Eastern peoples the most illiterate" and it was not by chance that the Acts of the Apostles (7:22) note that Moses was brought up by Egyptians. Not only was there no uniquely Hebrew wisdom distinct from other forms of wisdom, but the thesis of the continuity of Hebrew history and Hebrew lineage was untenable. Both the notion of the *isolation* of the Hebrew people and that of the *continuity* of their history met with Toland's energetic denial:

> The Hebrews in reality had mixed in with the Egyptians, no matter what some, more superstitious than the Hebrews themselves, might go dreaming or murmuring to the contrary. . . . Those who pertinaciously insist that all the Hebrews are certainly sons of Abraham or of Jacob without the least corruption of their blood, are blind. It is in fact clear to all that the Egyptians' slaves were not all of one stock, but were a promiscuous hodgepodge of all the nearby peoples.[15]

It often happens, Toland continues, that "the vulgar attribute to the First Founder, because of the reverence and the authority of his name, absolutely all innovations." How could this have failed to have happened concerning Moses? Were there not many examples, even among Christian sects, of attributing to Christ and to the Apostles acts never mentioned either in the New Testament or in any other text?

Toland insinuates, citing a famous passage in Strabo (*Geographica* 17:2, 3), that Mosaic philosophy was a form of pantheism and of Spinozism:

> Strabo affirms without hesitation that Moses was pantheist or, to speak according to more recent usage, Spinozist: he shows this, in fact, while he teaches that there is no God distinct from matter and from the structure of the world, and that nature itself, or the totality of things, is the one and supreme God whose several parts can be called creatures and the whole, if you like, creator. . . . This is the philosophy today of learned Chinese scholars and other oriental [scholars], and it was also sustained by many among the ancient Greeks and Romans. . . . I leave it to those who are competent to examine the extent to which this theology is sound and orthodox and to what extent Strabo's narration conforms with the Pentateuch.[16]

At the end of his *Origines judaicae*, Toland expresses his impious conclusions concerning the Hebrew people and an "Egyptian" and pantheist Moses in the form of questions: Were the Hebrews descendants of natives of Egypt? Had their leader and legislator, Moses, formerly "been an Egyptian priest and king?" Had he perhaps "left his country because of dissent about the public state of religion"? Was Nature "the one and supreme God" for him? And, finally, "Was Moses like Minos and Lycurgus and Zalmoxis and others of the same sort?"

In the earliest times, Toland wrote in the first pages of the third letter to Serena, the ancient Egyptians, Persians, Romans, Hebrews, and many other populous nations had neither sacred images nor statues, nor "peculiar Places or costly fashions of Worship": "The plain Easiness of their Religion being most agreeable to the Simplicity of the Divine Nature." Much later, men would arrive at picturing God, basing their vision of him on their earthly sovereigns—that is, they approached a changeable, jealous, and vengeful God with the same means used to seek the favors of their earthly kings. Idolatry and superstition had become so widespread among all eastern Christians and many western Christians that the situation could be summed up in the style of the popular ballad:

> Natural Religion was easy first and plain,
> Tales made it Mystery, Offrings made it Gain
> Sacrifices and Showes were at length prepar'd
> The Priests ate Roast-meat and the People Star'd.[17]

Many Christians, Toland wrote elsewhere, believe and observe things never mentioned in Scripture. They take them, they say, from tradition, "as if there were not ancient errors, ancient frauds, and ancient lies." They ought to remember, he warns, that the older a falsehood is, "the more dangerous, because its roots are deeper."[18]

The pro-Egyptian theses of the English scholars of the later seventeenth century did not necessarily lead to these conclusions, of course. They did lead, however, toward a comparison of civilizations and religions, toward putting the culture and the religion of the ancient Hebrews on the same plane as those of other peoples. They led to a denial of two points on which any defense of orthodoxy would continue to insist until well into the nineteenth century: the *isolation* of Hebraic civilization and the *continuity* of Hebraic history. When Pietro Giannone, in his *Triregno*, later contrasted the roughness of the ancient Hebrews with the refined civilization of the Egyptians, he paired Toland's name with Marsham's and Spencer's.

Thanks to studies by Margaret C. Jacob and Chiara Giuntini, Toland's thought has been amply explored in recent years, and Giuseppe Ricuperati has thrown much light on Giannone's work, so closely connected with this question. Let us turn, then, to a less well-known author to see how, as late as the middle of the eighteenth century, theses regarding the superiority of the Egyptians fused with those questioning the authority of the Bible and viewing it as merely one of many "fables" that mankind has created. During the same years in which Vico was rewriting and expanding the *Scienza Nuova*, Conyers Middleton was studying the ancient Germans and Egypt, studying the biographies of the great classical authors, and mixing his research on the functions of the Roman senate with an investigation of the history of languages.[19]

Between 1724 and 1750 Middleton reconsidered a whole series of familiar ideas and formulated and defended four theses:

1) That the Jews borrowed some of their customs from Ægypt.
2) That the Ægyptians were possess'd of arts and learning in Moses's time.
3) That the primitive writers, in vindicating Scripture, found it necessary sometimes to recur to allegory.
4) That the Scriptures are not of absolute and universal inspiration.[20]

Middleton's works have justly been compared to Hume's and Gibbon's. He starts from a precise conviction: the canons of historical research and the "rules" that Newton had applied with so much success to the natural world must be applied to the Bible as well. He speaks of the "incumbrance" of

> the notion, which is generally inculcated by our Divines, concerning the *perpetual inspiration and infallibility of the Apostles and Evangelists:* a notion, which has imported such difficulties and perplexities into the system of the Christian religion, as all the wit of man has not been able to explane: which yet will all be easily solved, and vanish at once, by admitting onely the contrary notion, that *the Apostles were fallible:* which is a sort of proof that generally passes with men of sense for demonstrative; being of the same kind, by which *Sir Isaac Newton* has convinced the world of the truth of his philosophical principles. . . . The case is the same in Theological, as in the natural inquiries: it is experience alone, and the observations of facts, which can illustrate the truth of principles. . . . Wherefore, as we learn from dayly experience that prejudice, passion, want of memory, knowledge or judgement naturally produce obscurity, inaccuracy and mistakes in

> all modern writings whatsoever; so when we see the same effects in antient writings, how sacred soever they may be deemed, we must necessarily impute them to the same causes.[21]

Middleton puts his thesis clearly: modern authors tended to follow a "double way of interpreting" the Bible. According to the circumstances, they take one sentence of the text as "literal" and the next as "allegorical": "one part as a fact, the next as a fable." Historical narration and allegorical narration "are compositions of quite different kinds, and serving to different ends." The first aim at representing the true state of things, the second at inculcating certain hidden verities. History and fable "naturally destroy each other and the same narration cannot be interpreted simultaneously as both historical and allegorical."[22]

Both the Old and the New Testaments, Middleton continues, are obviously full of allegories. In them, "certain religious duties and doctrines . . . are represented as it were to our senses, by a fiction of persons and facts, which had had real existence." But historical criticism cannot surrender before the sacred text, nor before statements that attempt to defend orthodoxy by doing violence to documented facts. If it is true, as Clement of Alexandria states, that *all* writers, both Greek and barbarian, who treated theological matters deliberately disguised their narration of the origin of things behind symbols and enigmas, then how can we refrain from applying this general rule in the particular case of the Bible as well? And if the Mosaic story of creation is also written "in this symbolical way of learning," should we not conclude that the Mosaic account is in many instances metaphorical? And that we must once again distinguish fact from fiction? How, for example, are we to doubt the testimony of Saint Stephen (Acts 7:22) that Moses "was educated in all the Lore of Egypt"?[23]

Middleton ardently defended the superiority of the Egyptians over the Jews, and he held that the civilization, the culture, and the customs of the Jews were of Egyptian origin. He greatly admired Newton's physics and Newtonian *regulae philosophandi*. He knew Newton's works on chronology, but thought that Newton's considerable work in this area had not sufficed to demolish that thesis. In a *Defence* (1732) of his earlier Letter to Dr. Waterland (1791), Middleton clarifies his position concerning the greatest of living scientists. He states that he had "asserted it to be more probable that the *Jews* should borrow from the Ægyptians, than the Ægyptians from the *Jews*" (which was the classic formula that Marsham had used more than half a century earlier), because the Egyptians, at the time of the encounter between the two peoples, were cultivated and civilized and the Jews were rough and illiterate. To the objection that it remained to be proved "if it were true" that at the time of Moses Egypt was a great kingdom and a highly civilized nation, Middleton answers:

> This then is the Fact that I undertake to make good; I will not say against you, who seem to know but little of the Matter, but against what you absurdly call the *Demonstrations and Discoveries of Sir Isaac*

> *Newton*. . . . You have heard much talk of Sir *Isaac's demonstrations* in mathematicks and his great *Discoveries* in Natural Knowledge; and imagined perhaps that those Words signified nothing more than *Conjecture* or *Opinion*. For had you reflected what a *Demonstration* meant, you could not have applied it to a *System of Chronology*, however probable, or preferable to all others, which from the Nature of Things can never reach Certainty or admit of *Demonstration*. But, pray Sir, after all what is it that Sir *Isaac* has discovered? has he brought to light any old Authors, which for Ages past had lain buried in Oblivion; or any Monuments of Antiquity unknown before to the Curious? Or has he done in the learned what he did in the natural World; invented a *new Telescope* to pry into remote and dark Antiquity with more Accuracy than had been practicable before? If he has done nothing of this, then *all his discoveries* can amount onely to conjecture; . . . for a thorough Knowledge of Antiquity, and the whole Compass of *Greek and Aegyptian* Learning, there have been, in my Opinion, and now are, many Men as far superior to him, as he within his proper Character is superior to every body else.[24]

The antireligious tenor of Middleton's theses was clear to his contemporaries, and his impiety was harshly criticized. One of the sharpest attacks came from John Rotheram, for whom Middleton's view of the Mosaic account ceased to be history: "We may call it an Apologue or Moral Fable." The most that could be got from it, Rotheram declares, were the verities that the deists had already stated: that the world was created by God and that man, once innocent, had lost his felicity in abandoning himself to sin. Sacred history became "a piece of history designed for the use of Jews alone." If the first part of the Mosaic books was fable, "where does the fiction end, and truth take place?" It is true, Rotheram admits, that the "histories of the remotest ages" are imperfectly known, and that the accounts of them are full of "a great mixture of fiction." The Mosaic account, however, cannot not be reduced to one of the many ancient "fables": it differs from them by its clarity, its certainty, its continuity, and its lack of fractures.[25]

One P. Williams expressed the opinion, in 1733, that Middleton's work should be burned and its author expelled from the university. He complained that Middleton's idea in his *Letter to Dr. Waterland* (1731) that a series of fables had been "interspersed" into the Mosaic account weakened the credit and authority of Holy Writ

> and therefore the certainty of every thing that depends upon them, which are no less than the truth of Christianity and the peace of society. . . . The author of Christianity as old [as the Creation, Matthew Tindal] is far more excusable: he endeavours to take away Revelation, but substitute Reason in its place; you divest us at once both of the sufficiency of Reason, and of the certainty of Revelation; and leave men to act and be governed by *Machivilian* schemes 'without God in the World.'[26]

Middleton defended himself from these accusations by repeating the theses we have already seen: the superiority of the civilized Egyptians over the uncultivated Hebrews, and the allegorical or fabulous character of many passages of Scripture. He points out to his adversaries that "M. de Fontenelle who is a Papist, still living in a Popish country" was enjoying, in his happy old age, the full respect owed to his merits, "notwithstanding his avowed *unbelief of the Heathen Oracles* and the fatal blow which he has given to their authority." He adds that if in his four theses he had stated something anti-Christian he was ready to retract it, but that the four theses remained true.

> What pity it is that *these inquisitors of ours* have not the power of the rack to extort what confession they please! All that this good man aims at, is to make me odious and detestable to every body; with a true *Popish Spirit*, he would draw me in to recant, and then proceed to burn; or with the old revenge of an *Italian*, first make me blaspheme, and then stab me.[27]

Until well into the eighteenth century, the evocation of John Marsham's theses was closely connected with a reconsideration of sacred history, with a comparison of sacred history with other histories, and with the defense of "free thought." In one of the basic texts for the freethinkers, *Christianity as Old as the Creation* (1730), Matthew Tindal returned to one of the themes then in fashion. Superstition had always held that when the body was shown no mercy, souls could better obtain divine mercy. Thus people thought, Tindal declares, that God delights in the pain and misery of his creatures, and that a man's best way to make himself acceptable to Him was to torment himself with "immoderate watchings, fastings, penances and mortifications of all sorts," and that the harsher these were, the more acceptable to God they made the sinner. Otherwise, Tindal argues, no one could ever have arrived at the notion that cutting the foreskin could be thought "a religious duty acceptable to a good and gracious God; who makes nothing in vain." If nature had called for an operation of the sort, nature, being "always the same," would have required it always. Thus circumcision was a historical institution: "This institution, as is prov'd by Marsham and others, seems to have been owing to the Egyptians, who thought all to be prophane who used it not; and it was after Abraham had been in Egypt, that *circumcision* was instituted [among the Hebrews]."[28]

18·History before Adam

THE controversy over the relationship between the Egyptians and the Jews concerned chronological *priority* as well as the *level* of their respective civilizations. Sacred history, from the orthodox point of view, was more than one model among others. It had to be defended from contamination because it was sacred, because it was the *unique* history of a people that had preserved its records intact from the very beginning of the world, with no mixture of myths and fables—because it was, as Vico was to say, a history "more ancient than all the most ancient profane histories that have come down to us." Thus the claim that there were histories *older* than sacred history could seem—from the time of Augustine's *City of God* to Vico's times and beyond—an attack on the truth of Christianity and a sacrilegious statement.

It was precisely by invoking the extraordinary antiquity of the Chaldeans, the Mexicans, the Peruvians, and the Chinese—and by comparing the chronologies of those peoples to that of the Jews—that Isaac Lapeyrère sustained, in 1655, the existence of the "preadamites," men who populated the earth before Adam, who was the first man to come from the hands of the Lord. In this alarming view, the Bible lost its standing as universal world history and was reduced to being a summary of the particular history of the Hebrew people. The Deluge was no longer a universal catastrophe, but became a particular episode in the history of one particular nation.

In his *Praeadamitae*, Lapeyrère gave an inhabitual interpretation of Romans 5:13–14: "usque ad legem peccatum non erat in mundo; peccatum vero non imputabatur, non existente lege" (for until the law sin was in the world: but sin is not imputed when there is no law). What, Lapeyrère asks, did Paul mean by *usque ad legem?* Was the *lex* of which he speaks that of Moses, as the traditional interpretation states, or was it instead Adam's? The text speaks not of a law in general, but of the law the breaking of which caused all men to sin; it speaks of the man who, having broken the law, brought death to reign among men. Surely, Lapeyrère argues, neither of these affirmations applies to Mosaic law. When he proclaimed the law to Moses, God was addressing only the Hebrew people, and only the Jews sinned when they broke that law. Their sin could surely not have had repercussions among peoples ignorant of that law. The text speaks of Adam; the *lex* of which it speaks is Adam's, and that anyone could have thought of Moses in connection with it seemed to Lapeyrère merely a sign of madness: such interpreters *allucinari mihi videntur*.

> The interpreters . . . were in every way streightened in the explaining of these words, until the law [*usque ad legem*]; for that law was either to be understood of the Law given to Moses or of the law given to Adam, if that law were understood of the law given to Moses, it must needs be affirmed that sin was in the world before Moses, and until Moses, but that sin was not imputed before Moses; if that law were understood of the law given to Adam, it must be held that sin was in the world before Adam and until Adam but that sin was not imputed before Adam. Therefore other men were to be allowed before Adam who had indeed sinn'd, but without imputation; because before the law sins were not imputed.[1]

Lapeyrère was well aware that the hypothesis that men had existed before Adam ran counter to current theological opinion and that it was apt to trouble the conscience and scandalize those overly steeped in tradition. But in antiquity, he argues, men were thought not to exist at the antipodes, and today this is admitted as an obvious truth. The controversy over terrestrial motion did not prevent day from following night and night day, and the seasons from succeeding one another—for followers of Ptolemy and those of Copernicus alike. Similarly, whether we accept Adam as the first man or admit other men before him, "stabit semper suo loco et suis misteriis religio omnis christiana." Indeed, the basis of the Christian religion was the one belief that all men were damned in Adam and redeemed in Christ.[2]

The hypothesis of the preadamites is based, according to Lapeyrère, on the Bible. Acceptance of this hypothesis, he argues, makes it possible to resolve two problems otherwise insoluble: (1) we can explain the great antiquity of some pagan peoples, which is documented by great monuments of profane history that date back to before Moses' time; (2) we can avoid all of the difficulties necessarily encountered by all those who argue the population of the entire earth by Adam's offspring alone.

> Moreover, from this Tenet, which asserts Men to have been before Adam, the History of *Genesis* appears much clearer, and agrees with itself. And it is wonderfully reconciled with all prophane Records whether ancient or new, to wit, those of the *Chaldeans*, *Egyptians*, *Scythians*, and *Chinensians;* that most ancient Creation on which is set down in the first of *Genesis*, is reconciled to those of Mexico, not long ago discovered by Columbus; It is likewise reconcil'd to those Northern and Southern Nations which are not known, All [of] whom . . . were, its probable, created with the Earth itself in all parts thereof, and not propagated from Adam.[3]

The idea that one single family could populate the world in six thousand years seemed to Lapeyrère inadmissible, even accepting a doctrine of Providence that denies autonomy to nature—a nature existing independent of a Being capable of contemplating it and enjoying the spectacle it offers. When Lapeyrère considered the assertion that plants and animals were present on an earth unpopulated by men, he asked the same sort of question that Kepler had asked, in 1611, when he considered Bruno's thesis of infinite worlds. Lapeyrère says:

> If we affirm Adam to be the first and the onely man by whom afterwards the colonies of men were drawn out and dispersed over the earth, to what purpose in that vast space of time [*immensissimum spatium temporis*] in which the whole earth must needs receive its people from one man: I say, to what purpose should the Countries of *Mesopotamia*, the *Antipodes*, bring forth grass and herbs?[4]

Many and, from an orthodox point of view, rather preoccupying consequences followed from these premises. Adamites and preadamites, Jews and Gentiles belonged to different species. Adam was no longer the first man: he became the progenitor of the Jews, who were the elect of God and, like Adam, were *filii Dei et genus Dei*, products of a "second creation." Sacred history was just the narration of the vicissitudes of the chosen people, and no longer coincided with universal history. The Pentateuch was confused chronologically, of mixed nature, and full of contradictions. It was not the work of Moses, but was of much later composition and the work of more than one hand. The biblical text narrates a particular history, and the episodes of this history did not regard the entire world, but only one particular area of the earth: the shadows that covered the world at the death of our Lord did not descend over all the earth, but only over the land of the Jews; the star that appeared to the magi at the birth of Jesus *facula fuit in aere, non stella in coelo;* the Noachian Deluge was not universal and did not strike the whole terrestrial globe, but only Hebrew lands, and God intended it to destroy the Hebrew people, not to eliminate all men.[5]

Sacred history, which was the particular history of the chosen people, had an enormous advantage over profane history: it was known in all of its details, from its inception, and it was continuous. The history of the Gentiles, on the contrary, was *confusa et incognita,* not only because it was lost in the night of time, but because our only sources for it were written down late and transmitted by the Greeks, who were "without history" and were relative latecomers to writing and to civilization. This was why the Greek chronologists divided time into "obscure" time (from the origin of things to the Deluge), "mythic and heroic" or "fabulous" time (from the Deluge to the first Olympiad), and "historic" time (from the first Olympiad on). The Greeks and their Latin successors and imitators are the only source we have. But for the Greeks, "everything was new": at the time of the Trojan War they most probably did not have writing, which they learned somewhat later from the Phoenicians. They did not have a *history* that could go back to before the first Olympiad. Only in fairly recent times did the Greeks learn of the ancient civilizations of the Egyptians, the Chaldeans, and the Phoenicians.[6]

If we turn to the thousands of years that constitute the past of mankind, Lapeyrère states, we arrive at the firm conclusion that even a small part of the past reaches far beyond the age usually set for the creation, and coinciding with Adam's creation. "Non dubitavi primorum hominum creatiionem, a principio rerum longissime ante Adami revocare tempora" (I doubted not to recall the creation of the first men to the beginnings of things, long before Adam's time).[7]

Lapeyrère reached an upsetting conclusion: the most ancient sources on which we can draw to reconstruct the history of nature and the history of man are not to be found in the Bible:

> Those which are scrupulously addicted to the books of *Moses*, use to referre the invention of all arts, sciences, and disciplines, either to *Adam* or his posterity; because in *Moses* there is no man read of

> before *Adam*. This they believe upon the same score, as they believe that all antiquities, both in natural and humane historie, are contained in holy Writ, especially in *Moses*.[8]

Lapeyrère was less interested in the world's first barbaric inhabitants than in its great civilizations, particularly in the progress that the Gentile peoples had made in astronomy, theology, and magic. Only "extremely long intervals," centuries long, could in any way explain those advances and account for the history and philosophy of ancient peoples or explain the perfection of the Chaldeans' astronomical observations, the Egyptians' calculations of chronology, or the theses developed by Chinese philosophers and historians. Did not the year 1594 of our era coincide, according to Scaliger, with the year 880,703 of Chinese chronology? Did not the computations of American civilizations also reach back innumerable centuries? And would not the same prove true as well of men of the austral lands, as yet unknown to us? And had not Salmasius shown that the Chaldeans computed time on such a large scale that they measured, not in years, but in terms that signified intervals of 60, 600, or 6,000 years? And the Mexicans and the Peruvians, did they not count time, according to Gomara and Garcilaso de la Vega, in "suns," each one of which referred to 860 years, so that their most recent sun began in the year A.D. 1403?[9] All peoples, Lapeyrère concludes, are in admirable agreement on the boundless time that separated the beginning of things from our days. Only contemporary chronologists were unaware of this:

> But as Geographers use to place Seas upon that place of the Globe which they know not: so Chronologers, who are near of kin to them, use to blot out ages past, which they know not. They drown those Countries which they know not: these with cruel pen kill the times they heard not of, and deny that which they know not.[10]

During those thousands of unknown centuries, Lapeyrère concludes, a pluralistic history took place, a history constructed by different peoples, which had found expression in different civilizations and had led men, already in the distant past, to the construction of grandiose monuments and the development of refined arts and abstract and difficult sciences. The enormously vast history Lapeyrère theorized extended like an endless, unexplored continent beyond the 5,617 years of traditional chronology. That ample and difficult terrain was soon to be populated not only by the wise Chaldeans, the mysterious Egyptians, and the refined Chinese, but also by barbaric men, ferocious beast-men, and by "apes" destined to become men.

19·The Backward March of the History of Nations

AN extremely dangerous mixture had been created. As Daniel Morhof saw clearly, the Hermetic notion of mankind's lost and original wisdom had been joined to attempts to compare Egyptian and Hebraic chronology, with the sacrilegious hypothesis of the preadamites, with Postel's dangerous notion of a Brahmin antediluvian wisdom, and with Spinoza's infamous denial of Moses as the author of the Pentateuch:

> Before the Deluge, human societies united in republics. The ancient Chaldeans in fact knew about the empires that had preceded the Deluge. . . . And there were antediluvian monarchs among the Egyptians and the Arabs, and Postel founds new and unheard-of affirmations on the books of Noah and Enoch. . . . It is the opinion of many that after the Deluge there was nothing more ancient than the Pentateuch, although the Hebrews tell of many things concerning a certain book of Abraham entitled *Liber creationis* which teaches in hidden fashion the fundamentals of the Cabala. And from this, Postel, I do not know with what reliability, infers many things concerning the schools of Abraham in Egypt and in India (which supposedly produced the Brahmin, as [the term is] close to Abrahamin) and teaching that among the Indians there were infinite hidden treasures of history and of books that preceded the Deluge. . . . In the Pentateuch there is reference to some books that seem to be older. And on this, that infamous propagator of atheism, Spinoza, founds his villainous calumny when, in the eighth chapter of the *Tractatus theologicus-politicus* he sustains, with many supporting arguments, that the Pentateuch was not written by Moses.[1]

Bayle was to repeat the notion of an antiquity or great age of the world vastly superior to what could be gotten from Genesis in the article "Caïn" in his *Dictionnaire* (1697). Cain, condemned to flee aimlessly throughout the world, was not to fear murder at the hands of those he met: "So the Lord set a mark on Cain, lest any finding him should kill him" (Genesis 4:15). Bayle notes that this language "seems to suppose that Cain was persuaded that there were inhabitants throughout the Earth." If all of mankind were limited to the family of Adam, Cain would only have had to *want* to get away from them. It was customary to argue Eve's fecundity in response to preadamite criticism and to calculate how many sons, daughters and grandchildren she could have generated in a hundred years. But it is difficult to believe, Bayle argues, that Cain could have been so afraid of his brothers and his nephews: "Thus it was the inhabitants of countries far away that he dreaded, persons unknown to him and with no blood connection with him." Furthermore, the Lord does not answer Cain to reassure him that those far-off countries are completely uninhabited.

Lapeyrère's work had provoked considerable reaction and was cause for grave scandal. It had also been sought avidly: "This impious and profane work should have remained buried in an eternal night: now it has seen the light in three languages it flies in an instant throughout the Christian world, and it is not only sold to the highest bidder, but is fought for among the buyers."

Within eleven years of the publication of the first edition of the *Praeadamitae*, no fewer than seventeen works had been published with the specific aim of refuting its impious hypothesis. As early as 1656, barely one year from its first appearance, nineteen refutations had been attempted. The title of a short work of Johannes Heinrich Ursin (Ursinus), published in Frankfurt in 1656, is enough to give an indication of the violence of this polemic: *Novus Prometheus Praeadamitarum plastes ad Caucasum relegatus et religatus*. Ursin says:

> A short time ago in Belgium there appeared a pestiferous work by a man who, by his own admission, leads a wandering life, and who fully merits that destiny. . . . The author claims to withhold his name through modesty, but it is a name as familiar to Frenchmen as the fingers of their hands. . . . He seeks to persuade the reader that his hypothesis is in no way harmful to faith in Christ . . . but instead he renews the damned heresies, the errors, and the wicked blasphemies of the Marcionites, the Manichaeans, and the Pelagians concerning the nature of man, the origin of evil, and original sin, and he mixes it all with the new art of selling one's goods.[2]

The attacks were aimed both at the author and at his ideas. Marten Schoock wrote in his *Fabula Hamelensis* (1662) that Lapeyrère was a man who had most shamefully abandoned the reformed religion, which he had formerly professed, and had turn apologist of Pope Alexander VII out of fear of punishment, and he added that he was a man who sustained impious and dangerous theses denying the universality of the Deluge and the descent of mankind from the sons of Noah. In his condemnation of the book, the bishop of Namur insisted on these points and he prohibited "the reading, holding, and selling of the book, as it contains heretical, erroneous, and rash affirmations." The many summaries of Lapeyrère's theses did much to spread the heresy—even when they were written, as in the case of Philippe Le Prieur, "so that all can understand the impiety, absurdity, and incoherency of his considerations." "The Book of the Preadamites", Richard Simon was to write, "at first made much noise in the world. There arose in Holland a sort of Sect under the name of Preadamites. But not only were its followers too few to form a body; they vanished immediately."[3] The disappearance of the sect was not enough to resolve the problems, and the doctrine of the preadamites continued to present an obligatory point of comparison until the middle of the eighteenth century.

It is undeniable that after the publication of Lapeyrère's book chronology, which before 1655 had been a field open to free speculation and controversy, became a sort of mined terrain over which movement was possible only with extreme caution. Reference to the early wisdom of the Egyptians and the Chaldeans took on different overtones after the hypothesis of the preadamites had been advanced. Discussion of the question no longer demonstrated simple curiosity, bizarre ideas, or "reactionary Hermeticism," nor did it remain within the context of an artificial attempt to prolong the life of a Renaissance and magical vision of reality. Now, invoking remote antiquity touched basic themes and furnished dangerous ideological weapons to the libertines and the *esprits forts*—to the naysayers and adversaries of religion.

> The Preadamites, the Libertines, and those who are called 'esprits forts' . . . claim to show that the first Empires, especially those of the Chaldeans, the Egyptians, and the Chinese, precede the age of Noah by many centuries, and that thus all that that holy Legislator said about the universal Deluge, about the confusion of tongues, and about the dispersion of peoples is not tenable.[4]

Anyone who discussed the oldest civilizations and the ages of human history risked the accusation of being a supporter of Lapeyrère, and attacks on Lapeyrère tended to join with those on the thesis of a wisdom more ancient than Mosaic wisdom. The many discussions that took place during those years on Manetho and Sanchuniathon were more than scholars' quarrels. Many ancient nations had left testimony of a history that seemed to go back to times immediately following the Deluge. As Shuckford wrote, Manetho "pretends to produce Antiquities of Egypt, that reach higher than the Creation by thousands of years," and the Egyptian kingdoms seemed to belong to an age that preceded the biblical date for the creation of the world.[5] Was it not true, it was argued, that such an attempt to lengthen world history was tantamount to smuggling in a denial of God and introducing the atheistic theory of the eternity of the physical world? Was it not precisely those who denied God and sustained the eternity of the world who used the age of Egyptian and Chinese civilization as an incontrovertible proof of their theses? Many, Isaac Vossius wrote in his *Dissertatio de vera aetate mundi* (1659), pushed back the moment of the creation by thousands of years: they do so, he declares, in imitation of those who conceive the machine of the world as eternal, and they present as proofs of their thesis the antiquity of the Babylonians, the Egyptians, and the Chinese. They attempt to "construct an immense antiquity for the world," destroying the truth of Mosaic history. These thinkers, Vossius adds, "did not lack success, and they drew many very famous men to their opinion."[6]

The polemics based on these presuppositions grew more and more inflamed. When Vossius used the Septuagint Bible (consequently giving the creation of the world as occurring in 5400 B.C. rather than 4000), Georg Horn immediately accused him of preadamitism, declaring that

> all of those stretches of time by means of which the pagans tend to have the better of the faith in the Scriptures were invented by a trick

> of the Demon . . . with the aim of establishing his impious errors in men's soul once the authority of the Divine Word was eliminated. . . . In his time, God will destroy the works of the Demon as well as of those who have been blinded by this vain antiquity. He will illuminate them with the light of his Word so they can see how vilely they have erred until now.[7]

As Vico was to see clearly, the problem of the great age of the Egyptians and the Chaldeans had become indissolubly tied, after the middle of the seventeenth century, to that of the immense age of China: China's antiquity also threatened the authority of the Bible and cast doubt on the universal Deluge. To accept the historical reality of that most ancient wisdom implied an acceptance of the data of Chinese chronology; to question biblical chronology would mean casting doubt on the priority of the Jews to the Gentiles. If it were true, as the Jesuit, Father Martini stated in his *Sinicae historiae deca prima* (1658), that Chinese history went back more than six hundred years beyond the time of the Deluge; if it was true that the Noachian deluge could be dated on the basis of Chinese annals during the third millennium before Christ; if it was true, as Martini declared, that the outer parts of Asia were most certainly inhabited before the Deluge, then the theses of that scholarly Jesuit could indeed be reconciled with the preadamite doctrines of the sacrilegious Lapeyrère. It is in this sense that Vico said that Martini's work had led many into atheism.[8] Why? Because it would be impossible to explain the presence of ancient Chinese chronicles previous to the Deluge if one had to believe, as the sacred texts state, that the entire human race, with the exception of Noah and his family, was destroyed in the Flood. Because the world would have to be declared older by at least a thousand years more than the Bible establishes, because we would have to turn to the chronicles and to the histories of the various peoples, not to the universal history in the Bible, for an understanding of the earliest epochs of world history, and, finally, because the Deluge, the construction of the Tower of Babel, and the dispersion of peoples would become episodes of local history, involving a minor people in a circumscribed region of the earth.

An author like John Webb, who declared in 1669 that Chinese was the first spoken language of mankind, was well aware that, according to Father Martini, the history of China "comprehendeth almost three thousand years before the birth of Christ." Webb strove to keep within the limits of orthodoxy, and he held that the primeval language was carried into China by Noah and his descendants. Webb states that China "is the most ancient and, in all probability, the first planted country in the world after the flood" but he goes on to say that China was untouched by the Deluge and the Chinese did not participate in the construction of the Tower of Babel (although they did colonize America). Here as in many other works, the idea of a Chinese history by far longer than Western history was associated (as in Martini) with the mythical idea of a simple, untouched China, fortunate to be closed within its frontiers, isolated from other nations. Martini, Webb declares, "very much enclineth to repose an assured confidence" in Chinese history, but that very isolation never gave the Chinese any "occasion to deliver untruths or report Fables." The Chinese

> preserve a continued History, compiled from their monuments and annual exploits of four thousand five hundred yeares. Writers they have more antient than even *Moses* himself. Ever since their beginning to be a Nation, they have never been corrupted by intercourse with strangers, nor even known what wars and contentions meant; but addicted only to a quietness, delight, and contemplation of Nature, have run through the space of *more than four thousand years*, unknown indeed to other Nations, but enjoying to themselves their own felicity at pleasure.[9]

During the second half of the seventeenth century, the libertines and the *esprits forts* made broad use of the data that the erudition and research of Jesuit missionaries and sinologists had made available. I have no intention of retracing here the vicissitudes of the controversy, for which the basic studies are still those of Pinot and Bontinck. The bitterness of that controversy can be well illustrated, however, by an Italo-French publication that has escaped the attention of those two scholars. It clearly illustrates the difficulties that the Jesuits' theses encountered in the beginning of the seventeenth century. The superiors and the directors of the Seminary for Foreign Missions of Paris write to Innocent XII in the following terms:

> We therefore do not fear, most Holy Father, to manifest to Your Holiness the error that some of the more powerful [the Jesuits] are presently scheming to maintain and that neither the malignity nor all of the consequences of which seem fully known. It is the following, Holy Father, and it pains us to overflowing to have to come to such a sad declaration:
> Proposition I: China had knowledge of the true God more than two thousand years before Jesus Christ.
> II: It had the honor of sacrificing to Him in the oldest temple in the world.
> III: It has had so much honor that it can serve as an example even to Christians.
> IV: It practiced a morality no less pure than its religion.
> V: It has had faith, humility, worship, both internal and external, priesthood, sacrifices, saintliness, miracles, the spirit of God, and most pure charity, which is the perfection and the characteristic of the true religion.
> VI: Thus the Chinese Nation has been the most constantly favored of God among all the others in the world.
> This is, Holy Father, the error that we denounce and accuse before Your Holiness, and for which, in the name of all the Church, we ask justice of the first Tribunal of the world, certain to obtain the condemnation both of the error and of the books that record it. God

> willing, we would conceal the authors. . . . The Jesuits have for long undertaken to justify and defend as innocent the Chinese idolatries and superstitions . . . and, to do things in the scholarly manner, not content to have said by chance some disparate propositions, they have made an entire system and a summum, so to speak, of doctrines that oppose the truth . . . .
> They have believed they observed in the ancient Chinese Books admirable principles and maxims favorable to the Christian religion, and marvelous conformity with the truths of Faith; and because nothing hindered them on their fair road, they have made a long voyage, and have gone towards the least known antiquity, as far as two or three thousand years before the Incarnation of the Son of God. Where they have placed the origin of the Chinese ceremonies, which, they will have you believe, were with most just and most holy intention ordained, and from there by sure channels, pure and sincere, were conducted to the birth of Jesus Christ. . . . All that is needed, they say, is to recall such ceremonies to their first origin and to correct the abuses that may have been introduced in them, and they will be good and holy. All that is needed is to give the names of *Heaven* and of *Sovereign Emperor* the primal sense that the ancient Chinese gave to them, and they will signify the true God; all that is needed is to tell the people that the tutelary Spirits that they venerate are the Guardian Angels and the Blessed Spirits, as known to the first Chinese . . . .
> From which it clearly appears that all of the system and the doctrine of the Jesuits concerning Chinese customs rests on this supposition, which is its foundation: that that vast and great Empire—that is, not some few like Job, but all the body of the Nation, and of the most numerous that there has ever been—has kept for two or three thousand years the true Religion, along with a pure Morality, and that there remain even today excellent traces of this. One deep calleth another and, in order to maintain the error of superstitious and idolatrous ceremonies they have recourse to an even more fatal error.[10]

On 18 October 1700, the Faculty of Theology of Paris subjected to censure the *Nouveaux mémoires sur l'état de la Chine* (1696) and the *Lettre des cérimonies de la Chine (1700)* of Father Le Comte, as well as the *Histoire de l'édit de l'empereur de la Chine* (1698) of Father Le Gobieu. The doctrines listed in the six propositions above were declared "false, rash, scandalous, impious, contrary to the word of God, heretical, such as to subvert the faith and the Christian religion and to render useless the virtue [*vertu*] of the Passion and of the Cross of Jesus Christ."[11]

In the later seventeenth century and in the first decades of the eighteenth, libertine thought made wide use of the notion of an immensely long past to support the idea, closely tied to the materialist and atomist tradition, of the eternity of the physical world. The imaginary voyages of Gabriel de Foigny (*L'antiquité du temps rétablie: la Terre Australe inconnue*, 1676) and of Denis Veiras (*L'histoire des Sévarambes*, 1677–79) speak of a history of lands in the southern hemisphere twelve thousand years old and of Chinese dynasties fifteen thousand years old. The *Espion turc* (1686) stated that the descendants of Panzon and Panzona (the progenitors of the human race) lived millions of years ago, that the Egyptians and the Chinese had preserved untouched the chronology of their long history, and, finally, that the Jews, who "do nothing that is not in view of making themselves important and raising up their race above all the other nations of the Earth," had only recorded "a limited and partial genealogy."[12] The Jesuits conceived of the myth of Egypt and the cult of China as efficacious aids to conversion and as means for the introduction of Christian truth. They had been used, however—as Tyssot de Patot declared in 1722, for example—to demonstrate that heaven and earth were not created in six days, but are of "inexpressible antiquity," and that the appearance of the first man on the earth is so far from our times that "the distance could scarcely be measured or expressed by any number."[13] One reaction to these theses was to deny the authenticity of these overextended chronologies and to identify the personages of Chinese history with their contemporaries in sacred history:

> The *Chinese* have been supposed to have Records that reach higher than the History of *Moses*: But we find by the best Accounts of their Antiquities that this is false. Their Antiquities reach no higher than the Times of *Noah*, for *Fohi* was their first King . . . and by all their Accounts, the Age of Fohi coincides with that of *Moses's Noah*. Their Writers in general agree, that *Fohi* lived about 2952 Years before Christ. . . . *Noah* was born, according to Arch-bishop *Usher*, 2948 Years, and died 2016 Years before Christ; so that all the several Computations about *Fohi*, fall pretty near within the Compass of *Noah*'s Life. But we shall hereafter see many Reasons to conclude *Moses's Noah* and the *Chinese Fohi* to be the same person.[14]

"Thus I see no harm," Dortous de Mairan wrote in the middle of the eighteenth century, "in setting back the infancy of the world." Even if the world were four times older than it is, even if it were eternal, that would not matter, provided we recognize "its true origin and its dependence even now on the hand that formed it."[15] This was a decidedly optimistic interpretation. During the first decades of the century, all the dangerous consequences that could be extracted from the "prodigious antiquity" of the Chinese and the Egyptians had been clearly brought to light: the age of the world, reduced by biblical chronology to little more than six thousand years, could be made to go back to a date a hundred times more remote, and Moses' account alone was inadequate to know about the history of origins. That account spoke, not of the history of the world, but of the history of *one* of the peoples in the world. These peoples were different from one another and they had different histories, and these histories—as is demonstrated by black-skinned men—do not all follow from Adam.

We need to call on a cultural conservative for a full picture of the situation, on some staunch defender of the Christian tradition that was threatened and offended by the doctrine of the immense age of the world. Eusèbe Renaudot announced (in 1718) that it was above all the libertines who had benefited from the excessive praise heaped on Chinese antiquities. They have profited, he states, to attack the authority of the Scriptures and of the Christian religion, to "combat the universality of the Deluge," and to prove that the world "was older than had been thought." The author of the system of the preadamites, Renaudot continues, having learned of China's long past from men more erudite than he, used this as a proof for his doctrines and asserted that the Assyrians, the Babylonians, and the Egyptians go back a nearly infinite number of years. Many, seduced by these dangerous ideas, even if they do not accept the doctrine of the preadamites, "spread opinions that tend to no less than the general Overthrow of all Religion."[16]

Father Perrenin, a Jesuit missionary at Peking, was a disillusioned man by the beginning of the 1740s. He describes with an open mind and a certain sense of humor the climate of resistance to novelty typical of the debate on the ages of world history, noting that it seemed easier to pass from the Ptolemaic system to the Copernican than to lengthened human history by even brief periods:

> I dare to hope that the Hebraicizing gentlemen will permit us to lengthen the duration of the world by a little. . . . It is much easier to persuade the astronomers than the chronologists. . . . There is not much hope that they will be touched by astronomical proofs, or historical proofs, or the proofs of physics. The scholars . . . have published large volumes on chronology and each one of them has done his best to prove himself right. They cannot agree among themselves, and if you dare to interfere in their disputes with arguments about far-off countries, they all jump on you and not a one of them will concede you a month of time or an inch of terrain to carry out your evolutions.[17]

"The precious monuments that remain to us of the antiquity of the world," Dortous de Mairan complained, are rejected and refused at the most tenuous suspicion that they do not fit into the epochs indicated by Scripture. You might think, he continues, that those who refuse them have available a sure, incontested chronology. Instead, from the moment that scholars began working to give us an accurate count of the ages that preceded the Christian era, we can count seventy or seventy-five different chronological systems. They place the age of the world somewhere between 3,700 and 7,000 years. In one system, the Deluge is supposed to have taken place in the year 1656, in another 2256, in still another in 3882. Father Riccioli, "a competent judge and above suspicion in these matters," has provided "the annotated catalog of seventy of these opinions, or hypotheses, or systems, of the age of the world." All of these systems are more or less valid, all are "equally championed and combatted." All claim to be founded on the most authentic biblical texts, all cite the same authorities and, more or less, the same facts. Each of the antagonists boasts that he has solved the problem and totally defeated *(terrassé)* his adversaries. Scaliger seems in contradiction with Petau, Pezron with Martianay, "and even the great Newton found some to contradict him, even in his own country and among his own disciples."[18]

The dispute over the ages of history was just as complicated as the dispute over fossils or the ages of nature—and perhaps even more so. "First get the more than seventy systems of your chronology to agree," Voltaire wrote not long after, "and then you can laugh at the Chaldeans."[19]

20·A Quarrel over Chronology

RATHER than linger over the many contending systems in the quarrel over chronology in the seventeenth century, we may gain a better idea of that quarrel through a somewhat more detailed examination of one dispute that was in many ways typical.

In 1655, the year of the publication of Lapeyrère's *Praeadamitae*, the learned scholar Georg Horn (or Hornius) published his *Historia Philosophica*, which began with a volume devoted to the *sapientia veterum*. "Si ullum saeculum, certe noster philosophicum appellari potest": since he believed he was living in a philosophical century, Horn took his stand firmly on the side of a rigid defense of orthodoxy. He chose to lash out at Poliziano, among others: Horn recognized Poliziano's merit as commentator of Aristotle's *Prior Analytics* and as a scholar of dialectics, but Poliziano, he asserts, had also been "the leader and the head" of a large faction of scholars who, "following the opinions of the pagans, scorn Christian philosophy." To the point, Horn reports, that once, when questioned on his reading of the Bible, Poliziano answered that he "had read that book only once, and had never wasted his time more than that once."[1]

Isaac Voss (Vossius), Horn's opponent in a long and somewhat labored polemic, had interests similar to Horn's. In 1659 he published a work that came to be widely known and provoked violent reactions and bitter controversy. Its title outlines his intentions: *Dissertatio de vera aetate mundi, qua ostenditur natale mundi tempus annis minimum 1440 vulgarem aeram anticipare*. By using the Septuagint Bible instead of the Masoretic Bible, the history of the world could be lengthened by 1,440 years, fixing the day of the creation at about 5400 B.C. Vossius takes a moderate position: the *principia temporis* ought not to be pushed too far back nor set too close to the present. He starts from two clearly orthodox premises: the world is not eternal; and the world is no older than what can be understood from the Mosaic account. That the world is not eternal, he argues, can be seen in the existence of the mountains, which would have been leveled to the ground if time were eternal. Vossius raises the problem of the antiquity of the Babylonians,

Egyptians, and Chinese in his very first pages, declaring that many have argued this antiquity to try to subvert Mosaic chronology, and that some have even made use of that antiquity to introduce the thesis of the eternity of the world. Taken as a whole, he admits, the voices of the champions of a longer history threaten to drown out those of the defenders of Moses.[2]

Vossius was in an ambiguous position, however. His professions of orthodoxy occasionally hid sympathies for libertine theses and he had broad-minded ideas on the biblical text (for which he would be bitterly reproached). Why, he asks, should the books of the prophets be denied the sort of reading that is unhesitatingly admitted for the Gospels? What man in his right mind "would bind the spirit of God to punctuation, letters, or words?" All that the Hebrews had to say "about the perpetual and uninterrupted conservation of the Sacred Books is nothing but imposture."[3]

Vossius later expounded on his reasons for preferring the Septuagint Bible in his *De Setuanginta interpretibus eorumque translatione* (1661). In the *Dissertatio* he tries to show that the Greek version should be preferred for the calculation of time. In traditional calculation, the age of the patriarchs lasted between 800 and 900 years, and they were supposed to have begotten their children in a period of about 150 years. If anyone examines the life of the patriarchs on this question, Vossius argues, he will see that "the longer they lived, the later they were supposed to have begotten children." The new chronology not only eliminated these absurdities; it also permitted a satisfactory account of the "antiquity" of the Chaldeans, Egyptians, and Chinese. From the creation of the world to the Flood 2,256 years passed; from the Flood to the death of Moses, 1,718. Hence, from the beginning of the world to the death of Moses there were 3,974 years. Thus 5,266 years had passed from the creation to the death of Alexander the Great, and in the present year (1659) the world was 7,048 years old. Vossius insists that he had no intention "to found a new chronology" on these calculations, but only "to recall to life an ancient chronology that has been abbreviated rashly." The difficulties raised by the age of China in particular—the most serious difficulties, in Vossius's eyes—seemed thus resolved. The appearance of the dynasties, which many, Father Martini among them, had placed long before the Flood, could now be placed 531 years after the Flood.[4]

Vossius had worked out an answer to Father Martini's proposal of a longer historical past for China. His direct opponent, however, the scholar he intended to take on face to face, was, once again, Lapeyrère. In chapter 12 of the *Dissertatio*, which examines the doctrine of the preadamites, it becomes clear what really motivated Vossius's chronological system. A simple negation of Lapeyrère would not do: the champions of orthodoxy had to be enabled to resolve the problems posed by the greater age of these peoples, not simply be obliged to make an out-of-hand rejection. Thus the annals of Gentile peoples should not be *opposed* to those of sacred history, but used to *confirm* the truth of that history. Even the most recent opinions could be used within the new "system," and the new chronology could, in the name of these new acquisitions to knowledge, differ to some extent from traditional computations:

> There is no lack of testimonies of other peoples to demonstrate this longer age of the world. We have with us the Siamese, the Persians, the Arabs, the Ethiopians, who all count seven thousand years more or less from the beginning of the creation to the present age. . . . Those who follow the usual computation have no defense but the opinions of the Rabbis. We, instead, are defended by the antiquity of the ancient Hebrew writing, by the authority of the Seventy Interpreters, by that of the Samaritan Codex and Josephus, by the consensus of all the Church Fathers and all the primitive Church and, furthermore, by the consensus of all ancient peoples, as well as by the weight of natural arguments and reasons. . . . Who is unable to see that to defeat the impudence of those who impugn the faith using the antiquity of heathen peoples, the one and most efficacious way is to show the truth of divine history on the basis of the annals of those very peoples?[5]

This position alone, Vossius contends, could successfully oppose the preadamite theses. He is forced to admit that the preadamites constituted a real sect, and that they had rallied to their positions all those who dissented from the Bible. Their members' various arguments served one aim alone: to give credit to the false opinion of the antiqity of pagan peoples:

> The opinion of the author of that book survives in the sect that he has left behind him, which is embraced avidly by those who are eager to dissent from the Sacred Scriptures. The arguments of the Preadamites may seem to be many, but it is easy to see that the opinion regarding the antiquity of peoples was the principal and, if I am not in error, the only reason for the writing of that work.[6]

Richard Simon—and his competence in this kind of discussion was undoubtedly very great—spoke on one occasion of "*le coglionerie* di Messer Vossio."[7] Nevertheless, the solutions that Vossius worked out had an undeniable subtlety. The Deluge, contrary to Lapeyrère's affirmations, was universal, but it still did not involve the entire earth. It was universal in the sense that it struck all men then living. Could there have been uninhabited lands, even in those times? And if so, what sense would there have been in submerging them? Many miracles would have to have taken place to inundate the entire globe, and God does not make miracles without reason. Voss calmly qualifies the traditional thesis that mankind at the time populated all of the earth as "foolish":

> Those who sustain that during the age of Noah men spread over the entire globe are quite far from the truth. Men went no farther than perhaps the confines of Syria and Mesopotamia. . . . Even if, on this account, we believe that only that portion of the Earth—perhaps but the hundredth part of the globe—was invaded by the waters, the

> Deluge still was universal because the destruction that fell on the entire then-inhabited world was universal. By accepting this thesis, all this trifling and these futile questions can come to an end.[8]

In reality, though, the questions were not trivial. Vossius had taken one of the theses characteristic of the preadamite sect—the thesis that the Deluge was not universal but limited to Palestine—and passed it off as a confutation of Lapeyrère.

Horn's response was swift, and it took the form of a direct attack. In the same year in which Vossius's *Dissertatio* was published (1659), Horn's *Dissertatio de vera aetate mundi qua sententia illorum refellitur qui statuunt Natale Mundi tempus annis minimum 1440 vulgarem aeram anticipare* was published in Lyon. Right from the start, in the preface, Vossius is presented as a secret follower of preadamite doctrine. The mad Lapeyrère ("delirus quidem Praeadamita") had tried "to destroy the authority of the Scripture, making use of the absurd antiquity of the pagans." His "impious" thesis, "which leads straight to atheism," had already been refuted by those who believed in creation and denied the eternity of the world. Horn's conclusion is somewhat menacing: "That folly was struck by the strongest of arguments, so that, scarcely born, that unhealthy opinion, which was supported by no proof, was snuffed out. If any of its followers remain, may they be punished with scorn or with Magistrates' sentences."[9]

Horn mingles praise of Vossius and threats in a passage that is so typically tortuous that it merits full citation:

> After [Lapeyrère] had already been defeated and that same Preadamite had abjured his opinion, there suddenly emerged a new sort of argument, by means of which that opinion could be more strongly defeated and one could, at the same time, vindicate the authority of the Scriptures against the false arguments of the impious. The argument is this: Do not restrict Mosaic history into as limited a span as the latterday Jews do, and the Christians that imitate them; reject the Hebrew codes, and follow instead the Greek version [of the Bible] as we now have it, consequently pushing back the beginning of the world by 1,440 years. Thus, it is thought, one can dispose of sufficient time to reconcile the antiquity of the pagans with faith in the Scriptures. The intention is undoubtedly good, and its principal aim is to reconcile more easily the chronology of Moses and the Gentile computation of years. Nevertheless, considering the matter attentively, we find that this assertion is not only false, but also dangerous, and such as to produce much harm and confusion to the Church.[10]

There are other, equally heavy-handed insinuations throughout the work. Horn argues that because the world is not eternal but was created by God, those who seek its true beginnings must seek for them in Moses. The books of Moses exist in different languages: one of them, Hebrew, is the original language; all others are translations. How could anyone doubt that the source is preferable to the rivulets that trickle from it, and the original to the translations? Only a curious person, bent on using any means to slash at a truth established for centuries, could proclaim the original source inferior or could accuse that original model of all sorts of crimes. Vossius asserts, Horn continues, that the Hebrew codex should not be preferred to the Greek texts. Have the Western churches, which have based their Bible on the Hebrew text and on the Latin version of the Hebrew text, labored for so many centuries to no purpose? Was their enormous effort to translate the Holy Writ then meaningless, and have they never held the authentic text of the Bible? If Vossius's theses are accepted, Horn protests, they lead to paradoxical conclusions: heathen writers who left no autograph text—Homer, for example, or Virgil—would enjoy more secure status than the Bible. Thus the Hebrew Bible alone, among all the world's books, would be so contaminated that we must prefer to it a more recent version that comes from the hands of obscure or unknown translators. The texts of Homer and Virgil have come down to us uncorrupted, then, and God did not take the trouble to have his word reach us intact through the centuries? If the Jews had really corrupted the text of the Bible, they would have done so for all the other verities of the faith, not just for chronology. The insult that had been directed at the Jews applies to Christians as well: "By supine negligence and vile stupidity" Christians would be guilty of having "conspired with the Jews in an abominable crime"—that of placing a corrupt edition of the Bible before the excellent, impeccable version of the Septuagint.[11]

As it often happens in this sort of polemic, the accusations became tangled. What sources did Vossius use to justify the superiority of the Septuagint, Horn asks. To what authorities did he appeal? The supposed adversary of the impious supporters of the world's immense antiquity was in reality their ally: in order to assert that the chronology of the Hebrew codices was corrupt, Vossius "depends on the authority of Josephus Flavius and on the annals of the Chaldeans, the Egyptians, and the Chinese." Horn's solution was instead very simple: anything in those annals that could not be immediately reconciled with Mosaic chronology must be declared "fabulous." From the creation to the Deluge 1,656 years had passed (2,256 according to Vossius). Christ was born 4,000 years after the creation. The current age of the world (in 1659) was therefore 5,695 years (for Vossius, 7,048).[12]

Horn devotes several chapters of his book to discussing once more the annals of the Chaldeans, Egyptians, Chinese, Ethiopians, and Arabs to offer new proofs that they could be made to agree fully with traditional chronology. This presented no difficulty and had the advantage of a venerable tradition: all the chronologies that differed from the Mosaic were "fabulous chronologies," expressing the desire, inherent in every people, of "prolonging its own history" and of "declaring its own origins to be very distant, hence more noble." We will return to the question of the distinction between *real history* and *imagined* or *fabulous history* more fully in chapter 22. For the moment, we should note that Horn—like many other writers of the seventeenth century—cites Varro in connection with Chaldean, Egyptian, and Chinese chronologies:

> Our first and fundamental premise is the division of time found in Varro. He gives three dividing lines in time. The first [period runs] from man's beginnings to the first cataclysm, and that time, because of our ignorance of it, is called *uncertain*. The second [runs] from the first cataclysm to the first Olympiad, and that time, since in it are found many things fabulous, is called *mythic*. The third [runs] from the first Olympiad up to our time, and this is called *historical* because the things that take place within it are contained within *true history*. This division can be applied not only to the Greeks but also and especially to the Chaldeans, the Egyptians, and the Chinese.[13]

The opponents in this polemic shared, among other things, a great prolixity and an extraordinary rapidity in drafting their texts. In the course of the same year, Isaac Vossius responded to Horn with the *Castigationes ad scriptum G. Hornii de aetate mundi* (1659). Many pages of the *Castigationes* concern the Septuagint and its relation to the Hebrew text, but many others discuss questions of chronology. How, for example, after the Flood and in the span of only one hundred years, could the three sons of Noah beget enough men to construct the Tower of Babel? No matter how long they lived or how frequently their wives gave birth, the maximum number admissible was three hundred fifty men and women. How was it possible for 350 persons to construct the Tower of Babel and found fifty kingdoms? As far as the Chinese were concerned, Voss accuses Horn of confusing Cathay and China. Furthermore, Horn had understood none of Father Martini's work, and had accumulated so much nonsense that he would have done much better to hold his peace.[14]

Horn was by no means willing to accept this advice, and in his *Defensio dissertationis de vera aetate mundi contra Castigationes Vossii* (we are still in 1659) he returns to the fray: Voss had no right to use Father Martini as an argument concerning China because Martini "limited himself to repeating what the Chinese believed to be true, but he does not by any means approve of their opinions." Martini could not, in fact, have even done so "without most grave danger," and was in reality convinced that "the antiquity of the Chinese was made up of fables." Martini thought many of the statements found in Chinese history and geography "ridiculous and fabulous." He held their annals to be "full of fables, both in the names of the Kings and in their acts." Martini, Horn insists, did not for one minute believe in the antiquity of China: in reality "he laughed at it."[15] At this point, Vossius held all the good cards. In his *Auctarium castigationum* (we are still in 1659) he repeats that Horn "was spreading a flood of nonsense." "You say," Voss writes, "that Martini thought in one way and wrote in another. Marvelous among mortals, you know what Father Martini thought better than he did himself!" Martini's work, however was available to all, and his history of China "takes place over 4,600 years: sure, constant, and uninterrupted."[16]

The conquest of time, or its discovery, was a fairly slow operation. In 1659 man and nature, which had come together from the hands of God, had a past of little more than seven thousand years (for Vossius) or of nearly 5,700 years (for Horn). The difference between the age of the world according to Vossius and its age according to Horn is exactly 1,353 years. To prove this difference admissible or inadmissible they discussed all the fundamental problems of chronology once again *ab imis fundamentis*, they debated the characteristics of the great ancient civilizations, they examined and evaluated sources concerning them, and they tackled the problems of biblical philology. The adversaries' countertheses were all subjected to minute analysis, every argument advanced met with a counterargument, invective was exchanged, veiled threats proffered, and the authority of the Fathers of the Church was called into cause. The disputants attacked openly or made insinuations; they appealed to the custodians of orthodoxy as well as to the solidity of the faith and the word of God, and they evoked temptations of the Devil. In more than one instance one scholar reached the point where he questioned the object of his own investigations. Horn, who devoted his entire life to the study of ancient civilizations, wrote later, in 1668: "It has always seemed contrary to Christian piety that many prefer to draw from the labyrinth of their own brain or from the stagnant waters of the pagans instead of from the crystalline fount of Moses and the Prophets."[17]

The choice no longer lay—as at the time of Augustine—between Christian truth and the pagan lies. The "Hermetic" authors who admired age-old Egypt, writers who cited the wisdom of the Chaldeans, the Jesuits who praised China, had raised a series of problems, set new ideas in circulation, awakened curiosities, and opened discussions. These problems called for a response, as did the questions that Lapeyrère had posed, the doubts that he had insinuated into men's minds, and the existence of the sect of his more or less faithful followers and admirers. This response could come in the intransigent and hard-hitting form that Georg Horn chose, taking on the role of champion of orthodoxy against the attacks of the Devil. Or a more flexible response was possible, one that made concessions to the opponents while refuting their theses on the eternity of the world and the immensity of historical time. Or, following the later "libertine" custom of many Enlightenment figures, there was also the possibility of letting "impious" theses filter through in works that seemed to all appearances neutral and inoffensive erudition and of discussing basic problems in the context of secondary figures or seemingly less important concepts.

Chronology was slippery terrain: it was easy to arouse bitter reactions and to be accused of atheism, materialism, and "preadamitism." A Benedictine of Saint Maur, a sworn enemy to the libertines, *esprits forts*, and preadamites, chose, for example, to adopt a much "softer" line with the new theses—in the interest of defeating them all the more effectively. In his *L'Antiquité des temps rétablie et défendue contre les Juifs et les nouveaux Chronologistes* (1687), Pezron reiterated the thesis of Isaac Vossius that had seemed to many to be tinged with libertinism.

Why not indeed accept the Septuagint? It had a more than respectable tradition behind it, and had been used by many of the Fathers. Some concession must be made to the thesis of the longer times of human history in order to defeat Lapeyrère's doctrines once and for all, to defend the faith, and to refute the claim of the greater age of the Chaldean, Egyptian, and Chinese peoples than Mosaic time allowed. "*If one does not give time its just extension*, it is difficult to give a solid answer to the arguments that Preadamites, Libertines, and the 'esprits forts' advance against the books of Moses and against what he says concerning the establishment of the world."[18]

For a long time an answer of a few thousand years seemed sufficient.

21·Moses as *Vir Archetypus*

WHAT lines of defense were open to orthodox Christianity to combat the idea of the immense age of the world, particularly when it served to spread the impious doctrine of the eternity of the world? How could one answer the doubts cast on biblical chronology and counter overly bold parallels between sacred and profane history? How to react to the affirmation of a superior morality and piety among pagan peoples, equal or superior to the ethics and the religion of Christian peoples? Or to the danger of seeing the Old Testament reduced to a chronicle of local history? Furthermore, how was it possible to avoid Spinozist or libertine positions? How could the sacred books be safeguarded as universal history and the Deluge as a universal catastrophe? How could one reinforce the tenet of the isolation and continuity of Hebrew history? How to prove the falsity of the Jesuits' declaration that civilization existed in China prior to the Flood, and how to refute their perilous statements about a Chinese morality and religion not only older than those of the Judeo-Christian tradition but equal or superior to it?

Untold pages were written to answer these questions. There were only two basic strategies, however (to which this chapter and the next will be devoted), and they moved in clearly divergent directions. To simplify to the extreme:

The first direction gave rise to a *process of identification* among different histories and aimed at reducing all different human histories, in more or less complicated ways, to sacred history.

The second direction consisted in a *denial*, declaring "imaginary" or "fabulous" all human histories that rivaled sacred history or deviated from it (as we have seen in Georg Horn).

Two other points should be made before we continue:

1. Both of these points of view arrived at a negative conclusion regarding the very earliest periods of human history: there was no *plurality in history*. For the

age that Varro had called "obscure" there existed one history and one only, which was sacred history.

2. Both points of view, born in defense of orthodoxy, *bore within them the dangers of heterodoxy*, or lent themselves to being used in "impious" ways. Thus, following the first direction to its logical consequences, all religions could be equated with and merged into one unique, "natural" religion. Following the second direction, not only the episodes of all of the "other" religions, but also those of the Judeo-Christian religion could be interpreted as "fabulous."

To support the idea of the greater antiquity of the Scriptures, then, one could start with calling on the testimony of all the peoples of the earth, and try to find in Egyptian, Chaldean, Greek, and Chinese history and mythology survivals or lost records of the truths the biblical text contained. This procedure had an extremely old tradition behind it, which took on new vigor and underwent a sort of rejuvenation in response to the sort of question we have listed. This was the direction followed by—among many others—Samuel Bochart, Gherard Voss, Daniel Huet (in his *Demonstratio evangelica*, 1679), and, concerning China, by Georg Horn in his *Dissertatio de vera aetate mundi* (1659). For these authors, the most ancient biblical wisdom lay hidden in the wisdom of the most ancient peoples. Samuel Bochart's monumental *Geographia sacra* gives a good example of this point of view:

> Tertullian justly states that all that is first is true, and all that follows after it is corrupt. It is in fact necessary that truth come before falsehood, given that falsehood is no other than the corruption of truth. For this reason antiquity is justly esteemed among the characteristics and the criteria of heavenly doctrine. . . . And the antiquity of doctrine cannot be better proven than by retracing all that is *more ancient* among the pagans: that is, what has been taken, or taken in a distorted manner, from our Holy Scriptures. Take the example of the ancient fable of Saturn and his three sons, who divide among themselves the empire of the world. Scholars have for some time seen this as concealing Noah and his three sons, among whom the Earth was divided. But to make this really evident, one would have to compare that fable and the Mosaic story of Noah more carefully. And this I believe no one has yet done with sufficient attention.[1]

The *myth* of Saturn hid the *truth* of Noah; the sons of Saturn were the sons of Noah and, among Noah's progeny, Canaan should be equated with Mercury, Nimrod with Bacchus, and Magog with Prometheus.[2]

The *De theologia et philosophia christiana sive de origine et progressu idolatrae* of Gherard Voss (Vossius) (1642) is a grandiose work, teeming with an endless quantity of gods. Pagan cults are classified according to their objects of veneration (listed in diminishing order of importance): spiritual creatures, celestial bodies, elements, meteors, men, quadrupeds, birds, fish, snakes, insects, fossils, plants, and symbols. One single archetype lies concealed behind the many Joves, Neptunes, and Bacchus: "Given that there were many Bacchus, let us speak first of the most ancient, who is no other than Noah."[3]

In this way Moses became the Thoth of the Egyptians, the Teutates of the Gauls, the Hermes of the Greeks, and the Mercury of the Romans. Saturn and Pluto were reminiscences of Adam and Ham. Hercules recalled Samson. The first Chinese emperors were identified as the biblical patriarchs: Fohi was Adam, Xu-nung was Cain, and so forth. Pythagoras and Plato, Aristotle and Cicero drew in various ways on Moses and the prophets.

Daniel Huet was a master at this sort of operation, and there is a passage in John Toland full of ferocious irony that gives a good idea of what an undertaking it was to draw all of profane history together into the single course of sacred history and to state that all peoples knew the teachings of the prophets:

> On the basis of a few similarities—which are necessarily present in all institutions and in all human comportment—Huet easily demonstrates: that all pagan theologies always derive from Moses; that all the gods and heroes of the Egyptians, Phoenicians, Arabs, Persians, Greeks, and Romans were none other than Moses; that even their kings, as, for example, Cecrops, Minos, and Romulus himself, and their legislators, like Zoroaster, and poets like Musaeus, Orpheus, Linus, Amphion, were no other than Moses, even though concealed in the fables and distributed among different masks and personages. According to Huet, Moses was known to all Indians and, in particular, to the Chinese and the Japanese, as well as to the Thracians, Germans, Gauls, Britons, Hispanics, and even the Americans. There is no reason for astonishment if we learn that the Greek infernal gods are graphic designations for Moses and that Bacchus, Typhon, Silenus, Priapus, Adonis, and other infamous and opprobrious tutelary spirits are identified with the divine Legislator of the Jews. When these premises are established, *more huetiano*, it follows immediately that, in the same manner, all of the fabulous goddesses are in reality only one.[4]

In defense of Huet and in sharp polemic with Toland, Jacobus Fayus wrote a *Defensio religionis necnon Mosis et gentis judaicae* (1709). Fayus raises an important point: if it were true—as indeed it was true—that Moses was *the oldest* of all the heroes; if it were true—as indeed it was true—that no heathen histories or fables were *older* than the history Moses narrates, then it made no sense to reproach Huet (as Toland had done) for his little respect for chronology. The princes, poets, legislators, and heroes whom Moses figures need not be his contemporaries. Moses was the first of all the heroes; he was a *vir archetypus*, and that was enough to explain how his person could inspire the many different "masks" or the various "fabulous personages" that appeared in different cultures and at different times. Huet did not "excogitate" the significance of the fables; he *found* the meaning that lay within the fables. Fayus, who had certainly read Samuel Bochart, shared one of Bochart's basic premises: what comes first is true; what comes after is corrupt.

After men "strayed from the truth and entered deeper into the multiple aspects of error . . . they never stopped adding fable to fable." The qualities, the "gifts" of Moses, became "distributed" in a series of different personages, and countless names were invented to indicate them.[5]

When Toland quite hypocritically accused Huet of equating the divine legislator with Priapus and other opprobrious tutelary deities, he hit home, as he often did. Even the decidedly and rigidly conservative thesis of Gherard Voss, Huet, Bochart, and Horn offered several dangers from an orthodox point of view. In the last analysis it brought the history of the Jews closer to heathen history and denied the isolation of the Hebrew people. Above all, it came dangerously close to the deist notion of a religion common to all peoples, above and beyond its historical forms in different cultures, and to the notion that elements of truth were present in all religions. It also showed worrisome affinites with the doctrine of a Christianity as old as creation and of a truth in some way present among all peoples because it had been transmitted to all the descendants of Noah in equal measure, starting from the time of Abraham.

Furthermore, if there were many Moseses as masks of the archetypical Moses, and many Adams and Eves who represented the first Adam and Eve, might not the very fables that hid Moses, Adam, and Eve refer to a time as far back as Moses or even as Adam and Eve? And would we not then have to admit for the Gentile nations that endless antiquity that ought to be reserved for the events of sacred history? Joseph François Lafitau noted that if the Greeks' Ceres, the Egyptians' Isis, and Phrygians' mother of God were none other than Eve, mother of all men, "then almost all fable in pagan mythology must refer to the ages that preceded the Deluge." The floods of Deucalion and Ogygus in this case would not have been limited floods, but the true Deluge, of which all nations have conserved "some sort of idea, but one that was very confused in the times of the profane authors who first wrote of it after Moses."[6]

But there was more: if the figures and the rites in Chinese religion were symbols or disguises for Christian figures and rites, did that not in some way place Chinese religion on the same plane with the true religion? When we extol the "affinities" between pagan religions and the true religion, are we not in the last analysis offering support to the libertines? Would this not put the ancient saints and prophets on the same plane as other exceptional men who lived during the earliest ages? Was not the comparison, finally, "an unworthy and insupportable insolence" toward the ancient saints?[7] If the Jesuits could see the Chinese mandarins as halfway Christian, why should the Greek and Roman philosophers be denied knowledge of God? Could we not say, like Dr. Cudworth, that peoples of multiform polytheism like the Egyptians nevertheless had knowledge of a supreme and universal God? Do we not risk stating that God was present even among peoples who knew neither the law nor grace? That God can be reached by natural reason alone?

In *Christianity as Old as the Creation* (1730), Matthew Tindal claimed there was no difference between natural religion and revealed religion: if God, from the very beginning of time, gave a religion to man, that religion must be perfect, not susceptible to augmentation or diminution, but as immutable as its Author. How could revelation add anything to a perfect and immutable religion, a religion "as old, and as extensive, as humane-nature"?[8] By insisting on similarities and on the perennial presence of true religion throughout *all* of the history of the human race, was it not inevitable to end up maintaining, like Tindal, that Christ's words were not a *new* message and did not differ in their essence from a natural knowledge of God? Could one not assert, like Toland, that even the most ancient Egyptians, Persians, and Romans, like the Hebrew patriarchs, had had an extremely simple religion without sacred images or particular places of worship? Did not an acceptance of Spinoza's affirmation of the superiority of the pagans over the Jews in matters of morality and relgion lie at the end of this road? And seeing the Chinese as "true worshippers of God"? And transforming Moses himself into a deist or a Spinozist?

These were dangers that did not escape the more attentive observers. Samuel Bochart, Gherard Voss, and Georg Horn were dedicated and impassioned champions of orthodoxy. Their comparative history aimed not only at defending sacred history from the attacks of the impious opponents of Christian truth but also at reconciling all histories with the one true history. Daniel Huet had also attempted to refute the theses of Spinoza's *Tractatus*. He considered that book an attempt to give credit to "the dangerous heresy of deism" and thought it the baleful fruit of the teachings of Ben Ezra, Hobbes, and Lapeyrère. But results often differ from intentions, even from pious intentions. Antoine Arnauld had some illuminating remarks to offer on Huet's *Demonstratio evangelica:*

> If the Protestant author [Arnauld is referring to a review of Huet's book that appeared in the *Histoire des ouvrages des savans*, June 1691] has not misconstrued what he speaks of . . . these are horrible things, capable of suggesting to young libertines that one must have a religion, but that all religions are good and that paganism itself can be considered equal to Christianity.[9]

The systematic, continuous, insistent, and minute comparison of the dogma and mysteries of the Judeo-Christian heritage with pagan myths and fables ended up giving the impression that even the verities of Christian faith could be placed on the plane of myth and fable. Citing Arnauld's judgment, Jean Racine declared Huet's book "abominable and full of impiety." True religion, Racine declared, became in Huet's work dissolved into the world of the "ancient fables":

> Moses did nothing miraculous when, throwing wood into the bitter waters he made them sweet, no more than Elisha in throwing in salt. We see in Pliny that the thing happened naturally. There was also nothing miraculous in the manna that fell in the desert. It usually falls in those countries. The same is true for the crows that nourished Elijah. Jupiter was nourished in his cradle by bees, Cyrus by a bitch, Darius by a mare, Romulus by a she-wolf and Moses made water

> flow from a rock. Darius dying of thirst with his army asked the gods for water and they accorded him a great rain. . . . The dead [were] resuscitated by Elijah, by Elisha, by Jesus Christ, and by Saint Peter: Theseus and Alcestis [were] resuscitated by Hercules, Semele by Bacchus, Hippolytus and Castor by Aesculapius. . . .[10]

Arnauld and Racine saw the problem clearly: that "pious" solution was full of dangers and orthodoxy's defense against the sect of the preadamites ended up offering weapons to the deists. John Toland offers proof of this, this time in positive terms. He is attacking Huet in this passage, but at the same time using him, citing him extensively for his own not exactly edifying purposes. With characteristic irony, ambiguity, and skill, Toland slips in the thesis of the equality of all religions as he is critically examining the theses of the "guileless" Huet:

> Huet, with his usual artlessness, affirms that Strabo records together three seers who can predict the future: Orpheus, Musaeus, and Moses, and he says that the three are one and the same person. As if Strabo had not also recorded together Amphiaraus and Trophonius and Tiresias and Zalmoxis and the Assyrian, Chaldean, and Etruscan deviners! All of these, and with equal right, could not only be compared to Moses, as the Geographer did [Strabo], but also identified with Moses, as the Demonstrator [Huet] would have it.[11]

This was, to Fayus's point of view, an odd way of arguing: not only did Toland tend to confuse all religions; he did not distinguish between religion and superstition. Someone had to "administer an antidote" for all who were unaware of the true purpose of Toland's arguments. "The young especially" must be warned about an author who should be compared with Bayle and Spinoza, a man "whom you could never persuade me should be tolerated by the Anglican Church."[12]

Pietro Giannone reacted just as strongly (in 1730) to the conclusions he found in Father Antonio Costantino's *Philosophia adamitico-noetica*. Costantino had found the Trinity in Chaldean oracles, Persian magic, and Egyptian wisdom. He had insisted on the analogies between Egyptian myths and the events of Christian history and he made wide use of the conclusions and the methods of Marsham, Bochart, and Huet. For Giannone, as for Arnauld and Racine, the idea of an original divine and earthly wisdom, transmitted from Adam and Noah to all the peoples of the earth and learned by the Jews through Abraham seemed downright dangerous:

> It is a very perilous undertaking to subject all to human discourse, to ratiocinations and deductions and conjectures taken from profane history, and to attempt to explain divine history with them. When it is suggested that the matter be examined in this way, I strongly suspect that the libertine writers, and especially the English, and a few Dutch and Germans, who have vomited many profanities on the question in several printed books, have won their cause.[13]

22·Real History and Fabulous History

AS we have seen at the beginning of the preceding chapter, another direction was available to anyone who wanted to rise to the defense of the unity and universality of revelation and the Holy Writ, or to anyone intent on combating the thesis of the immense age or the eternity of the world. He could distinguish between *facts* and *fable*, between the real historical facts characteristic of sacred history and of more recent history and the myths and fables proper to more remote, "obscure," and "fabulous" history, as Varro had called it.

John Webb, the champion of Chinese as the primeval language, defined clearly (in 1669) the dilemma confronting those who did not want to accept the "impious" theses. There were only two possibilities: to accept it as true that the first sovereign of China reigned 3,000 years before Christ (consequently accepting, as Father Martini claimed, "that the historical time of the Chinese began 2,847 years before Christ") *or else* to take the Chinese narration of their "history" to be legend and those times to be imagined, expressed in fabulous terms, and without correspondence to a real time.[1]

Ten years earlier, in 1659, Georg Horn had been equally explicit: either you admit that the various chronologies are all the work of the Devil and you refuse to take them into consideration, *or* you try to discern what, and how much, is fabulous in them. Inachus, for example, was held to be the earliest king of the Argives, and many illustrious scholars thought that he could be identified with Enoch. According to many commentators, Inachus lived before the Flood, but since we must identify the floods that the Greeks speak of with the Noachian Deluge, it follows "either that Inachus, who reigned before those floods, really lived before the Deluge of Noah, *or* that he was a fabulous personage."[2]

Thus the area of the "fabulous" tended to broaden out until it included all of the accounts and all of the traditions that seemed not to agree with biblical chronology. To defend the truth of sacred history, then, one could deny that the accounts and chronicles of the earliest periods of history documented a genuine historical reality; one could deny that they were proofs or documents of time *really lived* by mankind. It could be maintained that the more we look back in the course of the millennia, the less the accounts and traditions are *historical* and the more they are *fabulous*—that is, without correspondence to real facts. All the dangers implicit in the affirmation of an immense age or a near eternity of the world were thus nipped in the bud. The immensely long time of the history of the world became an imaginary or storied time, not real time. In this point of view that continent of earliest history, discovered by Lapeyrère, appeared populated not by men who had really lived thousands of years ago, but only by personages invented in the more recent times of human history.

The bitter attacks mounted against Lapeyrère had insisted, in fact, that knowledge of history before the Deluge was impossible independent of the Mosaic source. Le Prieur, for example, argued that we have no chronology of that history *(rerum gestarum series)* and for that very reason "all of that time was unknown and concealed." Varro, that "outstanding stimulator and excellent defender of the truth," was right to speak of an "obscure" or "uncertain" age of history. In that uncertain age, there were no historical *facts* but only fables and prodigies; men became heroes or giants, and the narratives took the form of myths or fables.[3]

It was also within the context of the attacks on Lapeyrère that the thesis that earliest antiquity was totally imaginary and sprang from the "vainglory" of the various peoples gradually emerged. Given that every people claimed to be the oldest, why should we not think that in every case this was a matter of unconscious vainglory or of conscious lies? How could anyone reject the biblical account—which is continuous and coherent—on the basis of a multiplicity of irreconcilable traditions, all in rivalry with one another?

Johannes Heinrich Ursin (Ursinus), as we have seen, violently attacked Lapeyrère's work the year after its publication. Soon after, in 1661, he published a work entitled *De Zoroastre bactriano, Hermete Trismegisto, Sanchonianthone phoenicio eorumque scriptis et aliis contra Mosaicae Scripturae antiquitatem*. In it Ursin indicts all scholars who drew their arguments from these three most ancient authors. Although he declares the falsity of the Hermetic texts and attacks Kircher, he credits the length of the histories of pagan nations to their desire for fame:

> If anyone can be persuaded that the things that are told about Thoth are true, then he must also believe true the things that were blathered about the kings, gods, and demigods who, before men [were created], are supposed to have reigned in Egypt for infinitely long periods. The dispute over antiquity is very ancient: not only among the Greeks, the Athenians, and the Argives, but also among the barbarians, Egyptians, Phrygians, Chaldeans, Ethiopians, and Scythians. Not only the Athenians in Greece have boasted to being autochthonous, but also the Aborigenes in Italy, the Turdetans in Spain, and the Burgundians in Gaul.[4]

The immense age of these nations, then, was documented by unreliable and fabulous sources. The "myriads of years" that preceded the history narrated in the Bible did not contain historical events: they were merely an expression of the vanity (Vico would call it the *boria*, or ostentation) of the people and the wise men of these nations. Arts, inventions, and sciences were not of remote origin, but were recent acquisitions. The ancient Egyptians and Chinese were not peoples of highest culture and refined civilization: behind their imagined wisdom hid rough and primitive nations who "fantasized" on their antiquity and their culture.

Many historians of ideas who treat the seventeenth and eighteenth centuries (as well as many Vico specialists) have failed to give adequate consideration to one point: the thesis that the most distant past was fabulous, that the long ages of human history had only been imagined or were legendary *necessarily led* to an acceptance of the thesis of an initial "roughness" in the first nations. If their age-old civilization was completely imaginary, if their "long" past was inexistent, then the traditional image of their wisdom must be replaced by one of their essential poverty of life-style, of ideas, and of customs. And the idea of their initial backwardness and lack of civilization could even slip into that of their "barbarous" origins. Since the theme is an important one, since much that is inexact has been written on the question, and since Vico's argument on the *boria* of ancient nations has been presented as an original "discovery" on his part, it is worth documenting the question more fully.

As early as the first attempts to confute Lapeyrère's heresy, we can begin to see the contrast between the pretense of reconciling Scripture and the histories of the Chaldeans, the Egyptians, and the Chinese and the recognition that the Bible was irreconcilable with "the ravings of those barbarous men." What need was there for this sort of operation, Philippe Le Prieur asked in 1656. Why seek support from the poets and the pagans?

> The Egyptians and the Chaldeans tried in vain to make the antiquity of their peoples go back beyond that time [of the Deluge] and they believed they could lie without restraint because they thought it very difficult to find out their lies. Among all peoples, in fact, Egypt had the reputation of a begetter of portents and master of lies. . . . Some serious writers reached the point of claiming that such a noble empire had existed before the Deluge. . . . And that, without any judgment whatsoever, is what Scaliger affirmed, when he made the kingdoms of the Egyptians and the Assyrians go back to before the Deluge.[5]

Marten Schoock as well, in his *Fabula Hamelensis* (1662), refers to the "presumption" of the Egyptians, which should be "held in suspicion no less than the antiquity of the Chinese, whom we are investigating."[6] Georg Horn (in 1659) was even more explicit: not only the Chinese, but many nations had wanted to extend the catalog of their kings back beyond the Deluge. Having learned from their own traditions that great empires and great heroes had existed before the Deluge, each people appropriated some of those heroes and identified them as their own founders: "singulae gentes hos sibi tanquam conditores illustres arrogarunt." It was neither legitimate nor possible, Horn concludes, to correct the Holy Writ on the basis of the false chronology of the Egyptians and the Chinese. One could not reject Moses on the basis of the "fabulous stories" of the ancient nations without passing over to the camp of the libertines and the preadamites. Furthermore, Horn continues, it was not in any way legitimate to heed the *insania figmenta* of Manetho, according to whom the dynasties of Egypt preceded Abraham by many centuries. If the extremely detailed Chinese chronologies were allowed, the beginning of the history of the Chinese people would fall seven or eight centuries before the Deluge or, as other scholars put it, 88,000 years would have passed from the beginning of the history of the world to the present. That apparent precision was only a sign of falsity: the Chinese traced their historical time back to 2847 B.C. but, Horn insists, "it is not really historical time, but fabulous time." To replace a period of real time that had lasted 640 years, the Chinese had imagined an infinitely longer period.[7]

Isaac Vossius, as we have seen, had something of a fondness for theses of libertine leanings, but he accepted (or seemed to accept) the distinction between real history and imagined history. The Chaldeans, he writes in his *Dissertatio* of 1659, exaggerated their age, and the Egyptians—*gens vanissima*—not to be outdone by the Chaldeans, added "their year-crammed list of gods and demons." Manetho, Vossius continues, based his entire construction on this imaginary antiquity, "and with malice aforethought extended his people's past to an endless age." The accounts of the antiquity of the Chinese, on the other hand, "have been rightly considered fables."[8] Gherard Vossius as well, in his *Chronologia Sacra* (1659), holds that the Egyptians "made such a fuss, boasting of the antiquity of their history, so they alone would be considered wise and all others would seem like children."[9] Sir Matthew Hale, in *The Primitive Origination of Mankind* (1677), is just as decisive:

> Notwithstanding these great pretentions of Antiquity, yet upon a true examination, their great pretended Antiquity is fabulous; and the Origination of their Monarchies began some Ages after the general Deluge; and so the truth of the Holy History concerning the Inception of Mankind, and the Inception of all the Monarchies in the World . . . is not at all weakened by those Fabulous Antiquities of the *Babylonians*, *Egyptians*, or *Grecians*.[10]

Lapeyrère's name comes up again in Hale, hardly by chance, as someone who put more faith in those fables than in Scripture; someone whose ideas, if true, would "not only weaken but overthrow the Authority and Infallibility of the Sacred Scriptures," and who, "*Judas* like while he seems to kiss these Sacred Oracles, . . . perfidiously betrays their Authority, and draws their Truth, as much as he can, into suspicion." When Lapeyrère gave more credit to the "Fabulous Traditions" of the Egyptians, the Chaldeans, and the Greeks than to the word of Moses, Hale asserts, he was really calling fables and fictions all of the first eleven chapters of Genesis.[11]

After the turn of the century, Melchior Leydekker undertook to write the history of the Hebraic state in twelve volumes. In his *De republica Hebraeorum* (1704), he describes the miracles of divine Providence in the constitution of the state, the theocracy, the political regime, and the public and private religion of the Hebrew people. As well as defending the verities of the creation and the Flood against the impieties of Thomas Burnet, the work aimed at correcting the false beliefs concerning the age of the world and at showing the "fabulous origins" of the Egyptians,

Phoenicians, Arabs, Chaldeans, Greeks, and Romans. Only the Hebrew nation knew its own distant past: for all other nations the past was uncertain, fabulous, and lay wrapped in thickest darkness. The historical time of the heathen nations began where sacred history ends. Leydekker wrote hundreds of pages to prove that earliest antiquity was fabulous. He used Bochart and Selden, Vossius and Stillingfleet, all of whom had tried, in various ways, to decipher the arcane contents of the fables and to find in them a corruption of the truth of sacred history. He criticizes Huet for going too far in this direction when he identified all manner of pagan heroes with Moses. Leydekker's conclusion, however, is the one we are already familiar with: "profanae gentium antiquitates meris fabulis constant."

> The ancient records of profane, Gentile peoples consist only of fables. And this appears from the fact that they invent years that total many centuries more than the real age of the world. Such are the ancient accounts of the Chinese, the Chaldeans, and the Egyptians, which are only dreams and invidious *kaukémata,* made to raise themselves arrogantly over other Peoples by means of an imaginary antiquity. Scaliger, who believed in the many dynasties invented by the Egyptians, labored in vain to digest Manetho's fable: the history of the peoples who glory over their antiquity is fabulous [*fabulose gentium de suis antiquitatibus gloriantium est historia*].[12]

To combat the libertines, Jam Martianay declares (in 1693) the ancient Egyptian histories to be fables. In his *Bibliothèque universelle des historiens* Louis Ellies du Pin mocks the supposed antiquity of the Phoenicians and the Chinese and declares that no people "boasted of a greater antiquity than the Chaldeans." Eusèbe Renaudot repeats, in 1718, that the arts and the sciences were recently produced, and he reduces the ancient wisdom of China to a wisdom typical of a primitive people (and adds that the libertines had used the Jesuits' statements to attack Scripture and to claim that the world is much older than it was thought). It is religion, Tyssot de Patot declares, that prevents us from granting the Chinese the past that they justly demand; it is Christianity that constrains us to consider as fables the perfectly documented accounts that testify to their real antiquity. Those who want at all costs to reconcile Chinese chronology with biblical chronology present the history of the earliest ages of China as "a mass of confused and unsure traditions" which only in recent times have been "tied together to form a whole."[13]

Only by accepting the idea of the *recent* appearance of the arts and civilization could faith in the word of God be maintained. Only by believing the boundless antiquity of the world and the nations to be the fruit of the vanity of those peoples and their philosophers could men be confirmed in the faith and biblical truth continue to be opposed to the myths of the heathen nations. In the *Voyages et aventures de Jacques Massé* (1710), Tyssot de Patot repeats an argument typical of the defense of orthodoxy against libertine theses. Egyptians and Chinese claim great antiquity, but "all of this is advanc'd without any Foundation, and from a

Principle of Vanity to challenge a Superiority over the other Nations of the Earth."[14]

In his attempt to "emend" the erroneous chronology of ancient kingdoms, Isaac Newton made wide use of this sort of argument. Every nation, Newton said, each claiming a more noble origin, pushed its history back in time. Thus the gods, kings, and deified princes of the Chaldeans, Assyrians, and Greeks were said to be more ancient than they really were. Thus the Egyptians built, out of vanity, the image of a monarchy older than the world by some thousands of years. Ancient histories are all uncertain, often imaginary, and always "full of poetical fictions." "The *Egyptians* anciently boasted of a very great and lasting Empire. . . . Out of vanity [they] have made this monarchy some thousands of years older than the world."[15]

In the text of *The Original of Monarchies* (1693–94), recently made available by Frank Manuel, we find the same assertions:

> Now all nations before they began to keep exact accompts of time have been prone to raise their antiquities & make the lives of their first fathers longer then they really were. And this humour has been promoted by the ancient contention between several nations about their antiquity. For this made the Egyptians & Chaldeans raise their antiquities higher than the truth by many thousands of years. And the seventy have added to the ages of the Patriarchs. And Ctesias has made the Assyrian Monarchy above 1400 years older then the truth. The Greeks & Latins are more modest in their own originals but yet have exceeded the truth.[16]

As Voltaire was to note in the seventeenth of his *Lettres philosophiques*, all of Newton's calculations aimed at shortening the history of the world: "Comparing the state of the sky as it is today, to the way it was then, we see that the expedition of the Argonauts should be placed about 900 years before Christ, not about 1400 years, and that consequently, the world is younger by about five hundred years than was thought. Thus all ages are brought closer, and all occurred later than it was thought."

The claims of the Egyptians, Armenians, Ethiopians, and other peoples regarding the antiquity of their language came, Augustin Calmet declared in 1723, from love of their country and their people.[17] Pietro Giannone echoed the same widely circulated ideas:

> It was the custom of almost all the ancient writers, when they wanted to give an origin to their own nation, to draw it back as far as they could in order to show the greater age and the preeminence of their own over all the nations of the world. . . . The Germans made Teutates, born of the Earth, the author of their nation . . . neither is there any nation that has not indulged in these dreams of showing off an unattainable and incomprehensible antiquity.[18]

Attacks on chronologies based on Manetho, the Egyptian, and on Berosus, the Chaldean, became a commonplace and can be found in many works. Even Antoine Yves Goguet, a level-headed author, repeated (in 1758) the themes of the "ridiculous pretentions" of the Egyptian "manufacturers of all those marvelous antiquities," who "instead of . . . giving us separately the catalog of the princes who had reigned at the same time over different parts of Egypt" brought them all together "in one and the same catalogue" and tried to make readers believe "that every one of those princes had successively reigned over all Egypt." Goguet as well is sure of one thing: "It is then provided, by the testimony of the highest and soundest antiquity, that it was only in modern times that the Babylonians and Egyptians began to made a parade of those thousands of ages."[19]

In the article "Histoire" of the *Encyclopédie* (1757) Voltaire too refers to the "absurd" origins of all these peoples, citing the "prodigies" of chronology of the Phoenicians and the Egyptians, but reaffirming the continuity of the history of the Chinese nation, "the most ancient of all the peoples who remain until today." Voltaire came to be reproached bitterly for these and other similar affirmations on China, among others, by François Nonotte (1772): "Only an imbecilic credulity could allow all of those supposed Chinese emperors." When he discussed infinity and chronology, Voltaire did not hesitate to use the argument of pride: "It is true that there is no family, no city, no nation, that does not try to move back its origin."[20]

The so-called reactionary literature of the eighteenth and the early nineteenth centuries is full of instances of denying to ancient nations a past of thousands of years and of qualifying such a past as imaginary and arising out of vainglory (or, as Goguet said, from the "weakness that almost all peoples have for their antiquity"). Three examples, chosen from a truly remarkable number of texts, may be enough: (1) an anonymous English pamphlet attacking "the most numerous sect of unbelievers: I mean the spinosists" (1734); (2) Laurent François's widely circulated work in defense of "the religion of Jesus Christ against the Spinozists and the deists" (1751); (3) Chateaubriand's masterpiece, *Le génie du christianisme* (1802).

The four essays that made up the *Free Thoughts concerning Souls* are devoted to the nature of the soul, to the comparison of the human soul and that of the beasts, to a presumed preexistent state of souls, and to the future state of souls. The theme of the periods of human history is completely marginal to this argument, but when the most dangerous and widespread heresies are discussed, the heresy of the eternity of the world is of course among them. Other than the Greeks, the anonymous author argues, there were many other nations who "have pretended to a much earlier time than even this of Moses." The Egyptians were a people "guilty of extreme vanity," and Chinese history could certainly not be older than 4,000 years. As later in Vico and in many other authors, the criterion of the relation between time and civilization was used in the *Free Thoughts* to put limits on the historical process. That the invention of writing and of all the other arts and sciences was recent offered further proof of the recent formation of the earth and against the great age of human history. Both from a "geological" point of view and from the "historical" one, six thousand years were more than enough to explain the changes in the earth and the "gradual improvement" of knowledge.[21]

The facts with which Manetho filled his "imaginary times," François writes in his *Preuves de la religion de Jésus-Christ*, are stories of gods and demigods, plus a long list of sovereigns who either never existed or who did not succeed one another in that order. Berosus certainly offered information no more reliable, and the Chinese discredit their own marvelous accounts when they tell of a usurper who some centuries before Christ burned all of their books and leveled all of the monuments of their previous ages. Chaldeans, Phoenicians, Egyptians, Assyrians, Arabs, Persians, and Chinese had nothing to offer against Moses: following the history of those peoples without the guidance of Mosaic history would be like "throwing oneself off a precipice with them into a dark hole, out of which one cannot escape except burdened with incertitudes, fables, and chimeras." How, François asks, can we keep from laughing at the Chaldeans' claims? "I know that some nations attempted to give themselves a vain antiquity . . . but the pretensions of those extravagant nations lack proofs and evidence."[22]

Chateaubriand certainly had no lack of talent for giving telling literary expression even to cultural clichés (or to ideas that had become clichés). In the *Génie du christianisme* (1:4) he too took on the "objections to the system of Moses." The opposition, visible in the title, between *logography* and *historical* fact summarizes perfectly the essence of a dispute that lasted for many centuries: "In vain shall we combine with imaginary ages, or conjure up fictitious shades of death; all this will not prevent mankind from being a creature of yesterday." All of the scholars who maintained that the natural world and the human world were of a great age had taken to citing Porphyry, Sanchuniathon, and the Sanskrit books, and referred to the tradition of the Theban priests who spoke of an Egyptian kingdom 18,000 years old. Human history, Chateaubriand argues, is instead recent, as are the inventions and the arts. This is shown by the fact that the names of their first inventors—Bacchus, Ceres, Noah—are "as familiar to us as those of a brother or a grandfather." History, geometry, medicine, the fine arts, and laws have not been in the world for long; we owe them to Herodotus, Hippocrates, Thales, Homer, Dedalus, and Minos. Are we not ourselves "a striking instance of the rapidity with which nations become civilized?" Twelve centuries ago our ancestors were as barbaric as the Hottentots, and today we surpass Greece in refinement of taste and the arts. If it were true that languages need a long time to reach completion, "why have the savages of Canada such subtle and such complicated dialects"? All of the arguments for the antiquity of history were "as unsatisfactory in themselves as their research is useless." Chateaubriand's work should also be cited in connection with the questions discussed in part 1 of the present work, as he too attacks all those who looked to "the history of the firmament" for proofs of the age of the world and the errors of Scripture, who spoke of the millions of years needed to form the solar system, or who theorized on "certain epochs" in history or claimed to read in the earth's "fossils, her marbles, her granites, her lavas" to discover in them "a series of innumerable years."[23]

To those who based their arguments for a past of numberless years on the antiquity of human history and the age of the history of the earth, the champions of shorter historical time periods invariably put the same two questions: If time is as long as you say, why are the arts and the forms of civilized life so recent instead of ancient? If the past extends infinitely, why is the earth today, minor variations aside, like the earth of yesterday?

Both the apparent immobility of nature and the correlation of time and civilization (as we have also seen in two of the three texts just considered) were thus used to limit historical time. If the world was eternal, if the past extended infinitely, why have the mountains, which are continually eroded by water and wind, not yet disappeared? Why were the arts not invented before our times? What are we to think when during infinite centuries the compass, printing, and gunpowder were unknown to men? Benoît de Maillet, in 1748, considered these questions "trivial difficulties, so often discussed and always with so little success." He thought them specious objections destined to "vanish like smoke, before the clear and resplendent lights of reason."[24]

> Can they who know nature, and have a reasonable idea of God, comprehend that matter, and things created, should be only six thousand years old? . . . [that] the human mind invents only little by little, and so slowly that, in order to produce the smallest novelty, it requires several centuries; that we have lost an infinite number of rare secrets, . . . that since the discoveries of all kinds made two centuries ago will certainly one day be buried in oblivion, the ancients had perhaps made a greater number of them which have not come down to us[?]. . . . The Romans . . . owed almost all their knowledge to the Greeks and to other nations; . . . the Greeks borrowed from the Egyptians all the skills which made them so famous; . . . for many centuries, the latter had reached perfection in all arts and sciences since . . . they were already famous in that respect when the Jewish nation was only in its infancy; . . . the Chaldeans did not yield to them in that matter; and . . . the Chinese dispute their title with both.[25]

The compass, printing, and gunpowder were new discoveries only for some peoples: they were of ancient origin as far as China was concerned, and they had perhaps been lost and fallen into disuse through carelessness. Phoenicians and Carthaginians may have had some sort of navigational instrument that served the purposes of our compass. Furthermore, in what way could emphasizing the ignorance of these peoples prove the brevity of history?

> Since, for so many centuries men have lived in ignorance of these discoveries, it is not impossible that such an ignorance may still be older; and that the world, having been without these inventions for

> 6,000 or 7,000 years, may have been as well without them for 50,000 or 60,000.[26]

Those who denied the long view of history always cited the exaggeratedly long life of the patriarchs. The life span of the earliest men, Maillet argued, was more or less what it is today. If there were a difference, it would be in favor of modern man, given that our ancestors "inhabited caves, lay upon the leaves of trees or upon the hard ground, and only lived on the herbs and fruits which the earth produced spontaneously, and without cultivation."[27]

The themes we find in Maillet's pages—occasionally thoroughly mixed together—had a long tradition. So it is with the idea of the slow progress of mankind, of the bond between natural history and human history, of the image of a pluralistic history in which the Hebrew people deserves no special place, of the conviction that regressions or losses occurred during the course of time, and with the idea of an original barbarity. De Maillet certainly did not lack imagination; still, at least concerning the problem of a long historical past, his faith in the "clear and resplendent light of reason" was excessive. The denial of a history longer than six thousand years was destined to continue, as is known, well into the nineteenth century. In his *Essai sur l'indifférence en matière de religion* (1817–24), Lamennais totaled up the "chimeras" that the eighteenth century had indulged in and found they led to bankruptcy:

> The philosophy of the last century spoke of nothing else but the prodigious antiquity of the Egyptians, the Chaldeans, the Indians, and the Chinese. Today even schoolchildren laugh at this chimerical antiquity, the falsity of which has been unveiled by . . . scholars of the first order. The more we investigate the history of these nations, the more we see it approach, in all it offers that is sure, Mosaic chronology.[28]

Much of the "prodigious" age claimed by so many authors of the seventeenth and eighteenth centuries did indeed prove to be "chimerical," and successive historiography did not leave much room to the fabulous dynasties of the Chinese and the Egyptians. But only a few decades later the modern concept of *prehistory* was to be born. Then the complicated process would be set off that would lead to imagining a history that was in some way unified—a history that begins with human paleontology and passes through prehistory and archeology to reach "history." The existence of a half-dozen major ancient civilizations, previously totally unknown, would slowly be revealed. Finally, men would come to accept the idea that there had been men who were contemporary with extinct species of vertebrates during the first ages of the history of the earth.

23·*La boria delle nazioni*

FIRST, beyond the six thousand years of biblical chronology, history—both natural and human history—had never existed. Second, the history of the Chaldeans, the Egyptians, and the Chinese prior to the starting point of biblical chronology is not real history, but an invented or imagined history (and for that reason, full of myths and fables). Third, the invention of that ancient, imaginary history was the fruit of the vainglory of those peoples; it arose, so to speak, spontaneously from the nature of man, who loves to give himself noble origins and who fantasizes about a glorious past even when he has none. These three assertions were certainly not entirely new when they were formulated in modern times to respond to the libertines. They were connected to ancient arguments that we can find as early as about 50 B.C. in a passage from Diodorus Siculus (*Bibliotheca* 1:9) often cited by Vico and by seventeenth-century historians and scholars:

> Again, with respect to the antiquity of the human race, not only do Greeks put forth their claims, but many of the barbarians as well, all holding that it is they who are autochthonous and the first of all men to discover the things which are of use in life, and that it was the events in their own history which were the earliest to have been held worthy of record.

In the *Contra Apionem* of Josephus (A.D. 94) there was another passage that was often cited: "In fact, each nation endeavors to trace its own institutions back to the remotest date in order to create the impression that, far from imitating others, it has been the one to set its neighbors an example of orderly life under law."[1]

And, finally, there was Augustine, who in the *City of God* (A.D. 416–23, 12:101; 18:39, 40) defends the count of six thousand years and inveighs against "the falseness of these writings" that speak of "many thousand years." "And the Egyptians themselves . . . are proved behindhand in comparison with our patriarchs." For this reason, "it is therefore a monstrous absurdity to say, as some do, that it is above 100,000 years since astronomy began in Egypt."

By the middle of the seventeenth century, the theme of the vainglory and the presumption of ancient peoples had become—as I have tried to document in the preceding pages—a true literary *topos*. This theme was present, in different forms and with different functions, in all of the authors who tried to limit the length of world history and refute the idea that man's presence on earth might date back to many thousands of years before the Flood. In any event—and this is what I was principally interested in documenting—Vico's concept of the *boria delle nazioni* was not a brilliant intuition and even less a *discoverta:*

> As for the conceit of the nations, we have heard that golden saying of Diodorus Siculus. Every nation, according to him, whether Greek or barbarian, has had the same conceit, that it before all other nations

> invented the comforts of human life and that its remembered history goes back to the very beginning of the world. . . . This axiom disposes at once of the proud claims of the Chaldeans, Scythians, Egyptians, Chinese, to have been the first founders of the humanity of the ancient world.[2]

"All the histories of the gentiles have their beginnings in fables": many pages of the *Diritto Universale* and of the first and second *Scienza Nuova* are devoted to attacks on the supporters of the "boundless" *(sterminata)* antiquity of human history.

In the *Diritto Universale,* Vico too begins with Varro's tripartite division of time into the obscure, the fabulous or heroic, and the historical epochs. Thus far it has proved a hopeless task to investigate the history of obscure times, Vico says, and the most careful critics are in violent disagreement concerning the reality of both the events and the heroes of fabulous history. Vico's preoccupation with the impious thesis that antediluivian wisdom had perhaps been preserved among the heathens and that the Deluge could thus be reduced to multiple local floods is clearly revealed in the *Diritto Universale,* particularly in the section devoted to refutation of the traditional scholarly error concerning the "intentional" or *ex electione* origin of poetry. How was it possible to believe that Homer, father of all poets and philosophers, born in an age of generally unpolished philosophy, "sprang up suddenly and spontaneously"?

> From this comes the problem that has disturbed religious souls: whether other peoples had not been submerged by other particular floods, and whether the survivors, remaining on the mountain heights, might not have preserved a wisdom that they possessed before the deluge. The Christian scholar must confront this problem, which would confirm the thesis of the eternity of the world.[3]

In spite of the magniloquent titles of the many books that have been published on the question, Vico continues, the beginnings of profane history are unknown, and those who have written on historical matters confess candidly that profane history "had neither a certain origin nor a certain course." If this were not so, "things of the obscure age would come to light, those of the heroic age would be stripped of the fables that envelop them, and we could know the causes out of which the events of the historical age were born."[4]

Vico realized clearly that the beginnings *(principia)* of history, as many had formulated them during the preceding century, constituted as many *incommoda* to the truth of sacred history. Those beginnings could lead to an acceptance of the enormous age of the world proclaimed by the Chinese. It was the duty of every Christian to refute that claim, and it was possible to do so by making use of Roman history:

> If that corresponded to the truth, the enormous antiquity of the world imagined by the Chinese would be confirmed, and that is something

> that the Christian scholar should not seek to hide, but criticize and refute. And this is easy to do on the basis of all the true things that we have learned from Roman history.[5]

A sure origin and a sure course, completely absent in profane history, could only be found in sacred history. Indeed, sacred history tells us of the events that occurred *during* the time that profane history *was passing through obscure and fabulous* or heroic times. This is why sacred history offers us the intermediary *(tramitem habemus)* by means of which we can join the history of obscure and fabulous times to the history of real times. Vico continued to hold firm on this point. It was a decisive point and, in my opinion, many and many serious misinterpretations have come of failing to grasp it.

> If Sacred History shows us the events that occurred *while* the obscure and fabulous age was taking place in profane history, we are in possession of an intermediary by means of which the history of the obscure and fabulous age can come to touch the history of real time. For this reason, what thus far has been said, with truth, about the dimensions of the terrestrial world, that is, that they draw their certainty from the certain measures of the heavens, should also be said concerning the truth of the civil world: the truth of profane events should be demanded of Sacred History alone.[6]

Sacred history was *true, perpetual,* and *ancient,* and all three of these characteristics differentiate it clearly from all other possible histories. Its greater antiquity is proved by the fact that in no profane history is there mention of theocracy in connection with *ex lege* mankind (either before or after the Deluge) in the ages before the foundation of the republics and the establishment of the laws. The Hebraic state, on the other hand, had theocratic forms from the beginning through the laws of the Decalogue, which do not specify punishments established by man, and this state of affairs lasted for five hundred years from the age of Moses to that of the first kings.[7]

In the *Diritto Universale* Vico presents the discovery of a ferine age as the result of an *inversion of chronology*. Orpheus and Amphion give evidence "that their Greek compatriots were wild beasts and rocks [*fiere e massi*]." Cecrops the Egyptian, Cadmus the Phoenician, Danaus the Phrygian are presented as carrying human characteristics *(umanità)* into Greece. But Orpheus and Amphion lived three hundred years after Cecrops and two hundred years after Danaus. None of these was correctly placed in time: the events of both ages were true, but they had been reversed. When we invert them, Vico says, the problem is solved:

> When I wondered at these chronological absurdities, this came to my mind: these events, even if they are reversed in time, are nevertheless true. After the fame of Egyptian, Syrian, and Asiatic power had spread among the Greek people, in order to make their own origin more

august they put it back in time. And this is the solution to the problem.[8]

The ambition of peoples had made ancient events more recent. When the chronology is reversed and the truth of how periods followed one another is reestablished, it is possible to state, according to Vico, (1) that the true religion of God the creator was maintained; (2) that mankind's fall into barbarism was limited to the Japhetites; (3) that the entire earth was submerged by the Deluge. The descendants of Shem preserved the true religion, human sociability, language (at least until the confusion of tongues of Babel), and remembrance of the arts of antediluvian humanity. Japheth's descendants "lived long, resembling Orpheus's wild beasts [*fiere*] and Amphion's rocks [*massi*]." They abandoned all religion and with it all humanity. The Hamites and the Japhetites were dispersed to wander alone through the earth's great forests; the men separated from the women and children from their mothers until, "repudiated and alone, they first forgot all religion, then all language, finally all humanity." This state of affairs lasted some thousand years. During the same period the Hamites, through contact with the Shemites, "were made aware of humanity" and had occasion to observe over the vast plains the rising and the setting of the stars. Impressed by stellar motion, "they pretended that the sky was a god." The Japhetites, on the other hand, were "reduced to a brutish stupor" and "had to be shaken by thunder to believe that the sky was a god and to see its will in Jove."[9]

The fact that the Assyrians had well-developed, refined cults while this savage way of life continued, even among the Greeks, for a thousand years after the Deluge could be explained only by remembering that the Shemites had kept the true religion. If the true religion had been maintained intact among the descendants of Japheth as well, or if Japheth, like Ham, had lived nearer a pious people, then the history of the West would have resembled that of the East.[10]

Moses, Vico continues, was antecedent to Lycurgus, Solon, and the decemvirs, and superior to them in wisdom. He was, at the same time, a consummate philosopher, historian, and legislator, and he lived in an age before "the heroic poets, the unjust legislators, the crude philosophers, and the fabulating historians" had appeared among men. Until now the reasons for the restoration of humanity to Western peoples has been unknown, Vico continues, and "one sole error, confirmed by the age of all the epochs, has occupied the soul of all the scholars." There was a truth that they did not see: that the first language among men was poetic, and that with this poetic language the first laws and the first religions were founded. When we unveil the origin of poetry, born not by intentional design, but by natural necessity, we reach a proof of Providence. Even the false religions served to insinuate the idea of divine omnipotence in men's minds, to help the fear of God to arise within them, and to lead them to live in society. God punished Adam's sin with shame: what was to have brought the ruin of mankind was instead the very source of its salvation and men's ability to live in association. This was what

Epicurus and Machiavelli had failed to see, and Hobbes, Spinoza, and Bayle as well.[11]

When in 1688 the physician Giambattista Menuzzi presented a denunciation to the delegation of the Inquisition in the Kingdom of Naples, he stated that atheists in Naples were claiming "that before Adam, there were men in the world."[12] In the *Scienza Nuova* Vico takes a quite clear position in opposition to the thesis of the preadamites and against the doctrine of the eternity of the world: "Sacred history assures us that the world is almost young in contrast to the antiquity with which it was credited by Chaldeans, Scythians, Egyptians, and, in our own day, by the Chinese."[13]

The fourth of the seven causes for the obscurity of the fables listed by Vico in the *Scienza Nuova Prima* refers to the distortions of the fables. Things heard or reported imprecisely became amplified, particularly when they concern events long passed and when the men who had word of them were rough and ignorant. For this reason, "word of ancient or distant things comes to us for the most part utterly false and always magnificent." This principle, which lay behind the distortions of the fables, was also at the root of the belief in the long ages of history and the nearly infinite age of the world, a belief that was contrary to faith:

> And this also is the reason for the appearance of the world, which seems ancient considerably beyond the merit of truth and faith, and which, in the dark until now concerning its origins, has seemed to those incredulous of Sacred history of an almost infinite antiquity, whereas in the light of this Science it is shown to be of very fresh date.[14]

Illuminating the darkness of those origins was one and the same, for Vico, as proving that the world was not as ancient as the incredulous would have it, and that it was instead extremely "fresh," or of recent origin: this demonstration was one task he set himself in the *Scienza Nuova*.

Men judge "distant and unknown things" by what is "familiar and at hand." For Vico, this aphorism described a property of the human mind, and when it was taken as a point of departure, as he says in the *Scienza Nuova Seconda*, the "proud claims" of those peoples "to have been the first founders of the humanity of the ancient world" is "disposed of at once." In like manner, this axiom "disposes of all the opinions of the scholars concerning the matchless wisdom of the ancients"; it "convicts of fraud" the oracles of Zoroaster the Chaldean, Anacharsis the Scythian, the *Pimander* of Hermes Trismegistus, the verses of Orpheus, and the *Carme aureo* of Pythagoras. All of the "mystic meanings" that scholars have read into ancient sources on "remote antiquities, unknown to us" derive from a characteristic illusion of the human mind, noted by Tacitus: "everything unknown is taken for something great" ("omne ignotum pro magnifico est"). These scholars have allowed themselves to be led by this and by hearsay, and they judge the origins of humanity as "small, crude, and quite obscure" in comparison with cultivated and enlightened times.[15]

Vico destroys at the root all possibility of allegorical interpretation of myths, he rejects the age-old tradition of an occult wisdom, and he interprets fables as the spontaneous expression of primitive mankind. At the very same time, however, he rejects the thesis of the world's great age, he reasserts that sacred history is older than any other history, and he contrasts the *boria* of ancient nations to the *verità* of sacred history: "This axiom proves the truth of sacred history as against the national conceit pointed out to us by Diodorus Siculus, for the Hebrews have preserved their memories in full detail from the very beginning of the world."[16]

Thus the unlimited antiquity of the nations, when it was presented not as a myth constructed by the vainglory of the peoples and of their wise men, but as an effective historical reality, was something that challenged the truth of sacred history. For Vico, all discussion of the real age of India, Egypt, and China was as meaningless as the use of allegories and "recondite wisdom." The Jews had "reckoned rightly the account of the time passed through by the world." Unlike the Chaldeans and the Scythians, and unlike the latter-day Chinese, the Jews did not distort (or distorted by "a very short period of time") the duration of real history.[17]

Vico was well aware—as we have seen in connection with the problems of Homeric wisdom—that behind the discussions of the world's great age lay the perennial threat to Christian tradition of the materialist and libertine thesis of the eternity of the world. Zoroaster, Mercury, Orpheus, and their like sprang from the beginnings of heathen mankind. "And if they did not, as they should not, wish to posit the eternity of the world, scholars should have meditated upon origins of this kind, in order to base the science of humanity, i.e., the science of the nature of nations."[18]

The philosophers who saw the founders of the ancient nations as adepts of a recondite wisdom had their first master in Plato, and it was to Plato that Vico attributed—even if in dubitive terms—that impious doctrine. He says in the *Scienza Nuova Prima:*

> We are unable, therefore, to consider the origins of the humanity of nations in terms of the reasons which philosophers have so far adduced. These reasons go back to Plato who, believing perhaps in the eternity of the world, took his start from the humanity of his own times, in which philosophers from other civilized nations must [already] have tamed the human race which remained elsewhere in a state of savagery.[19]

In the section that opens the *Notes to the Chronological Table* and with which he begins his discussion of Egypt, Vico clearly indicates his adversaries: Marsham, Spencer, and van Heurne shared the opinion that Egyptian civilization could be contrasted and systematically compared to Hebraic civilization, and from this comparison they had concluded that Egyptian wisdom was anterior to the Hebraic and a source for it:

> [The Chronological Table] takes a position quite opposed to that of the Chronological Canon [*Canon chronicon, aegyptiacus, hebraicus, graecus*] of John Marsham, in which he tries to prove that the Egyptians preceded all the nations of the world in government and religion, and that their sacred rites and civil ordinances, transported to other peoples, were received with some emendation by the Hebrews. In this opinion he was followed by [John] Spencer in his dissertation *De Urim et Tummim,* in which he expresses the opinion that the Israelites had taken from the Egyptians all their science of divine things by means of the sacred Cabala. Finally Marsham was acclaimed by van Heurn in his *Antiquitates philosophiae barbaricae,* in which, in the part entitled *Chaldaicus,* he writes that Moses, instructed in the knowledge of them by the Egyptians, had brought divine things to the Hebrews in his laws.[20]

In the same *Notes,* after he traces the portrait of the ancient Egyptians as rough, primitive, and conceited, incapable of rational thought, boastful of their imaginary antiquity over all other nations, and after a brief mention of the Chaldeans (who "did not fail to enter the lists in this contest of antiquity"), Vico takes up the problem of China. The Chinese, like the Egyptians, "grew into such a great nation having remained closed from contact with foreign nations." This isolation from "other nations by whom they might have been informed concerning the real antiquity of the world" lasted "many thousands of years." The Chinese, "in the obscurity of their chronology" and who "boast in vain of enormously ancient origins" (and with them the Egyptians and the Chaldeans) are "just as a man confined while asleep in a very small dark room, in horror of darkness [who, on waking] believes it certainly much larger than groping with his hands will show it to be."[21] To counter the theses that attributed many tens of thousands of years to Chinese history, Vico shortens historical periods: Chinese civilization began "some four thousand years ago and not before"; Japanese civilization, "some three thousand years ago"; that of the Americans, "some thousand five hundred years ago."[22]

Vico's references to a whole series of texts and authors take on a much clearer meaning in light of all this. He says, for example:

> It is true that Father Michele Ruggieri, a Jesuit, declares that he has himself read books printed before the coming of Jesus Christ. It is true further that Father Martini, another Jesuit, in his *Sinica historia* ascribes a great antiquity to Confucius, which has led many into atheism, as we are informed by Martin Schoock in his [*Demonstratione diluvii universalis*], in which he says that Isaac de la Peyrère, author of the *Praeadamitae,* perhaps for that reason abandoned the Catholic faith and then wrote that the flood spread over the lands of the Hebrews only. Nevertheless, Nicolas Trigault, better informed than

> Ruggieri or Martini, writes in his *De christiana expeditione apud Sinas* that printing was in use in China not more than two centuries earlier than in Europe, and that Confucius flourished not more than five hundred years before Christ. And Confucian philosophy, like the priestly books of the Egyptians, in its few references to physical nature is crude and clumsy, and it turns almost entirely on a vulgar moral code, that is to say on morals commanded to the people by laws.[23]

Aside from a few inaccuracies in his references, which Fausto Nicolini has noted with his habitual diligence, Vico's basic judgment remains totally valid—from his chosen point of view. The theses that the Jesuits (Martini in particular) had advanced concerning Chinese chronology could furnish ammunition to the libertines and lead to atheism, and they were irremediably connected with Lapeyrère's doctrines. What was important, faced with the challenge of these atheistic theses, was to demonstrate that this much-vaunted antiquity was illusory, to show that the invention of printing and the age of reason were recent, to see the "moral" age as unrefined, and to compress the ultra-long periods of that chronology into a rather shorter time, a time that agreed with the chronology of Father Petau.

Nicolini has already described the difficulties and the "contradictions" in which Vico became involved when he pushed human life on earth to the end of the third millennium before Christ and when he compressed into little more than a thousand years the reign of the beast-men *(bestioni)* and the two ages of the gods and the heroes.[24] During that thousand years, men, who had already become uncivilized by the time of the Tower of Babel, were supposed to have regressed to bestiality and then reemerged to rational thought. These difficulties are undeniable. But they really do not provide a basis for saying that Vico's discourse is the expression of a "confusion-prone" mind or is shaped by expedients of a "political" nature. Nor is there cause to wonder if Vico presents his "reflections on the vain opinion of their own antiquity held by these gentile nations," using human reasons "in support of Christian faith" ("tutto il credibile cristiano") and to confirm the thesis that the Hebrews, "whose prince was Adam, created by the true God at the time of the creation of the world,"[25] were the most ancient people in the world.

From this point of view sacred history becomes for Vico the criterion and the focal point of the problems raised by profane history. To oppose the view of the beginnings of profane history that was brandished in the pompous titles of so many books (and we really should not need to explain that the term *principi* in this context is synonymous with *inizi*, beginning), Vico found it necessary to recognize that profane history lacked clearly defined origins and a clearly defined course. Furthermore, fabled history was uncertain in origin—*incerta*—because knowledge of the obscure ages that preceded the fabulous and the historical ages (following Varro) is nearly impossible—*disperata*. Sacred history was *more ancient* than all of the various profane histories, and it was the only one to have certainty concerning its inception and its successive course. It tells of events that occurred while the obscure age and the fabulous or heroic age were taking place: "res nobis exponit actas, dum in historia profana tempus obscurum et fabulosum sive heroicum excurrit." The *boria* of nations, and the *boria* of scholars were an "inexhaustible source of all the errors about the beginnings of humanity that have been adopted by entire nations and by all the scholars."[26] The debate on Chinese chronology and on the antiquity of the Egyptians or the Chaldeans—questions that had shaken European culture for nearly a century—lost all meaning. Behind those quarrels, behind that difficult crisis, lay nothing but the illusion born from the darkness and the isolation of the ancient nations; lay nothing but the vainglory of invented, fictional antiquities, all of which had taken place in a time imagined by those peoples and by their philosophers, and not in real time.

24·History Sacred and Profane

VICO, as we have seen, did not limit himself to a rejection of the theses that viewed Hebraic civilization as *deriving from* or *depending on* profane civilizations (and in this his position contrasts with that of John Marsham, John Spencer, and Otto van Heurne). He also forcefully rejected any attempt to "illustrate" sacred history by means of profane history (and here he took a stand counter to that of John Selden, Daniel Huet, and Samuel Bochart). There is no doubt that Vico, as Arnaldo Momigliano has clearly seen, was intent on "defending the priority and the verity of sacred history against any attempt at contamination."[1] In light of these considerations—which are securely founded in the texts—it does not seem to me to make much sense to insist (as some have done) on verities that are obvious to any reader of Vico's works. There is little point, for example, in stating that (1) in the *Scienza Nuova* the two parallel histories never seem to meet; (2) in the *Scienza Nuova* we can find more a faith in Providence in general than in Christian Providence; (3) Vico's meditations are much more concerned with the world of the nations than with sacred history and, consequently, the biblical datum is often given parenthetically; (4) as soon as it is possible to "delineate a human history with documents and references to events of the history of nations, sacred history passes to the background."[2]

All of these affirmations are true, and no one has ever taken Vico for an ingenuous apologist for orthodoxy or for someone who echoes inoffensive commonplaces. Vico had precise adversaries in mind: Bayle, Hobbes, Grotius, and all of the irreligious tradition of those who denied the superiority of sacred history. These are the opponents he fought, and he posed problems, offered explanations, advanced theories, formulated explanations, and even discovered "verities" that were to have a decisive influence on modern thought. The cultural context in which he moved was complicated and difficult. The theses of the antiquity of the world, of the roughness of primitive man, even the thesis of "barbarism," could *serve different ends;* they could evoke suspicions and condemnations and, at the same time, could be welcomed as efficacious weapons for the defense of Christian truth. One point should be kept in mind: the lines of defense that the Christian tradition drew up against the impious theses of libertinism, against the thought of Hobbes and Spinoza, and against an emerging science of history that "implicitly or explicitly cast doubt on the validity of the Bible"[3] were contorted and broken at more than one point.

In a situation of this sort (abundantly documented in the present work) it is just a sign of ingenuousness to wonder at the fact that "popular wisdom, conceived as a highly radical means to defend orthodoxy, immediately turns back on itself." We seem to be back in the times of a historiography founded on the conviction of the progressive forward march of the human race, or on the ineluctable triumph of the spirit. Is it not clear, for example, that Galileo took a stand against the "true" doctrine of the moon's influence on the tides in the name of a totally valid polemic against occultism? Is there anyone who has not realized that Diderot conceived of his defense of the "false" doctrine of spontaneous generation as a decisive argument in favor of the clearly "progressive" thesis of the emergence of life out of matter? It has been seen as cause for great astonishment that Vico "pursued, with tenacious coherence, an instrument that was not only incongruous, but clearly counterproductive to the aims for which it had been imagined."[4] But this is just what we find ourselves recognizing (even without believing in a Viconian-Christian or a Hegelian Providence) both in historical research and in daily life.

It has been said that Vico's novelty consists in the theorization of a "complete parallelism" between two different historical developments (the Hebraic and the Egyptian), which "are not placed in a relation of reciprocal exclusion" but are "historically and logically co-possible."[5] We need to agree on the use of words. For Vico, the Hebrew people (1) was the first, or the oldest, in the world, and it originated in Adam, who was created by God; (2) was the only people to maintain its records, in its sacred history, from the beginning of the world; (3) was the only people to have had a notion of the time that had passed since the origin of the world that corresponded to real time and who, unlike other peoples, had only slightly misrepresented the duration of real time; (4) was free of the vainglory and the *boria* found in all other peoples, who boasted to having "founded humanity" and of having kept its records from the beginning of the world; (5) had lived hidden from all other peoples, both their Mediterranean neighbors and very distant nations; (6) unhesitatingly admitted this isolation and had no cultural contact with the Egyptians, not even during the period of the Egyptian captivity, both because the Egyptians, like all the other peoples, considered the peoples they vanquished to be men without religion, and because it is unthinkable that the prophets could have profaned their doctrines by revealing them to strangers and new arrivals; (7) came of a different origin from the pagan peoples, and were thus biologically different, to the point where the "entire original race" was "divided into two species," one of giants and the other of men of normal stature; (8) were of normal stature because they received proper upbringing ("pulita educazione")—a "human education"—whereas the pagan peoples had a "ferine" upbringing (the latter had also been generated of proper stature, but after the impious races of the three sons of Noah dispersed throughout the earth, wandering in savagery, they were found again as giants at the time when the sky thundered for the first time after the Deluge); (9) did not fall into "bestial wandering," were not dispersed over all the earth, but "persisted in humanity"; (10) had a "natural law" that must be distinguished both from the natural law of nations and from the natural law invented by the philosophers, and "on account of their failure to observe these differences between the three [natural laws], the three systems of Grotius, Selden, and Pufendorf must fall."[6]

Given that "historically co-possible" means (if I am not mistaken) possible at the same time, it is by no means legitimate to use this expression and speak of "parallelism" concerning histories of *different lengths,* where one of them contains a coherent narration of real events while the other gives legends and confused traditions that are the fruit of the vanity of ancient nations. The history of the Hebrew people, we ought not to forget, was for Vico "more ancient than all the most ancient profane histories that have come down to us." It alone, in a historical past of approximately 5,000 years, narrates in detail "the state of nature under the patriarchs; that is, the state of the families, out of which, by general agreement of political theorists, the peoples and cities later arose." Profane history says nothing, or says little and that little confusedly, of that familial stage; it constructs fables or imagines ages that are the fruit of vainglory.[7] Eight hundred years are almost one-fifth of a total of 4,500 to 5,000 years: thus, according to Vico, we have one and only one account of a fifth of the human history that has already occurred, and it is the sacred history of the Hebrew people. Sacred history (1) speaks of events that occurred *while the obscure and fabulous ages were taking place,* and is thus the only reliable source for those ages; (2) has one unique characteristic that distinguishes it from all profane histories: it is the *one* history that was *written as the events it narrates were taking place*. There can be no chronological "co-possibility" between the two kinds of history, and in no way have they a "qualitative homogeneity."[8]

In a recent and important work Mario Sina also expressed doubts concerning Vico's "decisive affirmation of two parallel developments in human history"—that of the Jews and that of the Gentiles. Sina says quite rightly that Vico used biblical arguments more and more parenthetically but, in my opinion, he has less reason to refuse Vico the least "desire for orthodoxy" or "apologetic intent." It is true, as Sina has emphasized, that Vico "gives the foreground to the history of the nations and of their advances, to the history of a humanity that gradually passes from the ferine stage to civil association," but it *not* in the least true that Vico consequently advocated "a chronology of human history of lay inspiration." The contrary is true: within a traditional and orthodox chronology, Vico puts the history of the nations in the foreground. Sina went quite far in his interpretation of a "lay"

Vico, so far as to state that Vico would like "to see as operative even in Sacred History those principles of humanity that constitute the new science."[9] Sina cites section 405 of the *Scienza Nuova Prima* to support this statement. In the chapter (5:3) entitled "Origins of This Science Found within Those of Sacred History," according to Sina, Vico tries to *insert* the beginnings of the New Science into sacred history. Vico says exactly the contrary. Vico says that the principles (or origins: *principi* has both meanings) of the New Science—"modesty," "curiosity," and "industry"—were *already contained within* sacred history. Vico was to return in his *Autobiography* to the question of the "indispensable and even human need to seek the first origins of this science in the beginnings of sacred history." But this same thesis had already been stated, fully spelled out, in the earlier section 25 of the *Scienza Nuova Prima* (which is "The human, as well as doctrinal, necessity that the principles of this Science be derived from holy scripture"). Retracing the origins of profane history in sacred history is not only a necessity dictated by faith; it is, for Vico, a "human necessity," *una necessità anco umana*. As far as the world's most distant past is concerned, the "first common principles of humanity" ("il primo comune principio dell' umanità") cannot be found either in the history of the Romans, which is too recent, nor in the history of the Greeks, which is fabulous, and not even in the obscure and discontinuous history of the Egyptians and the Chinese. That origin, that *incominciamento*, must be found in sacred history:

> Since all gentile histories have similar fabulous origins . . . and since we have abandoned all hope of discovering the first common principles of humanity from anything Roman, which, in relation to the world's great age, is of comparatively recent origin, or from the vanities of the Greeks, the remains of the Egyptians, such as their pyramids, and even from the total obscurities of the East, let us seek them among the principles of sacred history.[10]

One common beginning of humanity lies at the root of human history. The man Vico speaks of is neither the man of positivism and Darwinism nor the man of the synthetic theory of evolution. Vico's civil and rational man comes from a brutish beast-man who *once had been a man*.

Sacred history, which is "older than the fabulous faith of the Greeks," therefore narrates in detail the "era of families" during which "fathers ruled under the government of God." That "original stage of nature" or "era of families" was the first the world knew, and (as Vico added in the definitive edition) it lasted a good eight hundred years[11]—an enormously long period of time in Vico's chronological scale. At this point we should ask several questions: Does it really make much sense to note that sacred history and profane history "seem never to meet" in the *Scienza Nuova?* Or to point out that sacred history "passes into the background" when Vico discusses the history of the nations? Does the statement that Vico's main interest is in the world of the nations constitute an interpretation of Vico or just the recognition of an obvious fact? Why on earth should the two histories ever have *met?* If "meet" means "touch at one end," the two sorts of history obviously do meet for Vico, given that the precise task he set to his new science was to search out the certain origins ("origine certa") of a universal profane history and the way in which it "attaches" ("si riattacca") to a *more ancient history*, which is sacred history.[12] If instead the term "meet" is intended to mean "interweave" (as Reale and Sina and a number of other Vico scholars use the term), it is clear that the two sorts of history do *not* interweave. With what could a beginning, a *principio*, an *inizio*, an *incominiciamento* possibly interweave?

Where and *when* did profane knowledge ("lo scibile gentilesco") have its first beginnings in the world? Vico holds that to answer this question is to reinforce what it is legitimate for Christians to believe: "by adducing human reasons thereby in support of Christian faith [*tutto il credibile cristiano*], which takes its start from the fact that the first people of the world were the Hebrews, whose prince was Adam, created by the true God at the time of the creation of the world." The origins of profane history and its *perpetuità* (which in this context means *continuity*) with sacred history, Vico says, "have therefore lain unknown until now."[13] Between sacred history, which is continuous, coherent, and narrates real facts, and the "certain origins of universal profane history," which begins with the Greeks, there was an intermediary "obscure and fabulous" age that was as if *empty of history*, a sort of no-man's-land that all thought it legitimate to appropriate for themselves. Unlike sacred history, which regards a preceding period of time as well and in which *that void does not exist*, the history of Gentile peoples has neither an ascertainable beginning nor a sure continuity. Furthermore, we know nothing of the way in which the history of these peoples "continues" or connects with sacred history. The task that Vico set himself was to clarify the origins of the history of Gentile peoples, which followed sacred history and seems "to connect" ("riattaccarsi") to sacred history only "by means of the start of Greek history, the source of all we possess about profane antiquity." The *Scienza Nuova* demonstrates the certain origins of all of universal profane history and its continuity with sacred history through Greek fabulous history and more certain Roman history:[14] this was what seemed to Vico one of his greatest "discoveries."

Vico's achievement—and all of his interpreters are in agreement on this—consisted in having made his way through that world of the fabulous and that obscure age in which it was difficult even to penetrate. Anyone who discusses that world has to refuse to project into it the categories of reason and must realize that that humanity was different from our own, that those minds were "not in the least abstract, refined, or spiritualized," but were instead "entirely immersed in the senses, buffeted by the passions, buried in the body." But we must not forget that, for Vico, that obscure and fabulous world stood, chronologically, *in between two histories:* sacred history (which tells us of true things) and the first histories of the Greeks and the Romans (which were the first reliable sources of true things). That obscure and fabulous world stands chronologically in between sacred history—

narrated by Moses 1,300 years before Homer and Numa, and in which there are no "foul corruptions of the first traditions of the facts"—and the beginning, in Assyria, in Phoenicia, and in Egypt, of those "human governments" that *followed* after the two ages of the gods and of the heroes.[15]

Vico speaks of his distinct sentiment of having discovered a new land: "Ought we to . . . make ourselves possessors of the things of the distant past which have hitherto belonged to nobody, ownership of which can consequently be conceded legitimately to whoever occupies them?" It was like a voyage of discovery in a new world, where the explorer restores their real meaning to the things he sees after myths and legends had charged them with improper meanings. That *historical void,* Vico says, was filled arbitrarily through the *boria* of the learned and the *boria* of the nations, and works of recondite wisdom were attributed to it.[16] In that no-man's-land, however, all was instead "most incertain" because, during all those years (1) events that took place in different ages had been narrated as if they had occurred at the same time; (2) events that took place at the same time were narrated as if they had occurred at different times; (3) ages crowded with events were narrated as if nothing had happened in them; and (4) ages with few events were narrated as if they had been crowded with happenings.[17]

The problem of the length of the various ages is certainly not foreign to Vico's thought: his investigation, as he repeats on several occasions, is "a continual meditation on how much time was needed [*sopra quanto vi volle*]" for men of the impious races of the sons of Noah, fallen into barbarism after the Flood and dispersed throughout the great forests of the earth, to arrive at the epoch of the Seven Sages of Greece, which signaled their entry into humanity. By the operation of Providence, men were brought into humanity (at least in the West) in 1,500 years, but *during the same time* the Jews had continued on in humanity and in "civil discipline."[18]

How much time was needed: the ages of Vico's chronology are well known and have been analyzed thoroughly, even if all too often with a tendency to disregard them as irrelevant. How could a "Hegelian" philosopher, a champion of historicism, and a precursor of Benedetto Croce have taken an interest in these absurdities? Many carefully labored pages of the *Scienza Nuova* have thus been considered by Fausto Nicolini as "extravagances." Before the Flood 1,656 years passed. For Vico that period represented "a great void of a thousand six hundred years" within profane histories.[19] In the *Scienza Nuova Prima* and in the 1730 edition as well, Vico held that time could be "made up for" ("supplito") with the hypothesis that the same effects were produced in the "lost race of giants" of Cain's progeny as those the Flood had produced (after more than 1,600 years) in the "lost races" of the sons of Noah, before God "was pleased to enter into a new alliance with Abraham" and to "preserve his own true religion in his race." This hypothesis would have kept intact the difference between the history of the Jews and the history of the Gentiles, but it would have *set back* the age of the fall into bestiality by a great deal. Vico abandoned it in the definitive edition. The fall into bestiality of the impious races of the sons of Noah began the year after the Deluge (in the year of the world 1566 or 2328 B.C.). It was a process that lasted a hundred years for the race of Shem, and two hundred years for those of Ham and Japheth. Obscure time—or the age of the gods, or the familial stage—lasted nine hundred years: fabulous time—or the age of the heroes—two hundred years.[20] Put on a comparative scale, real historical time begins in each nation with the constitution of the "familiar" forms of society. As Sergio Landucci has pointed out in pages that are essential to the comprehension of Vico's concept of historical time, "the only chronological variable in Vico's theory is the length of the period of wandering in the wild, before the respective beginnings of the different nations."[21]

Historical time, for Vico, is neither unified nor uniform. Universal law is "uniform in all nations wherever, notwithstanding their differences in times." All the nations were propagated from Mesopotamia. About a thousand years after the Deluge, in Assyria, Egypt, and Phoenecia, "the age of the gods and that of the heroes have already gone by" and some sort of "human governments" have begun. When in the East, in Egypt and in Syria, the nations "had already gone under human governments," the Greeks and the Italic peoples *still* "lived under divine governments," and the Greeks reached human governance sooner than the Italics "proportional to the greater proximity of Greece to the East." Nearly two and a half millennia passed between the beginnings of the nations in the Old World and those of the Americas: the ferine stage of the Americans lasted much longer and their familiar stage lasted about 1,300 years instead of 900 years. The Chinese nation has existed for not more than 4,000 years, the Japanese for only 3,000 years, and that of the *Patacones,* or the giants of Tierra del Fuego only 1,000 years. Chinese, Japanese, and Americans have little history behind them: this is why the Chinese still have "few articulated words"—since they have had them only for 300 years—and why they write in "hieroglyphics." This is why the Japanese are "also a ferocious" people and their humanity "retains much of the heroic." This is why the Americans are governed "by terrible religions *still* in the familial stage."[22] Many nations still retain traces of barbarism, Vico continues, and these traces reveal the brevity of "human" history. Only one history is *umana* for its entire span, and that is Hebraic history. All the other histories *became* "human" in different, but in any event more recent ages. The length of the ages is calculated by the length of that one history that has *always* been "human." Behind the ages that biblical history narrates, there are no other ages and there is no other history. The territory that Vico explores does not coincide, at least in his intentions, with what we now call prehistory. It is enclosed within the clear confines of traditional chronology. But it is well known that at least in the world of ideas we can travel in countries that are different from those we thought we were passing through.

25·Rough Peoples and Barbarism

WHEN he advanced the thesis of the *boria* of the nations, Vico necessarily refused the first alternative direction referred to at the beginning of chapter 21. The traditions of the pagan nations were not accounts of a history that had really happened, but neither were they cryptograms for truths, for wisdom, or for theories. Even less could they be interpreted as symbols of Hebraic or Christian doctrine.

When he rejected all comparison between sacred history and profane history, Vico also rejected the theses that declared biblical truth to be present in the various historical forms of culture and religion. His polemic against Marsham's and Spencer's doctrines, his refusal of the "succession of schools" that supposedly led from the Chaldeans to the Greeks, and his ironic comments on van Heurne's "barbarian philosophy," which "the vanity of scholars has so much applauded" all seem to be based on a reaffirmation of the sacred and continuous character of Hebraic history, on the superiority of the Hebrew religion, "more ancient than those by which the nations were founded," and on the theses that Hebraic history preserved a detailed account *(spiegatamente)* that reached back to earliest times and had "kept its records from the very beginning of the world."[1] Vico cites Lactantius, who denied that Pythagoras was a disciple of Isaiah, and Josephus, who claimed that the Jews at the times of Homer and Pythagoras lived in complete isolation. Taking this as a base, Vico declares that he can "overturn" and prove false what Bochart, Selden, and most particularly Daniel Huet had said on the spread of Hebraic wisdom among the pagan peoples and on the covert or implicit presence of biblical truth among all the peoples of the earth. Daniel Huet, Vico says, was one of those Christian critics "who would have the founders of the gentile nations issue in a state of complete learning from the school of Noah." In the *Demostratio evangelica*, Vico continues, that most erudite man tries to make his readers believe that "the fables are sacred stories corrupted by the gentile nations." But it was improper, even impossible, to compare the two histories: "sacred" history was synonymous with "secret" history and its truth, "free of the filth contained in profane history," without the "foul corruptions of the first traditions of events," provided a "most luminous proof of the truth of the Christian religion" and a "weighty confutation of the errors of recent mythologists."[2]

To attribute civility, decorous customs, and culture to the founding fathers of the pagan nations, to conceive of the ancient gentile peoples as gifted with wisdom, was for Vico a doctrine that led to the denial of Providence. The systematic connection that Vico set up between the acceptance of the idea of occult wisdom and the denial of Providence is particularly significant from this point of view:

> Plato lost sight of providence when he perpetrated a common error of the human mind, that of judging the scarcely known natures of others

> according to oneself. For he elevated the barbaric and rough origins of gentile humanity to that perfect state of exalted, divine and recondite knowledge which he himself possessed. . . . Hence, through this scholarly blunder, which is repeated to this day, he found it necessary to prove that the first authors of gentile humanity were the possessors of recondite wisdom in the highest degree, whereas, in fact, as races of impious and uncivilized men, such as those of Ham and of Japheth must once have been, they could only have been beasts consumed by wonder and ferocity. . . . On this last point Grotius, Selden and Pufendorf have erred together. For lack of a critical method applicable to the founders of the nations, they believed them to be wise in esoteric wisdom and did not see that for the gentiles providence was the divine teacher of a common wisdom, out of which among them after the lapse of centuries the esoteric wisdom [of the philosophers] emerged. Thus our three authorities have failed to distinguish the natural law of the nations, which was coeval with their customs, from the natural law of the philosophers, which the latter grasped by force of reasoning, without ascribing any privilege to a people chosen by God for [the preservation of] his true cult [when it was] lost by all the other nations. . . . And through the arrogance of scholars [*la boria dei dotti*] all three of them have thought that the peoples, lost in the error of guilt, had observed in their customs a natural law in common with the Hebrews, who were illuminated by the true God.[3]

What Vico is saying is that if we start from the hypothesis of a wisdom present even at the origins of society, we will arrive at recognizing the existence of a natural law and a natural morality that ruled even the pagan nations. To admit the existence of this law and this morality, practiced and known by all the peoples of the earth, would once again imply seeing Jews and Gentiles as comparable. It would open the way to the dangerous and impious theses of Grotius and Bayle, according to which the system of natural law could "stand" even if "all knowledge of God be left out of account," and that ancient peoples could have "lived in justice" in a condition of atheism.[4]

By using "a really golden passage of Iamblichus, *De mysterii aegyptiorum*" to reduce Hermes Trismegistus to a "poetic character" that sprang from the fantasy of the earliest Egyptians—childlike men of vulgar wisdom—Vico could proclaim false the thesis that Moses, who lived later than Hermes, had learned the sublime theology of the Hebrews from the Egyptians.[5]

The rise of the "commonwealths" out of the "vulgar wisdom" of Solon (who was falsely believed to possess recondite wisdom) and the fact that those commonwealths preceded the laws and that the laws preceded philosophy showed the falsity of Polybius' statement (which "some acclaim") that "if the world had philosophers there would be no need of religions."[6]

The statement that "the early days" of all the nations were "quite barbaric" and that those beginnings were not "magnificent and enlightened," as Gherard Voss and "recent mythologists" had imagined; the thesis that the nations had been founded by "theological poets"—that is, by "the vulgar who founded their nations with false religions" and who "created deities with their imaginations"—proved how unreliable were the "monstrous hodgepodges" ("mostruosi ragguagli") of the travelers who had described societies devoid of the knowledge of God "to promote the sale of their books." This also made it clear how false Bayle's statements were, since Bayle had made use of these travelers' reports to claim that such peoples "can live in justice without the light of God."[7]

The inability to reason on humanity's origins "in terms of the reasons which philosophers have so far adduced" showed the error of the "Christian critics" who ended up giving equal weight to the verities of the faith and the legends and myths of antiquity. It also pointed out the limits of Selden's work. Selden, searching vainly to prove that the law of the Gentiles came out of Hebraic law, had failed to recognize the existence of a law particular to the Semitic peoples, and he had proposed "origins common to the gentile and Jewish nations, without distinguishing between a divinely assisted people and others completely lost."[8]

Vico's insistence on the irrational and fantastic origin of human society, on the savage and appalling ("fiere ed immani") natures of the bestial men, on giants devoid of reason and possessed of vigorous imagination had permitted him to consider meaningless the problem of the precedence of Egyptian and Chinese chronology over biblical chronology and to reject all attempts to bring pagan myths closer to Judeo-Christian verities. That solution also allowed him to stress the indispensable function of a Providence capable of changing vice into virtue and to direct the particular ends that motivate men to a universal purpose. This was the way to dismiss and defeat the libertine ideas implicit in the discussions of the immense age of the world and in the parallels between sacred and profane history and those doctrines of Spinoza and Bayle that were ever-present in Vico's thoughts: "Hence one cannot imagine more than two kinds of mode from which the world of gentile nations might have originated: either a few sages gave it order through reflection, or some bestial men were brought together by a certain human sense or instinct."[9]

"Having rejected the sages [of the philosophers], it remains to think about the brutes," Vico said elsewhere.[10] On many other occasions as well, Vico speaks in quite different and clearly opposed alternatives: for him the choices were *due e non più*.

Intellectual navigation could be facilitated by ceasing to defend the murky shadows of the obscure age and the fables of the heroic age:

> Hence we must make a choice. Ought we, in this dense night, in these rough seas, surrounded by so many dangerous reefs, to continue to sail in this merciless storm, which leads to the total subversion of all human reasoning, in order to defend the shadows of the obscure age and the fables of the heroic age, which were invented later rather than born thus from the very start? Or ought we to apply our reason to

> the fables, whose every interpretation has hitherto been quite arbitrary, to give them those senses which reason demands, and make ourselves possessors of those things of the distant past which have hitherto belonged to nobody, ownership of which can consequently be conceded legitimately to whoever occupies them, so that, in such fashion and by means of the principles of heroic nature proposed above, we may illuminate these dense nights, calm these storms, and escape these dangerous reefs? We refer here to heroic nature. . . . [11]

Those storms had not really been calmed, however, nor the reefs avoided. Declaring origins to be "fabulous" and insisting on the rough and primitive character of the earliest epochs, on imagination and irrationality, on the dull wits and the ferocity of the first inhabitants of the earth led Vico in the direction of a doctrine that was certainly no less impious and dangerous than the preadamite and deistic theses: it led him to the doctrine that Lucretius had advanced in the ancient world and Hobbes in the modern to explain the origins of human history.

For nearly a century, from Lapeyrère's day to Vico's, the view of the Chinese, the Egyptians, and the Chaldeans as sources of the original wisdom of the human race, as peoples of highest civilization and wisdom, had intertwined with a "Spinozist" sort of criticism of the Bible, with the antireligious thesis of the eternity of the world, and with a radical questioning of biblical chronology. As we have seen, however, the Chinese, the Chaldeans, and the Egyptians could be viewed as rough, ignorant peoples, and the earliest history of pagan peoples as an era of fables and of barbarity. The heathen peoples, once barbarous and crude, emerged laboriously into history: the same "impious" thesis of the initial barbarism of humankind could be used in defense of orthodoxy. To combat the thesis of a recondite wisdom and of the endless antiquity of the world, the oldest accounts of the oldest history of mankind had been declared energetically not to be histories but fables—except, of course, for sacred history. Was it not true that anyone who makes up fables and myths to boast of an imaginary antiquity was necessarily "barbarous," "rough," or "childlike"? One of Lapeyrère's fiercest adversaries asked himself how Moses' words could possibly be put on the same plane as the dreams and the "delerious ravings of barbarians" like the Chaldeans, the Chinese, and the Egyptians.[12] As we have already had occasion to note, the thesis that the time preceding the six thousand years of sacred history was not real time, but only an imagined, fabulous time, led *necessarily* to the thesis of a primitive barbarity of the nations.

It was indeed a question of "avoiding many reefs," to repeat Vico's image: Vico had to keep away from the dangers of pro-Chinese and pro-Egyptian views while making use of the thesis of "barbarism" and setting it into a Christian vision of history. At one and the same time he had to use Hobbes's thesis of beastlike men and accuse Hobbes of atheism and Epicureanism. While rejecting any slightest notion that man is "cast into this world bereft of divine care and assistance,"[13] he had to insert the history of the *bestioni* into the framework of biblical history and to shore up the idea of a stage of savagery that coincided with man's appearance on the earth—after the Flood. Above all he had to safeguard the idea of a *natural sociability* of man—contrary to Hobbes.[14]

There are several questions we need to try to answer to get to the bottom of what Vico's complicated undertaking meant: To what extent was Vico's attempt to use the thesis of the beast-men "original"? Was it not a matter (as with his rejection of the long ages of history and his insistence that they were legendary) of an idea that was common to a number of thinkers in Europe? Wasn't his solution to the problem presented again and again, in different guises, by those who picked their way across the scorching hot, difficult terrain of the origins of mankind and its primal—but postdiluvian—barbaric stage? And wasn't the question of barbarism and man's emergence from barbarism inextricably bound up with that of the origin of language? And how, in Vico's times and in Vico, was this connection made?

26·Vico and Fréret

BEFORE trying to answer some of these questions (part 3 will be devoted to an attempt to do so) we must return to at least two basic themes: (1) History could possibly be a *science* and take as its object not only the "easy" periods of human events—those for which we have sure records, contemporary to or close to the events narrated—but also the periods that are more remote, enveloped in fable and legend, and for which all that is left is uncertain and fragmentary evidence relating events much earlier than the time of the narration. (2) The course of history is unified, yet different civilizations intertwined from the beginning of human history. The Greeks and Romans had a decisive influence on this unified course of history, but also peoples distant in space and time: it is a history in which there were extremely ancient "high" civilizations and not just rough, uncouth peoples. Giambattista Vico and Nicolas Fréret both responded to these two great themes. Their responses were different and, at least in part, mutually exclusive.

For Vico there were only two hypotheses that could explain the origin of the pagan nations: either they were born of the conscious reflection of learned men, or else from human instinct buried in some beastlike men. "Rimossi i sapienti, ci rimangono i bestioni": Vico pursued, tenaciously and consistently, the alternative of taking away the sages, of a systematic demolition of the supposed wisdom of the Egyptians, the Chinese, and the Chaldeans. In order to "dispose of the proud claims" of those peoples "at once," he needed to insist on a real primitiveness that lay concealed behind their imagined wisdom. Vico, as is known, had made more than a few concessions to that wisdom in the first phases of his philosophic activities. In the *Liber Physicus* (as he explains in the *Autobiography*) he had made use of Hermetic themes from Egyptian wisdom. He had also drawn a comparison between the wedge ("the instrument with which nature makes everything") and the pyramids, attributed to the Egyptians the "distinctive mechanical medicine" of "the slack and the tight," and cited the *Medicina Aegyptiorum* (1591) of Prospero Albino. In the *Scienza Nuova,* on the other hand, Vico contrasted the idea of a highly developed philosophy hidden within the hieroglyphics to the image of an Egypt closed within its confines, barbarous, superstitious, incapable of contacts with other peoples, and not endowed with "human" qualities (as its admirers would have it) but inhospitable, proud of its imagined antiquity, and unaware of the true age of the world.[1]

The Egyptian priestly books contained "the greatest errors in philosophy and astronomy"; Egyptian medicine came down to "obvious nonsense and mere quackery"; the Egyptians' morality was dissolute and they "not only tolerated or permitted harlots but made them respectable"; their theology was "full of superstitions, magic, and witchcraft"; the arts of sculpture and metal casting, since "delicacy is the fruit of philosophy," were "extremely crude" in Egypt. Vico does not stop at half-measures: even the obelisks and the pyramids "might well have sprung from barbarism" since barbarism "accords well with hugeness." Since the Egyptians were unable to understand that "uniform ideas of gods and of heroes were born among the gentile peoples without their having any knowledge of each other," they believed that all the divinities of the nations with which they came into contact had originated in Egypt. They saw their own Jove as the oldest Jove and their Hercules as the earliest Hercules. When they took themselves as a rule for the universe, the Egyptians acted just like the Chinese, who

> grew into such a great nation . . . closed from contact with all foreign nations . . . and for many thousands of years they had no commerce with other nations by whom they might have been informed concerning the real antiquity of the world. Just as a man confined while asleep in a very small dark room, in horror of darkness [on waking] believes it certainly much larger than groping with his hands will show it to be, so, in the darkness of their chronology, the Chinese and the Egyptians have done, and the Chaldeans likewise.[2]

Vico's attack on the myth of Egypt and on the religion of the hieroglyphics did not come just from an opposition to the fantastic conjectures present in sixteenth- and seventeenth-century Hermetic literature (or literature with Hermetic tendencies). Many other authors (as I have shown elsewhere) contributed during those same years and quite independently of Vico to the destruction of that myth.[3] Vico's insistence on the backwardness of the Egyptians and his denial (in substance) that ancient Egypt had known a great civilization should not be opposed to the "foolishness" of Father Athanasius Kircher (as in Nicolini).[4] It should instead be considered as an element in a more general vision that attributed roughness and barbarity to all the peoples who had been seen theoretically as having had a wisdom that existed before Hebraic civilization or that could in some way rival Hebraic wisdom.

Vico stressed the superiority of Greece, which "alone . . . shone with all the fine arts that human genius has ever discovered," but he gives a judgment of Chinese civilization that is not unlike his judgment of ancient Egypt. Closed off until recently from the rest of the world, the Chinese "vainly boast an antiquity greater than that of the world"; they "boast in vain of enormously ancient origins." That their antiquity was imaginary could be proved by the fact that they "are still found writing with hieroglyphs" and cannot write alphabetically. They do not know how to paint shadows, and "since it has neither relief nor depth, their painting is most crude." Even Chinese porcelain statuettes seemed to Vico to show crudeness. Even Confucian philosophy—in reality "vulgar wisdom" and not philosophy—was "crude and clumsy" when it treats physical nature and it "turns almost entirely on a vulgar moral code, that is to say, a morals commanded to the people by laws." The invention of printing, finally, came only two centuries earlier in China than in Europe.[5]

While Vico rejected any mixing of or systematic parallel between sacred history and profane histories and saw pagan civilizations as recent, he qualified as "crude" the civilizations which, in a "vain rivalry" for antiquity, competed with Hebraic civilization as a source of a wisdom more ancient than sacred wisdom or as in any way comparable to Hebraic civilization. Pietro Giannone's treatment of the greatness of Egyptian civilization at the time of Abraham may suffice to give an idea of how a scholar who was working on the same problems and, in part, using the same texts, could reach a radically different interpretation:

> The kingdom of Egypt, at the time of Abraham even more than at the time of Moses, was large, populous, and famous for the many magnificent cities that adorned it, where the sciences and the mechanical arts were not only taught but perfectly exercised, where culture and civilization were at their highest point, and where foreigners came from the most distant parts to conduct their trade and commerce.[6]

In the pages of Giannone's *Triregno* the roles of the Egyptians and the Jews were totally reversed from their roles in the *Scienza Nuova*. The theologians who condemned as impious and perverse all those who held the Egyptian rites to enjoy precedence seem to Giannone "miserable and fantastic visionaries."

> What was so scandalous about it, as these hair-splitters claim, if God wanted, by means of Moses, to furnish the Hebrew religion with sacerdotal vestments, rites, and ceremonies, and chose to describe them to Moses as similar to those of which he had some idea from having long lived in Egypt so that Moses could understand them easily and teach them to as primitive a people as the Jews, so that they could more easily put them into execution and into effect? . . . The Jewish race had not yet become a people; they did not have their own

> republic, but were condemned to go wandering here and there, and after the death of Joseph they suffered miserable servitude.[7]

The choice between accepting an immense age for the pagan nations or presenting the ancient Egyptians and Chinese as primitive peoples without high civilization did not, however, indicate the only directions available at the time Vico was working on the different drafts of his *Scienza Nuova*. During those same years a different solution had been worked out that eliminated the choice itself as false and irrelevant—a choice typical of the seventeenth century, closely tied to the controversy over Marsham and Spencer, and conditioned by the debate on the wise Egyptians as opposed to the primitive Jews, or vice versa. Did not the denial of wisdom and civilization to the most ancient peoples in fact mean allowing oneself to be completely conditioned by the *use* that the Hermetics, the libertines, even the Jesuits and the supporters of the eternity of the world had made of that ancient wisdom? Was it not possible to take up these problems anew from a "critical" position that considered all past human history? Why could not the distinction be drawn between historical evidence and the evidence of fables (data that referred to an imagined history, not to real history), distinguishing the validity of each piece of information in turn and seeing each in relation to all civilizations? Did the fact that much of this evidence was fabulous mean it all was so? Did not the chronologists's task lie precisely in drawing this distinction and in comparing the testimony of different civilizations?

Many traditional ways of looking at the problem needed to be modified if these questions were to be answered in the affirmative: (1) It would be necessary to agree that historical criticism must follow the same rules that had been applied successfully in the study of nature. Even if the criteria of mathematics were admittedly not applicable to the only "relative certitudes" of history, rigorous reasoning "applies to all sorts of facts; it is not limited to the phenomena of nature alone."[8] (2) The working methods, the methodological procedures, the very image of chronology and of history that the great scholars and mythographers of the seventeenth century had constructed had to be rejected. Their same problems needed to be taken up, but their approach to the problems and the preconceived ideas that had made them incapable of a dispassionate examination of the sources must be put aside. Chronology and history ought no longer to be seen as a *means*, as ways to support philosophical opinions or religious opinions—as Scaliger, Petau, Vossius, Marsham, and Pezron had done. (3) Finally, a decisive step had to be taken: "to examine what are the proofs of the chronology of the Chinese annals, and what are the foundations on which it rests, to enable us to judge if it must carry the day against Judaic history, consulting only the rules of criticism, and leaving aside the religious respect that the books of Scripture in which it is reported inspire in us."[9]

Between 1718 and 1750 Nicolas Fréret took his stand in conscious opposition to the methods of study and the interpretations of the scholars, chronologists, and mythographers of the preceding century. He began by recognizing that although his epoch talked a good deal about the history of the earliest ages, it knew practically nothing about it. He rejected the methods of Scaliger, Vossius, and Marsham, whom he considered unable to distinguish between the true and the false or the probable and the improbable. He took up the problem of the methodology of the historical sciences, taking Leibniz as his base, and he states in a paper given in 1739 that even the history of the Hebrew people must be subjected to normal rules of criticism.[10] He worked for the full integration of Egyptian and Chinese history within the general history of mankind. As Jean Pierre de Bougainville said of him in 1749, he was

> born in a century in which the esteem owed to great men is not confused with a servile respect for their sentiments, . . . he dared to tread once more the paths on which the steps of [men like] Scaliger, Marsham, and so many other authors lay printed. Filled with their genius and taking their aims for his own, he followed a completely different method. Without prejudice, without preconceived plan, he carefully gathered citations, passages, vestiges of traditions, in a word, all of the fragments of the annals of the world scattered in ancient authors. He separated them from later glosses, weighed these different testimonies and, comparing them one to the other, he had the pleasure of remarking an agreement among them by which he himself was astonished.[11]

The traditions of ancient peoples shared one fundamental characteristic, Bougainville noted: searching back in time one "always finds an epoch beyond which these traditions no longer contain anything that is historical."[12] The inhabitants of the earth at this point are no longer men, but genii, monsters, and giants; nature no longer follows its own laws, and all events are presented as prodigious. In the history of some peoples—in Greek history in particular—these "fictions" were not linked by any chronology. Among the Chaldeans, the Egyptians, and the Oriental peoples they were connected within some sort of temporal "system." For Fréret also, the earliest history began several centuries after the postdiluvian dispersal of peoples, but his picture of Egypt is notably different from the one Vico or Newton had traced in their attempts to shorten the ages of Egyptian history. For Newton, who agreed in this with Marsham, Sesostris was none other than the biblical Shishak. For Vico, Sesotris "made the world seem older than it really is": Moses in Hebraic history and Danaus in Greek history were much more ancient than the mythical Egyptian king. For Fréret, Sesostris, Moses, and Danaus were contemporaries, and this contemporaneity "illuminates both the history of the Nations of Asia, that of the inhabitants of Palestine, that of the Phoenicians, and that of the Greeks." Egypt was a crossroads between a fabulous past and a present that was already historical. Even many centuries before that great conqueror, an "immense population" lived in Egypt, "cultivated the Arts, knew the Sciences, had a Religion, wise Governance, wise Laws, and flourishing commerce."[13]

To make a science of history, Fréret saw it as necessary to reject the choice that had underlain many of the works of seventeenth-century scholars. Only two roads had seemed open to resolve the difficulties that the chronologies of the early peoples had raised: some scholars had tried to *identify* the personages of Egyptian and Chinese history with the personages of the story of Noah and his ancestors; others had *rejected* the earliest history of both Egypt and China, seeing it as merely a confused mass of legends and fables. Fréret considered both of these positions false and partial. That Egyptian and Chinese history contained fabulous traditions was no reason to eliminate Egyptian and Chinese civilizations from history.[14]

To write a history of Egypt and China one must make use of archeology and epigraphy and must know and compare languages, writing systems, inscriptions, and fables. Scholars must give up the myth of one single people who distributed wisdom to all of mankind, and flee the temptation to deny multiplicity of civilizations in order to exalt an exceptional Hebrew history. Extolling and denigrating were equally harmful: "histories" as objects of critical and scientific investigation were equally valuable and all worthy of attentive consideration. This view made it possible to move more freely within a prospective broader than that of Western history and to feel oneself—as Bougainville says of Fréret—in some sense "compatriot and contemporary" not only of the Greeks and the Romans, but also of the Celts, the Chinese, and the Peruvians.[15]

Only in the "pluralistic" vision that Fréret sketches would the theme of barbarism and the "roughness" of the ancestors of civilized man take on productive meaning. It was the men of prehistory who would appear barbaric, not the ancient inhabitants of lands filled with the vestiges of high and highly refined civilizations.[16]

3
BARBARISM AND LANGUAGE

27·The Two Systems

THE seventeenth- and eighteenth-century theories on the human origins of language, on its slow historical formation, on its primitive simplicity and crudeness, and on its initially figurative and imagistic nature (as well as the consideration of language as a process tied to human history and not as the representation of a predetermined rational order) did not arise out of "linguistics." They were instead solidly linked to a series of affirmations and doctrines that provided them with a premise and a theoretical base. These affirmations, these doctrines concerned the origins of the physical universe and the appearance of man, the meaning of history and the notion of progress, the "bestial" or at least rough and sluggish beginnings of human life, the concept of perfectibility, and both the similarity and the distinction between man and the beasts.

Historians of language are often inclined to forget that the so-called pagan hypothesis (or the Epicurean and Lucretian hypothesis) of a spontaneous and physiological genesis of the different languages is found organically inserted within a broader discussion of the formation of the earth, the origin of life, and the birth of mankind. The connections between the history of a single discipline and the history of ideas are admittedly difficult to delineate. But when they are absent, the so-called special histories threaten to become completely imaginary histories. This is what happened to a leading contemporary Egyptologist, who has written a well-documented book on the myth of Egypt and its hieroglyphics in European culture taking into consideration only the authors who could be immediately qualified as "Egyptologists." In this way, he happened to eliminate from the debate on hieroglyphics that took place in Europe during the sixteenth and seventeenth centuries Della Porta and Pico, Conring and Fludd, Cudworth and Bacon, Wilkins and Vico.[1]

Lucretius, as is known, speaks in *De rerum natura* 5 of living things that spring from the earth, of the beginning of a discourse that uses different sounds to designate different things, and of the birth of the fear of the gods (vv. 69–73). The thesis that language derived from utility and from need, the parallel with the gestural language of children, the rejection of the hypothesis that one man could have given rise to language, the parallel with the language of animals—all this combines to form a "history" that begins with the origin of life out of the earth, continues with the appearance of the monsters that first populated the earth and with the struggle between the species (among which those who survived were those most capable of defending themselves from nature's insidious perils), and finds its completion in the arrival of man and in the construction of civilization. This "history" is based in presuppositions, and it is tied to a conception of nature and of time:

> Mutat enim mundi naturam totius aetas
> Ex alioque alius status excipere omnia debet

> Nec manet ulla sui similis res: omnia migrant
> Omnia commutat natura et verter cogit. [828–31]
>
> For time can change the nature of the universe,
> and one class of things must change into another class,
> and nothing remains the same, but all things are in flux,
> and Nature compels all things to be transmuted and changed.
>
> [trans. James H. Martinband (New York: Frederick Ungar, 1965)]

Language here is born of a feral mankind that feeds on what it finds and practices neither agriculture nor technology, that knows no law and has no interest in the common good—a mankind strong and agile, but that lives in terror of the wild animals. This mankind slowly learns to build huts and to cover its body. Families arise during this stage, along with the first friendships and the first pacts, thanks to which men, who remain faithful to them, survive.

Diodorus Siculus (1:7–8) also describes the passage from confused sounds to words, the construction of an intelligible language, and the formation of different languages in the broader context of the birth of the physical universe, the formation of the earth and its "fermentation," the appearance of the first forms of life and of the living species, and the miserable existence and savage life of the first inhabitants of the world. Groups of men, living naked and without agriculture, are driven by terror to join together here and there; they learn to use fire, they invent the arts, and slowly they learn to use their hands and their minds.

When these ideas were revived and discussed anew in modern times, they clashed, often dramatically, with a fidelity to the biblical account of a wise Adam, divine onomathete, in whose mouth God placed a primigenial language capable of reflecting the nature of things. In all cases, this opposition did not remain within the confines of a discussion on language, and these ideas were always projected against the more general background of a vision of the world based in presuppositions that for many centuries were considered in radical opposition to the Christian vision.

When the pagans reflected on the origin of languages, Calmet wrote in 1723, they explained it by two different systems. The first was summarized by the myth of an age of Saturn, a happy reign before the chastisement of God in which men and beasts all spoke the same language. The second system, which found typical expression in Diodorus and in Horace's and Lucretius' poetry, "attributed to chance, or to the Earth, humid and warmed by the Sun's rays, the creation of man." For those who supported this second system, Nature furnished the sounds of language and need brought forth, during an era of barbarism, the names that were given to things. Both of these systems, according to Calmet, had been elaborated only to compensate for the pagans' ignorance of the true origin of man: they did not know that all of mankind derives from one man alone, created by God, gifted with wisdom, who imposed names on the things and the animals. Adam's tongue remained unchanged until the Flood among the first men, or at least among the descendants of the just, up to Noah. Noah, the new Adam, repopulated the world and disseminated the very same language that Adam had received from God. About a hundred years after the Deluge (around the year of the world 1757) men, who had become too numerous to live together, founded colonies in different regions of the earth. At that time, as Genesis states (11:1), "erat terra labii unius et sermonum eorundem" (And the whole earth was of one language, and of one speech).[2]

Augustin Calmet was not an obtuse champion of tradition. He was a learned and brilliant writer, an innovator in biblical studies who combined an application of Richard Simon's methodology with an orthodox point of departure and who even included Protestant writers in his considerations.[3] For this reason, and also for the calm tone he brought to his discussions, his analyses and the positions he took—especially when they were quite blunt—are extremely useful as indications of general trends. Philastrius, bishop of Brescia, Calmet notes, declared that a plurality of languages had existed before Babel, and he interpreted the punishment of God as a forced forgetting of that plurality of languages, which was followed by the labored beginning of a more limited number of languages. But, Calmet charged, the Genesis text had been totally emptied of meaning "by three systems that amount more or less to one and the same," which had attempted to destroy the very idea of a miracle in the change in languages that occurred at Babel: the three systems were those proposed by Gregory of Nyssa, Jean Leclerc, and Richard Simon.

For Gregory of Nyssa, the mutation of languages came about as a result of the natural dispersal of men throughout the earth. God merely let nature work, and languages were born spontaneously from the needs of men confronted by things. Leclerc too declared linguistic differences to have a natural origin, and he interpreted the biblical passage as the expression of a community of sentiment among the men who devoted themselves to the construction of the Tower. Richard Simon, finally, quoting Gregory of Nyssa, interpreted the division at Babel as the expression of the divine will for a dispersion of men throughout the earth, and he too saw the "ordinary course of nature" in the formation of languages.[4]

Hypotheses like these, Calmet explains, are due solely to a refusal to recognize miracles. To deny that God was the direct cause of the confusion of tongues only served to "turn the historical and the literal meaning of Scripture upside down." Just as impious and equally dangerous, in Calmet's eyes, was the thesis of a natural language from which all other languages subsequently derived as dialects of the same idiom. The thesis of a natural language, initially similar to animal cries and that later evolved, also took away the meaning of the biblical verity according to which Adam and Eve, created with reason and speech, "as in a most perfect age," put names to things immediately after the creation. This hypothesis (which Herodotus attributed to the Egyptian king, Psamtik) was contradicted by the fact that to learn a language requires great effort and that—as recent cases of wild children grown to maturity in isolation had shown—there exists no natural inclination to speech.

When Calmet excluded the existence of a natural language, he was left with the problem of which among the known languages was the one Adam spoke. Here controversies and opposing hypotheses abounded: Hebrew, Syriac, Chaldaic, Ethiopian, and Armenian all claimed primacy. Becanus had recently claimed that ancient Flemish was the original language. There were many opinions, then, that converged to counter the idea of Hebrew as the original language. Gregory of Nyssa's statement that the first language had been lost had been repeated by many among the moderns, Hugo Grotius, Daniel Huet, and Heinrich Kipping among them. Each nation's claim of primacy for its own language and the combinations of compromise hypotheses had created a situation of almost inextricable confusion. The "babble" and the "pretentions" of the Arabs, Armenians, Egyptians, and Chinese must be combated by the thesis of the primacy of Hebrew, the language spoken by Adam and by Noah and the language in which—unlike other languages—"the names of men, animals, trees, places, and metals express their nature, their properties, the defects and the reason for their denominations." Successive alterations, the intermixing of languages, and the vicissitudes of history had slowly changed languages, however, and Hebrew among the others. It cannot be proved, Calmet concludes cautiously, either that the language of Adam exists to this day, or that it can be identified with Hebrew. Nevertheless, the arguments of the opposing camp are equally inconclusive:

> The sentiment of the profane, who believe there was a language natural to man, or who claimed that men, produced by chance out of the earth in different places in the world, had after many trials formed articulated sounds, and finally idioms different from one another—such systems have nothing in them, I will not say of the true or the existent, but not even the slightest sign of the stable or the probable. The making of men cannot have been the effect of chance, and man, created by God, has never been without the use of speech. The pretentions of the Egyptians, the Armenians, the Ethiopians, and the other peoples who want their language to be first among all, must in no way surprise us after the paradox of Goropius Becanus, who declares the Flemish, that is, the Dutch language to be so. Everyone loves his own land and his native language. . . . Finally, the confusion of tongues that occurred at Babel and was recorded by Moses was not the natural effect of the division that occurred among peoples and of the dispersal of peoples that followed: it was a miracle of God's omnipotence.[5]

A number of questions emerge from Calmet's work, directly or indirectly; questions that had been at the center of scholarly, historical, philosophical, and theological activity during the seventeenth century and the beginning of the eighteenth. The revival of Epicurean and Lucretian themes implied abandoning or questioning the idea that language was revealed by God and preceded society and history. The denial that the differentiation of languages was miraculous necessarily led to an investigation of how such a diversity among languages of different peoples did come about; it was also connected to the doctrine of the different and autonomous origins of the various peoples and races of man. Accepting the idea that Hebrew might not be the residue—more or less distorted—of the original linguistic patrimony transmitted by God to Adam also meant opening the way to quarrels over one language or another as the original language. This involved the danger of then having to recognize that the people whose language went back to the origins was superior to and older than the Hebrew people. In this way the traditions and the history of that people would be open to possible comparison with the history of the Jews as narrated in the Holy Scriptures. The question of the antiquity of the pagan nations was closely linked to the problem of the migration of peoples, to discussion regarding the American "savages" and their origin, and to the tradition of Hermetic wisdom. The question of the history of the ancient peoples could not be separated from that of their myths or of the "fables" that held the narration of their most ancient history. Among these myths that of a flood out of which one couple alone had been saved posed worrisome questions about the universality of the biblical account. During the latter half of the seventeenth century the notion of a great natural catastrophe that had taken place at the origins of history had become one with the problem of the formation of the earth and the modifications that had occurred on its surface during the course of time. What relation could be established between the time-honored thesis of a world that issued perfect from the hands of God, a world of perfect structures that man should celebrate and glorify, and the image of the slow formation of the physical world during the course of time? How could that image of the slow formation of the earth, of the growth of the mountains, of a slow sinking of the valleys be reconciled with the creation story? What position among the accounts of the origins of the world should be granted to the biblical account? How was its truth and universality to be defended in face of the other histories of nature and civilization? How could the image of a perfect Adam, a creature of divine wisdom, be put into relation with the concept of a primitive mankind, living in terror in a state of barbarism and emerging slowly, groping its way toward speech, the making of tools, and civilization? How were the ages of this history to be calculated? How could one combat the claims of many nations to an immense age, to millennia that reached far beyond the time allotted to God's chosen people? How could these processes—which concerned both natural history and human history—be made to fit within the six thousand years allowed by biblical chronology? And, to return to the first language, was Hebrew a postdiluvian language or had it existed before the Deluge? And if it was the very language of Adam—as Calmet and Vico and many other authors would have it—did Adam's speech result from a learning process or was it a miraculous and sudden illumination?

Calmet's discussion of the origins of language is closely connected with his discussions of the first giants, the origins of writing, chronology, the ceremonies of the Egyptians, and true and false miracles. This sort of connection between problems which seem to our modern mentality so very far from one another is typical of a large number of authors. This is why it does not seem to me to make much sense—to take an example from the case of Vico—to oppose Vico's "conception of language and languages" and his "conception of history," or to claim that Vico's linguistic ideas deserve the "central role" in his "philosophical construction." Vico, Tullio De Mauro has written, "is a thinker who elaborates a certain vision of language and of languages and who then *as a consequence* conceives of reality and the history of nations in a certain way."[6]

To take up the question of the variety of languages or the origin of language, between the mid-seventeenth and the mid-eighteenth centuries, implied being faced with the sort of problems mentioned above. The disputes were not carried on among "linguists" or professors of the history of language, but within positions of enormously broader theoretical—and practical—implications. To discuss the passage from an age dominated by sense and by fantasy, incapable of abstraction, to the age of science and reason; to discuss the problem of the advent of civilized man and the historical beginnings of language, writing, the arts, and political institutions—all directly involved themes central to the culture of the time: chronology, ancient history, and the history of the earth. Above all, it involved decisive questions of theology.

28·A Triumvirate of Demons

THE most radical denial of the miraculous birth of languages as the work of God that the late seventeenth century produced can be found in a text of the sort that historians of language generally have no occasion to mention and in an author whom it is certainly difficult to count as a student of linguistics. Isaac de la Peyrère, or Lapeyrère, as I have said, joined with Hobbes and Spinoza to form that "triumvirate of demons" that men who lived in an age of solid convictions concerning the existence of devils regarded with fascination and terror.

Lapeyrère's *Praeadamitae* has already been discussed in some detail (chapter 18). "The men of the first Creation were created long before Adam," and the sources of that older history cannot be found in the Bible. In this view, the entire problem of Adam's divine wisdom and of a first language of divine origin came to lose all meaning. In the third volume of the *Praeadamitae*, in fact, the attribution of divine onomathesia to Adam, as described in the second chapter of Genesis, is clearly presented as a legend. How, in only one day, could Adam have given a name to all of the animals of the earth and all of the birds of the air? How could the elephants, whose step is so grave and slow, ever have arrived from the distant regions of Asia and Africa to the terrestrial paradise in one day? How could anyone think that animal species living in the other hemisphere could have swum endless oceans to reach Eden and appear before Adam? Lapeyrère insists on one point with particular force: language is founded in operations of the mind, and mental operations *require time*. Divine knowledge is immediate, but human knowledge is born of seeds, the growth of which requires thought and ratiocination. And thought and ratiocination *grow through time:*

> God . . . brought unto [Adam] all the creatures of the earth, and all the Fowls of heaven, that he might see what he would call them: for every thing whatsoever Adam called it, that is the name thereof. . . . "Vidit Adamus quid vocaret ea. Vidit:" Adam made use of his reason in the proper and determinate Names which he gave to all the living creatures and fowls. He saw what he would call them. He saw; that is, that . . . he had those operations of mind which are distinguishable, discursively, and in time. Nor is it possible that Adam could in one half of a day see all the creatures of the earth, and the fowls of heaven, then premeditately give every one of them a convenient name; which neither in pronouncing of their names, nor in a continual rehearsing of them, he could have perform'd in so short a time. . . . They that say the world was created with Adam, say that Adam, the first hour he was made, had all sciences, arts, and disciplines perfectly, [but] they doe not seriously consider what they say. Truly, in the beginning, the prime causes and means of all sciences were in God, but the seeds of them were only sown in *Adam*, which could not arise, but by meditation, reasoning with himself, by cultivating, and time.[1]

Lapeyrère proposes no theories concerning the rise and development of language among the preadamites. He limits himself to showing that traditional solutions to the problem were untenable, to stating that the biblical account must not be taken literally, and to accepting the idea that thousands of years must have passed between the age of the first and the second creation.

In his *De Quatuor Linguis Commentario*, published in London in 1650, Méric Casaubon (son of the more famous Isaac) adopted a compromise solution to the problem of the divine origin of language: God was the source of the words Adam pronounced; man is the source of the subsequent developments of language.[2]

Hobbes, during those same years, denied privilege to Adamic language and, in the long run, proved the entire question of a mythical, presocial language to be meaningless. But Hobbes's importance, his "novelty," his continued influence during the centuries that followed, and his position at the center of discussions on language as the one great subject of modern anthropology, certainly do not lie only in this rejection of traditional theses, and even less in a generic "regard for the variety and the historicity" of the world of languages.[3] They lie instead in his analytic discussion of a series of firmly linked themes, connected (as later in Vico and in Rousseau) into a single problematic cluster. The problems involved a certain number of paired concepts: animal nature/human nature, nature/culture, barbarism/civilization, natural language/human discourse, natural repetition/human progress. It would certainly be inappropriate to take up here either questions of Hobbes's anthropology and politics, or those more strictly related to his linguistic doctrines. Both have been amply studied.[4] It does seem appropriate, on the other hand, to try to enumerate a few of Hobbes's fundamental theses concerning these concepts. If they were to be ignored we might risk losing sight of some essential reference points in the discussions that took place after the middle of the seventeenth century.

1. Sense, imagination, understanding, prudence (as the ability to profit from experience), deliberation, and will were, for Hobbes, common to man and the animals. Understanding is the imagination "raised in man" by "words, or other voluntary signs": the dog, like many other animals, is able to understand the call or the correction of his master. Prudence, which is prediction of the future on the basis of past experience, is not only common to man and the animals, but more highly developed in some animals than in man: "There be beasts, that at a year old observe more, and pursue that which is for their good, more prudently, than a child can do at ten." Deliberation, as the succession of alternating appetites, aversions, hopes, and fears, is present in all living creatures no less than in man, and for that reason "beasts also deliberate." Since will is the last appetite or aversion immediately adherent to action or lack of action, "and beasts that have *deliberation*, must necessarily also have *will*." In general, experience of fact (which at its highest level is called prudence) characterizes man and the animals alike, but only man is capable of the *evidence of truth* (which at its highest level is called science).[5]
2. Aside from senses, imagination, memory, prudence, and deliberation, which are characteristic of all animals, consequently also of man, there is no "act of man's mind" that is "naturally planted in him," the use of which requires only being born human (being a member of the human species) and living with the aid of the five senses.[6]
3. Thus as man is not by nature a political animal, he is not by nature a rational animal. Men "receive not their education and use of reason from nature." Unlike sense and memory, which are "born with us," unlike prudence, which is "gotten by Experience only," reason is artificially acquired; it is "attained by Industry." The faculties that are specific to man and that distinguish him from the animals "are acquired, and encreased by study and industry."[7]
4. The specifically human faculties, those that signal the passage from the natural world of the animal to the cultural and political world of man, "proceed all from the invention of words, and speech." This is why children do not possess reason until they have acquired the use of speech, and they are called reasonable creatures only because it appears evident that they can acquire the use of reason in the future. Furthermore, given that instincts, which come from nature, are not in themselves bad (although the actions that sometimes spring out of them can be bad) children are neither culpable nor bad and, since they do not have use of reason, they are exempt from any sort of duty.[8]

5. Since the faculty of reason is a consequence of the use of speech, reason is

> attained by industry; first in apt imposing of names; and secondly by getting a good and orderly method in proceeding from the elements, which are names, to assertions made by connexion of one of them to another; and so to syllogisms, which are the connexions of one assertion to another, till we come to a knowledge of all the consequences of names appertaining to the subject in hand; and that is it, men call SCIENCE.[9]

6. The sort of "understanding" that characterizes man and that is peculiar to him is the ability to understand conceptions and thoughts by means of the succession and connection of the names of things in affirmation, negations, and other forms of discourse. Man's superiority over the animals consists in the fact that man, when he thinks something, is able to investigate its consequences and effects. While sense and memory are nothing but the knowledge of a fact (which is past and irrevocable), science is knowledge of the dependence of one fact on another and knowledge of the consequences. By means of science—unlike the animals and unlike many men who live out their lives at the level of children—we know more than what we are doing in the present; we know how to do it again: "when we see how any thing comes about, upon what causes, and by what manner; when the like causes come into our power, we see how to make it produce the like effects." Thanks to words, man can reason, that is, he can reduce to general rules the consequences that he notes. Thanks to words and to method, the "human" faculties can be elevated to such a high degree that man becomes clearly distinguished from all the other creatures.[10]
7. As the single word is to the idea of a single thing, so discourse is to the *discursus animi*. Reasoning and discourse are equivalent terms: they both signify a context made of words established by man's will to signify a series of concepts. In this sense, animals have no speech because they understand what we command or what we desire not through words as words, but through words as signs. They do not know the significance that has been imposed on words. In the animal species, finally, there is no discourse because the calls—the *voces* that express their feelings of fear or of joy—are not caused by the animal's will, but arise out of the necessities of his nature. Animals give different signs to each other to invite another to feed, to excite to song, to solicit love, still, those signs do not constitute discourse because they spring by a natural impulse out of the passions proper to each animal. Those signs do not constitute a language: this can be seen in the fact that the vocal signs of animals of the same species are identical in every part of the world, whereas human language varies.[11]
8. Since man is the only animal able to use the universal significance of individual words to construct general rules, man is also the only animal able to construct

false rules and transmit them to others for their use. Thus he can teach what he knows to be false, and he can turn others' minds against society and peace. This cannot happen in the society of animals, since they evaluate with their senses what is useful or harmful to them. Only when we refer to beings who make use of discourse can we speak of truth and falsehood: an animal who takes the image of a man reflected in a mirror for a real man does not apprehend that appearance as true or false, but only as "singular" and thus he is not fooled. Given that truth and falsehood regard propositions and language, the animals also cannot multiply one falsehood with another as man can. Being superior to animals in understanding also means being superior to them in error: absurdities are born of linguistic capacities and of the related exercise of reason. An animal cannot fool himself. Only man can. Discourse does not make man *better* than the animals, but only *more powerful* than they are.[12]

9. Hobbes describes the origin of mankind in the first chapter of his *De homine*, following Diodorus Siculus: once again, we meet the original confusion of heaven and earth, the separation of one mass from another in the upheaval of the universe, the constitution of a humid, muddy, revolving mass in which the land becomes separated from the sea, the action of the sun and the consequent formation, in swampy places, of membranous bubbles out of which animals and men are born. This account, according to Hobbes, is close to what is said in the first book of Genesis ("Propinqua quidem haec sunt iis quae traduntur in capite primo *Geneseos*, sed non eadem"). There are two differences, however: first, according to the biblical account, the separation and the gathering together of the bodies was produced "by the movement of God's spirit" ("propter molum spiritus Dei"); second, the earth produced every sort of animal, but by virtue of the divine Word and with the exception of man, who was created in the image of God after all the other animals. Hobbes draws two conclusions from this comparison: first, the ancient authors understood all that it was possible to understand without the aid of revelation; second, the first generation of the universe cannot be known by anyone: it is known only to God, who was its author.[13]

10. Just as he is not by nature a rational animal, man is not by nature a political animal. At birth, man is not able to associate with others, no matter what political writers might suppose, claim, or postulate. The world of the political, like that of culture, arose out of the invention of language. Without language, neither society, peace, nor discipline would exist—only savagery, solitude, and lairs to live in. Without language, society, contractual agreements, and peace would no more exist among men than among the lions, bears, and wolves. The principles by which we know the just and the equitable from the unjust and the inequitable have been laid down by men. The world of pacts and laws, tied to the existence of language, is an artificial construction; before it came into being neither justice nor injustice existed, neither good nor evil—as it is among the animals. Man's superiority over the animals is, once again,

connected to a series of negative possibilities: wolves, bears, and snakes are not rapacious unless they are hungry, and they become cruel only when provoked. Thus man can exceed the animals in rapacity and cruelty.[14]

11. At the roots of human history (which is the history of artificial products such as language and politics) there consequently lies a period of savagery and barbarity. Such was the life during past centuries of peoples now civilized and flourishing, and such is presently the life of the American savages. Then, as in America now, Hobbes says, those peoples were "few, fierce, short-lived, poor, nasty, and deprived of all that pleasure and beauty of life, which peace and society are wont to bring with them." When men live without a common power to hold them in subjection, they live in a state of war, with no other source of security than their strength and wit. In that state (as with American natives now, Hobbes adds), there is no industry, no agriculture, no navigation, no use of imported goods, no buildings, no machines, no knowledge of the earth's surface, no understanding of time, no arts, no letters, no society: "Continual fear, and danger of violent death; and the life of man, solitary, poor, nasty, brutish, and short." This barbarous state never occurred simultaneously throughout the world, but its existence is certified by our experience of savage nations and by the histories of our ancestors.[15]

12. The invention of language thus coincides with the specificity of man and with his emergence out of animality and out of barbarism. When Hobbes establishes language and rationality as constructions, as produced artificially, he destroys the myth of an original, perfect language and contradicts the notion that God revealed a complete language to the first man before the establishment of society. In the fourth chapter of *Leviathan*, Hobbes accepts the biblical account to the extent that "the first author of *speech* was God himself, that instructed Adam how to name such creatures as he presented to his sight." Adam was thus enabled to increase the number of words and to acquire, as time passed, that small amount of language necessary to his life and his experience. Instead of being perfect, the language of Adam appears in this view extremely impoverished and rudimentary: "for I do not find any thing in the Scripture, out of which, directly or by consequence can be gathered, that Adam was taught the names of all figures, numbers, measures, colours, sounds, fancies, relations; much less the names of words and speech." Adam is no longer seen as one of the great sources of a God-given, original wisdom: he becomes a *potential man* who possesses only the simplest elements of language, and who fashions the language he needs ("so much language might be gotten as he had found use for"). Hobbes returned to these same ideas in the ninth chapter of *De Homine*, where his views are expressed in even more radical terms. In *Leviathan*, God had *taught* Adam the names of the creatures who presented themselves to his sight; in *De Homine*, Adam imposes names ("imposuit") according to his will ("suo arbitrio") on only some animals—those that God has placed before him. Subsequently, and in the order in which they

come before his senses, Adam puts names to other things. The communication between God and Adam is not linguistic, but takes place in supernatural ways: how could Adam have understood God's commands, when he still did not know the meaning of fruit, eat, tree, know, good, or evil? How could he have understood the serpent when it spoke of death, of which he could not have had the slightest notion?[16]

13. For Hobbes, then, Adamic wisdom came down to the use of names that designate objects and animals. In denying that Adam's speech contained the names of figures, measures, relations, or names of names ("names of words and speech"), Hobbes reduced the language of the first man to an extraordinarily rudimentary level, the level of "notes" and "marks." The things that Adam can name are just a few perceptible objects. When he denies Adam names "of second intention," Hobbes places Adam's knowledge on the level of the everyday practicality that precedes scientific knowledge:

    "Of the *first intention* are the names of things, a man, stone, etc.: of *second* are the names of names and speeches, as *universal, particular, genus, species, syllogism,* and the like. But it is hard to say why those are called names of the *first,* and these of the *second intention,* unless perhaps it was first intended by us to give names to those things which are of daily use in this life, and afterwards to such things as appertain to science, that is, that our second intention was to give names to names."

    In most cases, for Hobbes, reasoning was connected with the use of abstract names. Unlike concrete names, which designate a thing supposed to be existent and "invented before propositions" (for example, heat). Abstract, on the other hand (as, for example, in "to be hot"), "is that which in any subject denotes the cause of the concrete name." Abstract names consequently "proceed from the proposition," given that "these could have no being till there were propositions, from whose *copula* they proceed." In denying the "names of numbers" to Adam, Hobbes arrives at a truly radical result:

    "By the advantage of *names* it is that we are capable of *science,* which beasts, for want of them are not; nor man, without the use of them: for as a beast misseth not one or two out of many her young ones, for want of those names of order, one, two and three, and which we call *number*; so neither would a man, without repeating orally or mentally the words of number, know how many pieces of money or other things lie before him."[17]

14. Adam, as we have seen, does not know the names of numbers and "it seems, there was a time" in which the names of numbers were not in use and men had to use their fingers to count. The ability to count as well as to tell clock strokes was thus an acquired skill. When this was achieved, it was possible to subject times, weights, motions, and varying degrees of qualities to cal-

culation and, consequently, develop the arts of the measurement of bodies, the calculation of time, the computation of celestial motion, and with them geography, navigation, and the construction of buildings and machines. All of this, Hobbes concludes, found its point of departure in numeration, which in turn found its point of departure in speech. Speech—which underlies our very ability to transmit knowledge—originates in man's will ("Origo sermonis naturaliter alia esse non potuit praeter ipsius hominis arbitrium"). There are no "natural meanings" and the thesis that names were imposed on things according to the nature of the things themselves is childish ("Quod autem dicunt aliqui, imposita esse singulis nomina juxta ipsarum rerum natura, puerile est"). There is no affinity between *voces* and *corpora*, and while the nature of bodies "is everywhere one, languages are diverse" ("Qui enim fieri potuit, cum una sit ubique natura rerum, diversae tamen essent linguae?"). If language derives from will and from institutions, from *whose* institutions does it derive? To answer this question, Hobbes calls once again on the idea of a slow historical formation of language, on its transition from a rudimentary phase, in which it was made up of few simple names of objects and of animals, to a later phase in which the names of numbers, the names of names, and the names of relations appear:

> It is incredible that at one time men agreed together to decide and establish what words and verbal contexts should mean; it is therefore credible that at the beginning, names were few and that they were the names of the most familiar things. [Nam convenisse quondam in consilium homines, ut verba verborumque contextus quid significarent decreto statuerent, incredibile est. Credibile ergo est nomina ab initio pauca fuisse, et earum rerum quae maxime erant familiares.][18]

Spinoza's reflections on the *impositio nominum*, given in the *Tractatus de intellectus emendatione* and repeated in the *Ethics*, are founded, as is known, on the opposition of imagination (which forms, *nobis invitis*, confused notions) to understanding (which, by means of its "force," forms clear and distinct ideas). In a letter to Balling dated 20 July 1644, after describing the face of a "certain black and hirsute Brazilian" who had appeared to him in a dream and whose face persisted, like a real object, even after he awoke, Spinoza defines the effects of the imagination, which "arise either from bodily or mental causes." The imagination, he continues,

> follows in the tracks of the understanding in every respect, and it arranges its images and words, just as the understanding its demonstrations and connects one with another, so that we are hardly at all able to say what will not serve the imagination as a basis for some image or other.[19]

Images and words stand opposed to demonstrations: in order to acquire the true method, which consists only "in the knowledge of the pure understanding, and its nature and laws," it seemed necessary first to distinguish between the true ideas of the mind and the false and dubious ideas of the imagination. The mind, unlike the body, is not "exposed to chance"; clear and distinct perceptions arise only out of other clear and distinct perceptions: "nor do they acknowledge any cause external to us." They depend on "our absolute power, not on fortune"; they do not depend, that is, on "causes . . . yet unknown to us, and alien to our nature and power."[20]

Unlike the ideas of the intellect, which arise from the active power of the mind and are "internally disposed," the ideas of the imagination reflect changes in the human body and alterations or states produced in it by other bodies. They arise from the "fortuitous concourse of things," they depend largely on chance, and reflect experience, but a vague and confused experience: "The human mind, as often as it perceives a thing in the common order of nature, has not an adequate cognition either of itself or of its body, or of external bodies, but only a confused and mutilated cognition." Transcendental terms (being, thing, something) and universal notions (man, horse, dog) belong to this level of experience: "The human body, being limited, is capable of distinctly forming in itself only a certain number of images at once. . . . If it exceeds this number, the images begin to be confused." The human mind can thus conceive, dintinctly and simultaneously, "as many bodies at once as there can be images formed in its own body." When images become confused, "the mind also will imagine all bodies confusedly without any distinction and will comprehend them under the attribute, namely under the attribute of being, thing, etc." In the same manner, when "so many images of men, for example, are formed in the human body at once, as that they surpass the power of imagining, not indeed greatly, but so far that the mind is unable to imagine the small differences of individuals," then the name *man* (which is then predicated on an infinite number of particular instances) is used to imagine "that alone in which all . . . agree." This is, however, a false universality, since men form notions in different manners: general notions vary for each person "according to the object by which the body has been oftenest affected, and which consequently the mind easily imagines or records." Some will use the term *man* to indicate an erect animal, others an animal capable of laughter, others a featherless biped, or a rational animal, and "each will form universal images of things according to the/his physical constitution."[21]

General notions can be formed from individual things "presented by the senses to the intellect in a mutilated, confused, unorderly manner" or by signs, as when "hearing or reading certain words we remember things and form ideas of them similar to those which the things themselves first produced in us." From the thought of a thing the mind can in fact pass to the thought of another thing that has no resemblance to the first one. A Roman would pass immediately from the word *pomum* to the thought of the fruit, "which has no resemblance to that articulate sound," and has nothing in common with it "except that the body of the same man was often simultaneously affected by these two things; i.e., that the same man often heard *pomum*, when he saw the fruit." The transition from one thought to another and from a word to a notion therefore depends on the familiarity or the habit of making particular associations: from the hoofprints of a horse, a soldier passes to the image of the horseman and to that of war; the rustic passes to that of the plow and the field. Unlike the linking of *ideas*, which "is formed according to the intellect, which is the same in all men," this sort of association involving things external to the body "arises out of the order and concatenation of the affections in the human body."[22]

At this point, several passages on language in the *De intellectus emendatione* may become clearer. In that type of *perceptio* that arises "when the essence of one thing is inferred from another thing, but not adequately," the imagination introduces an element of confusion:

> When things are conceived thus abstractly, and not through their true essence, they are apt to be confused by the imagination. For that which is in itself one, men imagine to be multiplex. To those things which are conceived abstractly, apart, and confusedly, terms are applied which are apt to become wrested from their strict meaning, and bestowed on things more familiar; whence it results that these latter are imagined in the same way as the former to which the terms were originally given.[23]

As Bacon wrote (*Novum organum*, 1:43), words "plainly force and overrule the understanding." Not only do they condition language, but also the very functioning of the intellect: "The faulty meaning of words cast their rays, or stamp their impression, on the mind itself; they do not only make discourse tedious, but they impair judgement and understanding."[24] We affirm or deny many things, Spinoza repeats, "because the nature of words allows us to do so, though the nature of things does not." Words, for Spinoza, are thus "part of the imagination" ("verba sunt pars imaginationis") and we construct many of our concepts "in accordance with confused arrangements of words in the memory, dependent on particular bodily conditions." *Verba* are *constituta ad libitum et captum vulgi* (words are formed according to popular fancy and intelligence); they are *signa rerum* (signs of things) "as existing in the imagination, not as existing in the understanding." This is why "all such things as exist only in the understanding, not in the imagination" have been given *nomina negativa* (incorporeal, infinite, etc.) and why many affirmative notions have been expressed negatively and by opposites (uncreate, independent, infinite, immortal, etc.). "So, also, many conceptions really affirmative are expressed negatively, and *vice versa*, such as uncreate, independent, infinite, immortal, etc., inasmuch as their contraries are much more easily imagined, and, therefore, occurred first to men, and usurped positive names."[25]

One of the fundamental properties of the intellect, for Spinoza, was the capacity to form positive ideas *before* negative ones. On the basis of the opposition *imaginatio/intellectus*, he links the *impositio nominum* and the formation of language to the imagination and he strongly emphasizes the differences between common language and scientific language. There are some singularly interesting aspects, from this point of view, to his observations on the original, "vulgar" meaning of certain philosophical terms, on the alteration in the original meaning of such terms, on the alteration in the original meaning as it passed to learned use, and on the persistence of the original prejudice (which was tied to the first, vulgar meaning) within scientific language. I have found two examples of this sort of analysis in Spinoza. The first is taken from the sixth chapter of the *Cogitata metaphysica:*

> That we may rightly perceive two things, the true and the false, we shall begin with the meaning of the words, from which it will appear that they are only extrinsic classifications of things and are attributed to things only rhetorically. But since the common use first discovered these words, which were only afterwards used by philosophers, it seems pertinent for anyone who inquires into the first meaning of a word to see what it first denoted in common use, especially in the absence of other causes, which might be drawn from the nature of language for the purposes of the investigation. The first meaning of true and false seems to have had its origin in narratives; a narrative was called true when it related a fact which had really occurred, and false when it related a fact which had nowhere occurred. Later, the philosophers used this to denote the agreement, and disagreement, of an idea with its object. For this reason an idea is said to be true which presents a thing as it is in itself; but false, which presents a thing to us other than it really is. For ideas are nothing other than narratives or mental histories of nature. Later, the notion was transferred metaphorically to mute things: for example, we call gold true or false, as though gold could be summoned to tell us about itself, whether something is or is not in it.[26]

The second example comes from the preface to part 4 of the *Ethics:*

> If we see a structure (which I suppose to be not yet complete) and know that the object of the architect is to build a house, we say the house is imperfect; and on the contrary we say it is perfect as soon as we see it carried out so as to fulfill the end for which the architect had designed it. But if we see a work unlike anything we have seen before, and are unacquainted with the mind of the author, then assuredly we cannot know whether the work be perfect or imperfect. And this seems to have been the *primary* signification of these words. But after men began to form universal ideas, and to conceive types of houses, edifices, towers, etc. and to prefer some types of things to others, the result was, that each called that perfect which appeared to be in accordance with the universal idea he had formed of any particular kind of thing, while each pronounced imperfect what appeared less in accordance with the type he had conceived, although according to the opinion of the author it might be thoroughly complete. And it is apparently for the same reason that natural things, such as are not

> made by human hands, are commonly called perfect or imperfect; for men are wont to form universal ideas of natural as well as of artificial things, and these universal ideas they regard as the archetypes or models of things. Moreover, supposing that Nature does nothing save for the sake of some end, they imagine that it contemplates those universal ideas and proposes them to itself as archetypes. When therefore they see anything come into existence in Nature, which is not in accordance with the type or model which they have conceived of that kind of object, they believe that Nature itself has failed or erred, and has left the thing in question imperfect. Thus we see that men are wont to call natural things perfect or imperfect more from prejudice than from a true knowledge of those thing.[27]

Spinoza was acutely aware of the force of these prejudices and of the tenacity and insidiousness of linguistic usage. It is extremely difficult, he writes in the *Theologico-Political Treatise* "to change the meaning of a common word" and to "keep up the change among posterity or in common practice or writing," given that both the common man and the learned join to maintain language as it is.[28]

The distinction between images of things and words, on the one hand, and ideas and conceptions of the mind, on the other, remains one of the fixed points of Spinoza's views: we must distinguish carefully—as the Scholium of proposition 49, part 2 of the *Ethics* says—between an "idea, or conception of the mind, and the images of things, formed by our imaginations." It is just as necessary to distinguish "between ideas and the *words* by which we signify things." Those who confuse ideas with images persuade themselves that "those ideas of things of which we can form no corresponding image are not ideas but only fictions, which we form by virtue of free will." Those who confuse words with an idea "suppose that they can *will* the contrary of what they *feel;* when they merely affirm, or deny in words alone." Anyone who reflects on "the nature of thought" will also understand that "an idea, since it is a mode of thought, consists neither in the image of anything nor in words. For the essence of words and images is constituted by corporeal motions which do not in the least involve the conception of thought."[29]

Notions like substance, eternity and the like cannot be acquired with the imagination, but only with the intellect. Unlike universal notions, "the true definition of anything includes and expresses nothing besides the nature of the thing defined." The definition of a triangle "expresses nothing else than the simple nature of a triangle, and not any particular number of triangles." The knowledge proper to common sense, the notions with which common mortals usually explain nature are just modes of imagination; they do not indicate the nature of things, but how the imagination is constituted. The names of those notions refer to entities of the imagination, and not to entities of reason. What is contained objectively in the understanding, whether in function finite or in function infinite, "must necessarily exist in nature."[30]

Spinoza's theses on the difference between common language and the language of demonstration and on the relation between names and things and between images and words were not destined to influence greatly the disputes over the birth of language. They are of central importance, however, to any investigation of how discussion on the rise of language is connected to the notion of a state of barbarism. The importance of Spinoza's ideas lies chiefly (1) in his affirmation that faith and theology have no relation to and no affinity with philosophy either from the point of view of their aims, their bases, or their method;[31] (2) in his affirmation that Scripture does not teach philosophical doctrines, but only piety, and that all that it contains seems adapted to the preconceived opinions of the heterogeneous and inconstant Judaic populace;[32] (3) finally, in his affirmation—in which he approaches Bacon and Hobbes—of the slow emergence of culture and science out of rough, imperfect origins as a progressive construction of increasingly perfected intellectual instruments.[33]

The first two of these affirmations were to have a decisive effect on the discussions on the origins of language, civilization, and culture. In Spinoza's broad perspective, the problems of the interpretation of the Bible and of the relations between sacred history and profane history were, in fact, destined to change meaning radically and, with them, the problem of the original language of mankind and of its relationship with Hebrew. If, as Spinoza would have it, the laws that God revealed to Moses were no other than the particular laws of the Hebraic state, not designed to be extended to other peoples; if the "election" of the Jews was limited to the political sphere alone; if the Jews' superiority over other peoples consisted solely in their material prosperity, in the well-being produced by their civil structure, and in the scorn that they showed toward other nations; if in every other way the Israelites were equal to the other peoples and, on the intellectual level, they "knew scarcely anything of God" and their scriptural doctrine contained "only very simple matters, such as could be understood by the slower intelligence"; if their prophecies did not spring from more perfect minds but only from more lively imaginations; if God manifested himself to other nations more frequently than to the Jews; if Moses treated the Jews "as parents treat irrational children" and they were "men accustomed to the superstitions of Egypt, uncultivated and sunk in most abject slavery" when the written law was handed down to them; if "God adapted revelation to the understanding and opinions of the prophets," who "taught nothing which is not very simple and easily grasped by all, and further, they clothed their teaching in the style, and confirmed it with the reasons, which would most deeply move the mind of the masses to devotion to God"[34]—if all of this was true, then all distinction between the sacred history of the chosen people and the history of heathen peoples collapsed and the idea of an incarnation of the *meaning* of universal history in the particular history of the Hebrew people was destroyed. The *Theologico-Political Treatise*—in both radical condemnations and cautious adherence, through both explicit or implicit references—was to remain for more than a century at the center of all discussion on mankind's earliest history.

The indifference that historians of language have shown toward Spinoza's theory of language is clearly totally unjustified. It is true that we would look in vain in Spinoza for passages devoted to the problem of the origins of language or to the life of the first men on the earth. But, as we can see in a famous passage from the *De intellectus emendatione*, Spinoza shared one firm conviction with many intellectuals and philosophers of his day: the transmission of culture and the creation of intellectual instruments and material instruments represent a broadening of possibilities that were, when human life began on the earth, poor, crude, and limited.

> But as men at first made use of the instruments supplied by nature to accomplish very easy pieces of workmanship, laboriously and inperfectly, and then, when these were finished, wrought other things more difficult with less labor and greater perfection; and so gradually mounted from the simplest operations to the making of tools, and from the making of tools to the making of more complex tools, and fresh feats of workmanship, till they arrived at making, with small expenditure of labour, the vast number of complicated mechanisms which they now possess. So, in like manner, the intellect, by its native strength, makes for itself intellectual instruments, whereby it acquires strength for performing other intellectual operations, and from these operations gets again fresh instruments, or the power of pushing its investigations further, and thus gradually proceeds till it reaches the summit of wisdom.[35]

Even in the context of the problem of the origin of language, an acceptance of Spinoza's theses, like those of Hobbes, necessarily led to positions difficult to reconcile with Christian tradition. In Pierre-Daniel Huet's attack on the "dangerous heresy of deism that is making such deplorable progress today" as in many other works, the names of the three philosophers we have been discussing are joined, by no means by chance:

> A Theological and Political Treatise has recently appeared whose author . . . is not content to undermine the bases of religion and of a solid theology, and goes as far as to shake political order and the notions of common sense. . . . The author has drawn his arguments in part from Aben Ezra, in part from the *Leviathan* of Hobbes, and from the *System of the Preadamites*.[36]

29·The Fabulous Poets

EDWARD Stillingfleet, whose *Origines Sacrae* was published in 1662, was well aware of the nature of the weapons that the atheists were using to attack religion in the modern age:

> As the tempers and genius's of ages and times alter, so do the arms and weapons which Atheists imploy against religion; the most popular pretences of the atheists of our age have been: 1) the irreconcileableness of the account of Times in Scripture with that of the learned and ancient heathen nations; 2) the inconsistency of the belief of the Scriptures with the principles of reason; 3) and the account which may be given of the origine of things from principles of philosophy without the Scriptures.[1]

Stillingfleet's chief interest—like Vico's later—was to compress and shorten the ages of the world's history in response to the impious claims of the Epicureans' and the libertines' new chronology. He aimed in particular at showing the ambiguous, uncertain, and fabulous character of all of the histories of the pagan nations, as opposed to the certitude and continuity of Hebraic history. He argues that the latter remained the only reliable source by which to measure the partial truth of all other histories. His denial of the boundless ages of pagan nations and of their claims to primacy in the history of the world led Stillingfleet to accentuate the stupidity, the barbarism, and the crudeness of the first peoples. The sons of Noah "did gradually degenerate into ignorance and barbarism."[2] The primitive world, in Stillingfleet's view as in Vico's, was dominated by the sluggishness and the ferocity of its first barbaric inhabitants.

The transition from this barbarism to civilization appears as a slow construction that had left its confused traces in the fabulous tales of the ancient peoples. To pass from a mythic and "fabulous" historiography to true accounts of the events of particular peoples had to wait for the foundation of the great empires. Before that time, during the obscure ages of history, men's only occupation was a struggle for survival. They knew no technical arts and no sciences. Stillingfleet, like Vico, respects the brief spans of orthodox chronology, and he poses the problem of the transition, among the pagan nations, from those obscure ages to the age of civilization, from the world of "fables" to the world of reason. Language, writing, the arts and the sciences, institutions and civilization were something that mankind had lost and had had to reconstruct slowly and laboriously during the course of time.

Stillingfleet supports his idea that the fabulous histories were unreliable, among other ways, by pointing out that men found it impossible to communicate with one another during their period of barbarism. The rise of writing was not instantaneous; it had a history and it took place within time. During the first ages, the transmission of knowledge was carried out through "representative symbols" or hieroglyphics:

> Now this defectiveness of giving testimony of ancient times by these Nations, will further appear by these two considerations: *First*, What ways there are for communicating knowledge to posterity. *Secondly*, How long it was ere these Nations came to be Masters of any way of certain communicating their conceptions to their Successors. Three general ways there are whereby knowledge may be propagated from one to another; by representative Symbols, by Speech, and by Letters. The first of these was most common in those elder times, for which purpose *Clemens Alexandrinus* [*Stromata* 5] produceth the testimony of an ancient Grammarian Dionysius Thrax: . . . *That some persons made a representation of their actions to others, not only by speech, but by symbols too*. Which any one who is any ways conversant in the Learning of those ancient times, will find to have been the chief way of propagating it (such as it was) from one to another: as is evident in the Hieroglyphicks of the *Ægyptians*, and the custom of Symbols from thence derived among the *Grecian* Philosophers, especially the *Pythagoreans*. It was the solemn custom of the *Ægyptians* to wrap up all the little knowledge they had, under such mystical representations, which were unavoidably clog'd with two inconveniences very unsuitable to the propagation of knowledge, which were Obscurity and Ambiguity.[3]

In order for humanity to arrive at the construction of symbols appropriate to the representation of concepts "a great deal of time" was needed. Those first symbols were uncertain and subject to a great variety of interpretation. Alphabetic letters brought a remedy to the ephemeral nature of speech and the weak and inadequate memory of man. With them, "mens Voices might be seen, and mens Fingers speak." Galileo had been right to call letters "the choicest of all humane inventions." The fact that nearly every nation attributed the invention of writing to one or more persons—the Jews to Adam or to Moses, the Egyptians to Thoth or to Mercury, the Greeks to Cadmus, the Phoenicians to Tauto, and the Latins to Saturn—served merely to show the obscure and fabulous character of the most ancient histories. Stillingfleet concludes with an expression that Giambattista Vico would have appreciated: "Thus it happened with most Nations: what was first among themselves, they thought to be the first in the world."[4]

The "fabulousness of the poets" lay behind the profound change undergone by the most ancient traditions, which contain "a strange confusion of things together" that makes the distant past an inextricable labyrinth. We can find our bearings in this labyrinth, to some extent, by throwing light on some of the constants inherent in the fabulousness of the poets, constants which can explain the changes to which the account found in sacred history had been subjected. One of these constants was the tendency to attribute to men and women of one's own nation the great actions of the founders of mankind. According to another "principle of mythology," a series of actions accomplished by many people comes to be attributed to one mythical personage:

> But yet a more prolifick Principle of *Mythology* was by attributing the Actions of several Persons to one who was the first or the chief of them. Thus it was in their stories of *Jupiter, Neptune, Mars, Mercury, Minerva, Juno, Bacchus* and *Hercules,* which were a collection of the Actions done by a multitude of Persons, which are all attributed to one Person. . . . Before the time of the *Trojan* Wars, most of their Kings, who were renowned and powerful, were called *Joves*. Now when the actions of all these were attributed to one *Jupiter* of *Creet,* they must needs swell his Story up with abundance of Fables. *Vossius* had taken a great deal of pains to digest in an Historical manner the Stories of the several *Jupiters,* whereof he reckons two *Argives,* a third the Father of *Hercules,* a fourth a King of *Phrygia,* and two more of *Creet,* to one of which, without any Distinction, the Actions of all the rest were ascribed, and who was worshipped under the name of Jupiter.[5]

In his lengthy discussion of the origins of the alphabet, Stillingfleet gives particular attention to the presumed invention of letters by Cadmus. Given his initial premises, one point is particularly important to him: Cadmus was not as "antique" as many had claimed and, consequently, the invention of alphabetic writing came into use "rather late" among the Greek nation, "which hath most vainly arrogated the most to itself in point of antiquity."[6]

The hypothesis of the preadamites, Stillingfleet calmly affirms when he reaches the end of his labors, was undoubtedly false. After the dispersal of the sons of Noah, the true history of the world was distorted and transformed by the barbarism of the pagan nations, by idolatry, and by the confusion of languages that prevented the preservation of the original patrimony of mankind. On the problem of the origin of linguistic differences, Stillingfleet sadly admits, "there is no agreement in men's minds."[7] Once again this champion of orthodoxy, who had theorized the barbarous state of the ancient nations in order to reaffirm the truth of sacred history, was put on the defensive: the positions of Méric Casaubon in his *De Quatuor Linguis Commentario* and of Jean Bodin in the ninth chapter of his *Methodus* were untenable. The truth concerning languages did not depend only on differences in climate and in the customs of the various peoples. It required the direct and miraculous intervention of God.

30·Moses, the Deluge, and Barbarism

JOHN Woodward, professor of physics at Gresham College and expelled from the Royal Society for his arrogant conduct, had a more specific interest than Stillingfleet's: he was interested in showing that the fabulous traditions of ancient peoples were unreliable in the particular context of the Flood. In his *Essay towards a Natural History of the Earth* (1695), Woodward rejects most of Burnet's hypotheses and qualifies Burnet's picture of the earth as "imaginary and fictitious." The Flood, for Woodward, was an enormous revolution, "the most horrible and portentous catastrophe that Nature ever yet saw." An originally "elegant, orderly, and habitable earth" was "quite unhinged, shattered all to pieces, and turned into a heap of ruins." What was involved was not simple alterations in the form of the earth or in its position in relation to the sun, but a true dissolution of the constitutive principles of matter, which were mixed together and then separated anew. The Flood, following the letter of the Genesis account, was a veritable *destruction* of the world.[1]

That terrible phenomenon did not affect mankind alone. It was not to be interpreted *only* as divine punishment for the sins of man. It did not serve only to punish one generation and to terrify its posterity. It was directed not only against men, but "principally against the Earth that then was": fertile and luxuriant, inhabited by happy men who had no need to work and no need of the skills that serve to overcome a hostile nature:

> The said Earth . . . was yet generally much more fertil than *ours* is. . . . The exteriour *Stratum* or Surface of it, consisted entirely of a kind of terrestrial Matter proper for the Nourishment and Formation of Plants. . . . Its Soil was more luxuriant . . . Earth requiring little or no Care or Culture; but yeilding its encrease freely, and without any considerable Labour and Toil, or assistance of Humane Industry; by this means allowing Mankind that time, which must otherwise have been spent in Agriculture, Plowing, Sowing, and the like, to Purposes more agreeable to the Design of their Creation.[2]

This world of primitive innocence was contaminated by sin. The very fertility of the soil and its exuberant production excited and fomented lust and vices. Passions were let loose: human creatures mated "without discretion or decency, without regard to the age or affinity, but promiscuously and with no better a guide than the impulses of a brutal appetite." At this point the Flood occurred. For Woodward, the Flood could not be explained by natural causes alone—as Burnet would have it—but required instead the presence of a supernatural power, guided by wisdom and acting according to a precise plan. The system of Epicurus—which envisioned the world as arising out of chaos, denied God or reduced him to a Being indifferent to the affairs of this world, and which attempted to explain the stupendous and harmonious edifice of the universe by a chance meeting of atoms—was absurd and meaningless.[3]

After the great catastrophe, Woodward explains, human history began. Once again, it is a history of obscurity and barbarism. Men, dispersed and shaken, lived without knowledge of their past and "have only dark and faint ideas." They were totally involved in a struggle for material survival, in the cultivation of the soil and the search for food. The effort required to guarantee simple survival, the laborious task of working the soil, and their involvement in manual arts left no room either for "works of the brain" or for the contemplation of nature. Thus it was impossible to believe the mythical and confused narrations of the Flood offered by the ancient pagan philosophers: all nations were once barbarous, and "philosophy was then again in its infancy." The few marks that then survived of the older tradition were "defaced by time." The ancients were without exact observations or reliable documentation of the state of the earth before their times; they had limited and insufficient ideas of Providence, and they formulated their hypotheses with the aid of imagination and fantasy alone.[4]

After the Flood, in an age that had no reflective thought and no science, no form of writing existed. Woodward rejects the myth of an original alphabet:

> There was no shew of *Learning*, or matters of Speculation among them; and we hear little or nothing of *Writing*, nay 'twas a very considerable time before *Letters* themselves were found out. I know very well, there are some who talk of Letters before the Deluge; but that is a matter of meer Conjecture, and I think nothing can be peremptorily determined either the one way or the other, though I shall shew, that 'tis highly probable they had none. But *that* how it will, I shall plainly make out, that the Ages which next succeeded the Deluge *had none;* so far from it, that they knew nothing at all of them; and the first Writing they used was only the simple Pictures, and Gravings of things they would represent, Beasts, Birds and the like; which way of Expression was afterwards called Hieroglyphick. But this fell into disuse, when Letters were afterward discovered; they being, in all respects, a far more excellent and noble Invention. We see therefore that there were several Reasons why those early Ages could not transmit Accounts of the state of the Earth and of these Marine Bodies, in their times, down to the succeeding Generations.[5]

Woodward is known particularly as an antiquarian, a geologist, and a fossil collector. But his interest in the history of barbarous ages and in the world of primitive man was neither marginal nor secondary. In his *Remarks upon the Ancient and Present State of London, Occasion'd by Some Roman Urns, Coins, and Other Antiquities Lately Discovered* he set the life of the primitive inhabitants of Britain into a framework that was immediately enlarged to discuss the generally barbarous conditions of all mankind:

> The *Britains* in those Days were barbarous, and wholey unciviliz'd. . . . They went naked, and painted their Bodies with the Figures of various Animals, after the Manner of other Savage Nations. . . . There was little or nothing that could claim the Name of Science among them. What they had was lodg'd with the Druids, who were the Devines and Philosophers of those Times. . . . So much of the Religion of the *Barbarous* nations is placed chiefly in Things realy mean and trivial. Nor will it be thought strange that our Progenitors should be, in those early Times, thus rude and unciviliz'd. . . . All Mankind quite round the Globe were once so, I mean at their first Original, in the Ages that ensu'd next after the Deluge. This the Histories and Accounts of the *Assyrians*, the *Ægyptians*, the *Chineses*, and all others, agree in. Even the Grecians, that became afterwards the most polite and refined People upon Earth, were once *barbarous*. They made as little scruple as the *Britains* of slaying men; and sacrificing them to the Deities which they worship'd. They liv'd upon *Leaves* and *Herbs*, or upon *Acorns*, 'till *Ceres* and *Triptolemus* taught them to *Plow* and *Sow Corn*. They had no other clothing than only the Skins of Beasts. Some of them dwelt in *Caves*, others in mean *Hutts;* others run *wild* in the *Woods*, like so many *Brutes* 'till *Pelasgus*, *Orpheus*, *Amphion* and some other great Men, found out Ways to tame them, and to deter and reclaim them from their *Rapine* and *Ferity*.[6]

At the end of his most famous work Woodward speaks of two projected *Discourses*. The first was to concern the migrations of peoples; the second, the tradition of the Flood among the Scythians, Persians, Babylonians, Phrygians, Africans, and so forth. Rites, ceremonies, and superstitious worship common to several peoples had their origins in the age immediately following the great diluvian catastrophe and arose from that common, convulsive experience. The first discourse was to take up the question of the repopulation of the world, America in particular. It was to discuss the traces of the Americans to be found in the most ancient traditions, the discovery of Atlantis, the reasons for the diversity of different peoples in external aspect, language, eating habits, customs, arts and sciences, religions, and the organization of civil and military life.[7]

This project was probably not carried out in this manner. The English translator of Woodward's *Naturalis historia telluris* noted among Woodward's unpublished writings a work of vast scope:

> A Representation of the State of Mankind in the first Ages after the Deluge; with an historical Discourse wherein the Manners, Customs, Opinions and Traditions, as also the Arts, Utensils, Instruments, and Weapons, of all the most Antient Nations, are carefully compared; in Order to the Discovery of the Origin of Nations, but more

> particularly of the Americans, Negroes and Indians. Tho' in the Compass I am confined to, [continues B. Halloway, the translator] it be not easily practicable to give an Idea of a Work of the Variety and Extent that this is, yet I cannot but take Notice that it makes out very plainly, from Reflexion on their Notions, and Practices, from their chief Customs Religious and Civil, from the Disposition of their Minds, and constitution of the Bodyes of *Americans, Negroes,* and *Indians,* that they, with the rest, came all originally from one and the same Stock: and that the present Difference as to Stature, Shape, Features, Hair and Complexion, is owing wholey to the Diversity of Heat, Climes, Soils and their various Productions, Diet, and the different methods of Living.

Holloway's summary, among other things, confirms that Woodward held that hieroglyphic writing preceded alphabetical writing: "As to the *Americans,* in particular, . . . they kept their Records and preserv'd the Memory of Things, by Hieroglyphic Representations; all which the most antient *Asiatic, African,* and *European* nations, the *Chineses,* the *Ægyptians,* and the rest, likewise did."[8]

Taking up the problem of origins or returning to the remote historical past involved taking a position on the Mosaic story. In this Woodward, when he had finished with his precautions, his anti-Epicureanism, and his continual and repeated professions of orthodoxy, took some decidedly modern positions. His interest in the history of the earth and the history of human origins led him to approach the biblical text in ways that many authors preferred to avoid carefully:

> For Moses, he having given an Account of some Things which I here treat of, I was bound to allow him the same Plea that I do other Writers, and to consider what he hath delivered. In order to this, I set aside everything that might bypass my Mind, over-awe, or mislead me in the Scrutiny, and therefore have regard to him here only as an Historian. . . . Wherein I do him but simply Right, and only the same that I would to a common Historian, to *Berosus* or Manetho, to Herodotus or Livy, on like occasion.[9]

Woodward, as is clear from his unpublished correspondence, was an impassioned reader of Lucretius, whom he considered, as he wrote to John Edwards on 26 June 1699, "a mighty master of the learning of the still earlier ages."[10] Indeed, it was by citing the authority of Lucretius over Ovid and Virgil that he had held that the cultivation of grain and the plow and other tools for cultivating the fields were unknown to the primitive Egyptians, Greeks, and Italians.[11] But he was also extremely interested in defending his positions as completely orthodox and perfectly reconcilable with Scripture—unlike those of Burnet. In another letter to Edwards dated 23 May 1699, he defends himself against a detailed accusation of heterodoxy that had been leveled against him.[12] Later, he passed on to more challenging and more compromising readings. Abbé Bignon thanked him from Paris on 12 August

1721 for having sent him John Toland's *Pantheisticon*, which had been published in London the preceding year.[13]

When he took the position that Moses should be considered a historian of the barbarous ages, it was almost fatal that Woodward be bracketed with Toland. Moses, according to a tradition that runs from Spinoza to Toland and Giannone, ended up perforce being portrayed as a writer of barbarous times. Moses became a "crude" writer. This transformation duly takes place in the second and enlarged edition of Woodward's *Natural History* (1726):

> *Moses*'s own Observations could give him little Light into this Affair [the Flood and transformations of the earth]. . . . The World was not then thorowly settled, Things sufficiently establish'd, or Arts so far advanc'd as to afford Leisure to Curiosity, or such Kinds of Speculation. . . . Tho' indeed, had *Moses* been ever so curious or inquisitive, it would have been to little Effect, as he must have wanted Assistance to carry his Enquiries on to a sufficient Extent. Navigation was then in its Infancy, . . . the Mariners Compass by which we are conducted on our long voyages being not found out. Indeed there was then only a small and very inconsiderable part of the World known; whereas *Moses* could not have Intelligence sufficient to found Propositions of so great Extent upon without Accounts and Observations procur'd from Countries the most distant, and even Antipodes to those he had seen, from the remotest Part of *Africa*, and *Europe*, from *China*, and even from *America* itself.[14]

Observations on the progress of knowledge and the superiority of the moderns derived from Bacon are applied here to the relationship between Mosaic philosophy and the modern world. Reconciling his cosmological history with the biblical account of creation was not an easy task for Woodward. Dangers and difficulties were inherent in it from the start—from the moment that a "scientific" account of the formation of the world was opposed to the Genesis story, thus posing the problem that Scripture might be reduced to the level of the "fabulous" accounts. Woodward's explicit and insistent polemic with Burnet was obviously not enough, in the eyes of the more rigorous champions of orthodoxy, to cancel out the dangerous possibility that he might slip into out-and-out impiety. As early as 1699 an anonymous letter (perhaps written by Thomas Baker, although the author is careful not to leave compromising documentation) posed some fundamental objections:

> By *natural*, I do not mean your theory is so in every thing, and particularly not in that instance I have objected, and tho' I do verily believe your design in that work was pious, yet I am afraid your account of the dissolution of the earth will upon examination be found not so reconciliable to the Text, for reasons your own thought will suggest to you; and to deal freely with you, I am of opinion that no theories I have yet seen (I will not except yours from that number) have done service to Religion, for if they fail in one thing, they loose their

> end. Not being good at counterfeiting a hand, nor willing my own should be known, I must be short.[15]

How is it possible, Thomas Baker was to write to Woodward on 15 April 1700, that such grandiose changes could have taken place on the earth without being mentioned in any in sacred history? And how can we know the details and the particulars of a history and of a past that Scripture does not discuss?[16]

31·Adam as Beast-Man

IN Woodward's works, which were published between 1695 and 1720, the origin of language and of writing represents a somewhat marginal theme. Samuel Shuckford, on the other hand, gives a broad and carefully articulated treatment of the question in his *Sacred and Profane History of the World Connected,* first published in London in 1728. The problem he set himself was to establish how man, the only "conversible creature" in the universe, achieved this extraordinary skill through time. According to Shuckford, two different doctrines disputed the field. The first, found as early as Diodorus Siculus and Vitruvius, states that the first men lived like the beasts in forests and caves. At this stage, men produced only strange and bizarre sounds, until fear forced them to join together. They then began to communicate with one another, first by means of signs, then by means of the names they gave to things, finally by the construction of languages that differed because the humans banded together in different places. The second doctrine, advanced by those who took the Mosaic account as true, states that Adam was created "not only a reasonable, but a speaking creature." Many in this camp went so far as to claim that Adam possessed an "innate set of words" in the same way that he possessed the power of thinking and the faculty of reasoning. In theory, this innate and original language would have been transmitted to all his descendants if they had not been taught another language in infancy.[1]

The first of these doctrines seemed to Shuckford an "ingenious conjecture" which could be received as probable if our knowledge of human origins were to spring—as in the view of Diodorus Siculus and Vitruvius—from our own imagination. Since we have the Mosaic history at hand to inform us of the origins of both men and languages, he argues, we, like Eusebius, can reject those inconsistent conjectures "as a mere Guess that has no manner of authority to support it."

Shuckford found the second doctrine to be equally untrue, because there was no need to think that Adam disposed of a series of words innately. It is not in the least true, he argues, that children would speak an original language "naturally" if their upbringing did not prevent them from doing so. This "very wild and extravagant fancy" was proven false by the fact that in infancy all languages can be learned with equal facility. When such learning is absent, as many experiments have shown, children were only able to emit inarticulate, accidental, and meaningless sounds.

After rejecting the more daringly heterodox thesis, which smacked of Lucretius, and after setting aside the traditional doctrine of divine onomathesia and of an innate, original, and natural language, Shuckford declares his intention to hold firm to the thesis of the divine origin of language, but he ends up by completely emptying it of meaning:

> The first Man was instructed to speak by God, who made him . . . his Descendants learnt to speak by Imitation from their Predecessors; and this I think is the very Truth, if we do not take it too strictly. The Original of our Speaking was from God; not that God put into Adam's Mouth the very Sounds which he designed he should use as the Names of Things, but God made *Adam* with the Powers of a Man. He had the Use of an Understanding, to form Notions in his Mind of the things about him; and he had a Power to utter Sounds, which should be to himself the Names of Things, according as he might think fit to call them. These he might teach *Eve*, and in time both of them teach their Children; and thus begin and spread the first Language of the World.[2]

In this clever "connection" of sacred history to profane history Shuckford repeated arguments that Richard Simon had advanced in his *Histoire critique du Vieux Testament*, which had been translated into English and published in London in 1682. Adam, Shuckford continues, was just a creature gifted with the faculties and the possibilities that characterize man. The traditional affirmations that Adam was a master of wisdom, that he listened to the true teachings of God, that he invented divine onomathesia thus collapsed. Adam's speech was not complex: he emitted sounds and set up a connection between such sounds and particular things. Those names for things did not reflect their nature: they were the names of those things *for Adam*. Adamic language embodied no relationship of names to things; it was not the natural language that mirrored the immaculate wisdom of the origins. It was the language that served the extremely crude and limited uses of the first inhabitants of the earth. Adam did no more than "name" the creatures so that he could tell them apart:

> The first language consisted chiefly of a few Names for the Creatures and Things that Mankind had to do with. *Adam* is introduced as making a Language, by his naming the Creatures that were about him. The chief occasion he had for Language was perhaps to distinguish them in his Speech from one another . . . so this might be all the Language he took care to provide for the Use of Life.[3]

Any attempt to attribute the sacred and the exemplary to Adamic language was here cut off at the root. Any discussion of the priority of Hebrew, or of any other particular language, became senseless. On this foundation Shuckford could build the thesis of a historical origin to the formation of language and of its slow and progressive formation through time. The first language was like baby talk: it consisted of "very simple and uncompounded sounds" of the sort that infants emit in their first uncertain attempts at speech. The transition to words of varied length and composed of separate sounds was the work of "Contrivance and Improvement" and it took place in response to the expressive insufficiency of monosyllabic language. Shuckford notes that we consider languages currently spoken that abound in "short and single" words to be poorer and less polished, hence the original, unrefined language was of this sort.[4]

Names for *things* or *persons* also characterized the first forms of language. Abstract concepts or notions were lacking, and terms like "lion-man" were used to indicate the abstract quality of courage. In a language made of nouns and which did not yet contain verbs, the names of beings who typically accomplish a certain action served to indicate the action. (The use in English of the terming "dogging," Shuckford notes, to indicate the action of observing and following a person wherever he goes is a holdover of this older use, as it uses the noun "dog" to indicate a behavior and an action like those typical of dogs.)[5]

In mankind's past, then, there was no natural, original, and perfect language. Languages were formed historically; they were a conquest that sprang from human art and human invention and from the new needs life put before men, a conquest that required time and that was realized within time:

> It was Time and Observation that taught Men to distinguish Language into Nouns and Verbs; and afterwards made Adjectives, and other Parts of Speech. It was Time and Contrivance that gave to Nouns their Numbers; and in some Languages, a Variety of Cases, that varied Verbs by Mood, and Tense, and Number, and Person, and Voice . . . and made Men thereby able to express more things by them, and in a better manner, and added to the Words in use new and different ones, to express new Things, as a further Acquaintance with the Things of the World gave Occasion for them.[6]

When it came to the origin of writing, Shuckford clearly could not fail, given these premises, to be decidedly opposed to the thesis that credited God with the origin of the alphabet and that saw alphabetic writing as the earliest form of writing. It seems strange, he states, that an invention as surprising as that of letters has been attributed by many to ages close to the beginning of the world. It might be easy to conceive that nature offered man the possibility of expressing himself and of speaking, but it was really not so easy to imagine that mankind, "amongst its first attempts," succeeded in creating a method appropriate to the expression of all that can be said or thought through a system of sixteen, twenty, or twenty-four characters. The idea of an original alphabet, learned immediately and directly, "exceeds the highest notion we can have of the capacity we are endued with."

Man was undoubtedly gifted with extraordinary capacities and was able to use and turn to his service all parts and all the creatures of this world, but human knowledge proceeds "by steps and degrees" and the first attempts in any invention or in any science have never achieved perfection. The first men, Shuckford insists, made a series of attempts in many arts that successive ages carried to perfection. Thus it happened for language, where the first men gave rise to a comprehensible language without arriving at elegant speech. Thus it happened in the arts of making glass and forging iron, in construction, in painting, and in sculpture, where "the first Tryals were only Attempts."

> Men arrived at Perfection by degrees; Time and Experience led them on from one thing to another, until by having try'd many ways, as their different Fancies at different Times happened to lead them, they came to form better Methods of executing what they aimed at, than at first they thought of. And thus, without doubt, as it happened in the Affair of Letters: Men did not at first hit upon a Method extremely artificial, but began with something easy and plain, simple, and of no great Contrivance, such as Nature might very readily suggest to them. And, if I may be allowed to make some Conjectures upon this Subject, I should offer, that it is not probable, that the first Inventers of Letters had any Alphabet, or set Number of Letters, or any Notion of describing a Word by such letters as should spell, and thereby express the Sound of it. The first letters were, more likely, Strokes, or Dashes, by which the Writers mark'd down, as their Fancies led them, the Things they had a mind to record; and one Stroke, or Dash, without any Notion of expressing a Sound or Word but it was the Mark of a whole Action, or perhaps of a Sentence.[7]

The first men had no form of writing capable of expressing sounds and words. They used characters in some ways comparable to Chinese characters, to astronomic symbols (which are intelligible to people who speak different languages and who "read" them in their own tongues), and, finally, to the signs still used as mnemonic devices by the many people in the world who can neither read nor write nor sound out letters:

> I have been told of a Country Farmer of very considerable Dealings who was able to keep no other Book, and yet carried on a Variety of Business in Buying and Selling, without Disorder or Confusion: He chalk'd upon the Walls of a large Room set apart for that Purpose, what he was obliged to remember of his Affairs with divers Persons; and if we but suppose, that some of his Family were instructed in his Marks, there is no Difficulty in conceiving, that he might this way, if he had died, have left a very clear State of his Concerns to them. Something of this sort is like the first Essay of Nature, and thus, without doubt, wrote the first Men. It was Time and Improvement that led them to

> consider the Nature of Words, to divide them into Syllables, and to form a Method of spelling them by a set of letters.[8]

In this work, Shuckford places the birth of language and writing against the background of a more general "philosophy of history" and a doctrine of progress that sees the infancy of mankind as analogous to that of the individual. He emphasizes "fancy" and draws a parallel between the first forms of writing and forms used by primitive peoples or still in use among "illiterates." Shuckford, like Stillingfleet, and like Vico and Warburton later, placed the discussion of language and writing within the broader problematics that had as its object human origins and the rise of the first forms of civil life, and this led him to take on the great subject of the "records" or the "vestiges" of the earliest nations.[9] Shuckford as well was an enemy to the irreligious theses of the libertines and of Lapeyrère: he denies that the civilization of the Chaldeans or of the Phoenicians could be made to precede the age of Moses; he rejects Manetho's declaration that Egypt's history reached back many centuries beyond the date accepted as that of the creation of the world, he identifies the figure of the Chinese emperor Fohi with Moses, and he judges much of the recent speculation on the origin and formation of the earth difficult to reconcile with the authentic record of the Mosaic account.[10] But his primary interest lay in the *connection* of sacred history and profane history: Although he insisted that the truth of the Mosaic account must be safeguarded, he emphasized the function and operation of time and the slow, gradual formation of both civilizations and the tools and ideas that were at work within them. His rejection of a "natural" language infused into Adam's mind, his insistence on a gradual apprenticeship in language and writing, which progressed and grew along with the growth and progress of knowledge and the arts, his thesis of an inarticulate speech that preceded articulate speech and of a writing made of images and hieroglyphics that preceded alphabetical writing, his image of a mankind of rough ways and few ideas, becoming adult and mature as the centuries pass—all of this led Shuckford to conceive of the first men as "without any stock of actual knowledge," as rising gradually into knowledge by observation of the natural world and by the sentiments that observation provoked in them. Thus the idea of a natural and original language—"philosophical" in that it expressed the true nature of things—which then was perpetuated in the language of the Jews seemed to him necessarily "romantic and irrational."[11] In Shuckford's discourse, the figure of Adam ended up coinciding with that of "potential man." It was precisely for having reduced the figure of Adam to that of a "savage" and of a "wild animal" that he was to be reproached bitterly:

> Oh horrible! So the salvation or reprobation of mankind to all eternity depended upon the behaviour of a poor simpleton. . . . Does not the Doctor see how he impeaches the divine justice, in resting an affair of eternal moment to the souls of so many millions, upon the trial of a person no ways prepared or fit to stand it, of one ignorant of God, the creatures and himself . . . a mere child, an ignoramus, an idiot?

> What shall we say to these things, and whither will this modern scheme of commenting from Ovid, Tully, Mr. Pope, the Ars critica, and our own imaginations, carry us? For heaven's sake let us return into the good old way of explaining Scripture by Scripture, and not deny two and two to be four, because the Hutchinsonians maintain it.[12]

When he mentions Pope and Hutchinson, the anonymous pamphleteer who attacks Shuckford so forcefully does not note a much more dangerous name: Hobbes. He was wrong not to do so, because in many of his pages on language Shuckford did little but amplify and comment on a brief passage in which Hobbes speaks of Adam's limitations and of the fashioning of the first language. Hobbes says:

> The first author of *speech* was God himself, that instructed Adam how to name such creatures as he presented to his sight; for the Scripture goeth no further in this matter. But this was sufficient to direct him to add more names, as the experience and use of the creatures should give him occasion; and to join them in such manner by degrees, as to make himself understood; and so by succession of time, so much language might be gotten, as he had found use for; though not so copious as an orator or philosopher has need of: for I do not find any thing in the Scripture, out of which, directly or by consequence can be gathered, that Adam was taught the names of all figures, numbers, measures, colours, sounds, fancies, relations; much less the names of words and speech, as *generall, speciall, affirmative, negative, interrogative, optative, infinitive,* all which are usefull; and least of all, of *entity, intentionality, quiddity,* and other insignificant words of the school.[13]

32 · Born with Fear

IN 1728, the year that saw the publication of Shuckford's book, the second part of Bernard de Mandeville's *Fable of the Bees* was published in London. The fourth, fifth, and sixth of Mandeville's *Dialogues between Horatio and Cleomenes* (spokesmen, respectively, for Shaftesbury's ideas and those of the author) gave more ample and detailed discussion of some of the ideas on the nature of language that had already been presented in *A Search into the Nature of Society* (1723) and offered a well-structured vision of the origin of languages and the connection between language and the transition from savage to civilized life.

In the *Search,* as is known, man's "bad and hateful qualities," imperfections, and "want of excellencies" (present in other creatures) were seen as the reason for the greater sociability of man as compared to the other animals:

> No societies could have sprung from the amiable virtues and loving qualities of man; but, on the contrary, . . . all of them must have had their origin from his wants, his imperfections, and the variety of his appetites; we shall find likewise that the more their pride and vanity are displayed and all their desires enlarged, the more capable they must be of being raised into large and numerous societies.[1]

Man, who lived in a state of innocence before the expulsion from Paradise, who was "endued with a consummate knowledge the moment he was formed," who was protected from the aggressions of the animals and of nature, who was "yet not conscious of guilt" and was "unrivalled lord of all" was not a social creature by nature: he was "wholly wrapped up in sublime meditations on the infinity of his Creator, who daily did vouchsafe intelligibly to speak to him and visit without mischief." Beyond this mythic "blessed state," in which arts, science, and commerce among men would be totally superfluous, real men were a bundle of desires, impulses, and passions. Independently of "whatever fine Notions we may flatter our Selves with," all human beings "are sway'd and wholly govern'd by their Passions." This was true, in exactly the same way, both for those who "strictly follow the Dictates of their Reason" and for those, acting in the opposite manner, "whom we call Slaves to their Passions."[2]

Society in Mandeville takes the form of an *artificial* and *unnatural* way to control and rationalize impulses that are part of human nature. It represents the denial of demands that are natural to man, and is at the same time able to make use of some of these demands. Since they are an integral part of human nature, the passions persist, and are *identical* in both the savage and the social and civilized man:

> There is no difference between the original nature of a savage and that of a civilized man: they are both born with fear; and neither of them, if they have their senses about them, can live many years, but an invisible Power will at one time or other become the object of that fear; and this will happen to every man, whether he be wild and alone or in society and under the best of discipline.[3]

Only moral ills (the passions: fear, the love of power, the innate tendency to oppress others, etc.) and natural ills (the hostility of the elements, the ferocity of the animals, etc.) make men sociable, that is, move men to accept becoming "subdued by superior force," and to becoming a disciplined creature "that can find his own ends in labouring for others." Finally, they create a situation in which "each member is rendered subservient to the whole, and all of them by cunning management are made to act as one."[4]

Mandeville sees politics as an art capable of inserting into the immutable passional makeup of man those elements of rationality and those values that make orderly cohabitation possible and that permit men to escape the perils of total anarchy. Values, virtues, and the elements of rationality are introduced into the body social by using the mechanism of the passions for leverage. Reason and virtue will be *experienced* by the majority of civilized men as liberty and as a ransom from the servitude of the passions. Anyone capable of analyzing human nature, however, will see reason and virtue as manifestations of the passions, at work within men without their knowledge:

> For we are ever pushing our reason which way soever we feel passion to draw it, and self-love pleads to all human creatures for their different views, still furnishing every individual with arguments to justify their inclinations. . . . Those who can examine Nature will always find, that what these People most pretend to is the least, and what they utterly deny their greatest Motive.[5]

Mandeville draws a first conclusion on the level of communication: given that "it is impossible that man . . . should act with any other view but to please himself," given that "we are always forced to do what we please, and at the same time our thoughts are free and uncontrolled," it is impossible to be sociable creatures without hypocrisy. In all civil societies men learn, unconsciously but necessarily, "to be hypocrites from their cradle." The very existence of social life—which always and in all cases involves the repression and deviation of the natural impulses from their object—gives rise to a necessary and irreducible fracture between thought and language:

> Since we cannot prevent the ideas that are continually arising within us, all civil commerce would be lost, if by art and prudent dissimulation we had not learned to hide and stifle them; and if all we think was to be laid open to others in the same manner as it is to ourselves, it is impossible that, endued with speech, we could be sufferable to one another.[6]

If society both denies and uses natural passions; if values are introduced artificially in the world by means of the promise of prestige to those who respect them and the threat of scorn to those who follow their natural antisocial tendencies; if human actions are always motivated by self-love even where this seems "mediated" by the demands of life in society; if hypocrisy is a constructive element in society and one of its essential supports; if, for example, "the sexton would be stoned should he wish openly for the death of the parishoners, though everybody knew that he had nothing else to live upon";[7] if all of this is true, then the principal end of language is not that of making our thoughts known to others, but of both communicating and concealing them:

> HORATIO: The design of Speech is to make our thoughts known to others.
> CLEOMENES: I do not think so.
> HORATIO: What! do not men speak to be understood?
> CLEOMENES: In one sense they do; but there is a double meaning in those words, which I believe you did not intend: if by man's speaking to be understood you mean, that when men speak, they

> desire that the purport of the sounds they utter should be known and apprehended by others, I answer in the affirmitive: but if you mean by it, that men speak in order that their thoughts may be known and their sentiments laid open and seen through by others, which likewise may be meant by speaking to be understood, I answer in the negative. The first sign or sound that ever man made, born of a woman, was made in behalf, and intended for the use, of him who made it; and I am of the opinion that the first design of speech was to persuade others, either to give credit to what the speaking person would have them believe, or else to act or suffer such things as he would compel them to act or suffer, if they were entirely in his power.[8]

In Mandeville, both the libertine thesis of the astuteness of politicians and the myth, connected with that thesis, of a society founded by the skill of wise legislators who introduced social values into the human world in order to assert and maintain their own power were included within much broader theories concerning the rise of politics and the communicative and social functions of language. The "prudent dissimulation" by which ideas are hidden or suppressed seemed to Mandeville to result from art only in the sense that it is not natural to man. Unlike cows or sheep, he argues, men are not made to "keep together out of a natural affection to their species or love of company." If "society" means something like "flock" or "herd," no creature in the world is less fit for society than man. But the fact that dissimulation is artificial does not at all imply that it is conscious. What Mandeville enjoys, what he qualifies as "a great pleasure," is precisely the possibility offered to the intellect of subjecting the diverse and contradictory attitudes of men to analysis, thus unmasking the reality of passion and impulse that is both revealed and concealed by noble discourses and high-sounding appeals to values.[9]

The libertine heritage and the influence of Hobbes joined in Mandeville's works with Epicurean and Lucretian theses concerning the slow formation of language and its emergence in connection with the first sentiments and the primary, fundamental passions. Among the latter, fear occupied a central place, as we have seen. Religion for Mandeville draws its life from fear, and since the idea of an invisible cause is one of the earliest ideas, the name given to it is one of the first names invented by the first men. To sustain that "fear made a God" would be "silly as well as impious." Nonetheless, it was not opposed either to good sense or to the Christian religion to state, concerning the savages, that "whilst such men are ignorant of the true Deity and yet very defective in the art of thinking and reasoning, fear is the passion that first gives them an opportunity of entertaining some glimmering notions of an invisible Power."[10] When mankind had grown in experience and practice and its intellectual capacities were more perfected, it was led to recognize an infinite and eternal Being. The idea of God as an abstract entity arose slowly, nourished by the primal, elementary terrors in the face of setbacks and disasters, excessive heat and cold, unsupportable humidity and aridity, thunder and lightning, noises in the dark, and the dark itself—all that is frightening and unknown. Even with his "weak understanding," primitive man soon realized that "fruits and other eatables" were not always and everywhere available, that fruit trees can be stricken with drought, that storms can destroy laboriously accumulated provisions, that babies can sicken and die for no apparent reason. Primitive men tended to attribute a different cause to each of these setbacks. His terrors multiplied and, with a mentality typical of young children, who "seem to imagine that everything thinks and feels in the same manner as they do themselves," he populated the world with animated beings and hostile spirits.[11]

The analogy between the primitive mentality and the child's psychology—which has so often been emphasized in connection with Vico and with romantic culture—seems particularly strong in Mandeville: he refers to children who project their reactions onto objects they imagine as animate and to the behavior of nurses, who imitate children, scolding and punishing an object that has hurt the child. For Mandeville, no savage will ever, during his lifetime, manage to liberate himself from this sort of "natural folly." His curiosity toward the world, his investigative attitude toward surrounding reality arise, once again, from "ills" and from the hostility of the world. In enjoyment and in pleasant sensations, man is relaxed and free of worries; he can "swallow a thousand delights, one after another, without asking questions; but the least ill makes him inquisitive whence it came in order to shun it." The knowledge of the causes of good "seems not to promise any addition to his happiness"; the knowledge of the causes of evil, on the other hand, seems indispensable in order to avoid it. Difficulties, adversities, and fear underlie religion: "the word *religion* itself and the fear of God are synonomous." Together with religion, they also underlie the activity of knowing and the first forms of language:[12]

> Whilst a man is but an imperfect thinker and wholly employed in furthering self-preservation in the most simple manner and removing the immediate obstacles he meets with in that pursuit, this affair perhaps affects him but little; but when he comes to be a tolerable reasoner and has leisure to reflect, it must produce strange chimeras and surmises; and a wild couple would not converse together long before they would endeavour to express their minds to one another concerning this matter; and as in time they would invent and agree upon certain sounds of distinction for several things of which the ideas would often occur, so I believe that this invisible cause would be one of the first which they would coin a name for.[13]

History, Mandeville continues, which had led from the first hesitating speech of a savage couple to the civilized nations, has a duration that it is impossible to determine, nor can we fix the number of the ages that mankind has had to go through. The invention of tools, the use of iron, division of labor, and the use of money are some of the fundamental moments in this history. The three "steps" or fundamental stages that led to the establishment of the first societies, however, were determined, Mandeville states, (1) by the "common danger" that arose from the presence of wild animals and that pushed men to unite in a common defense; (2) by the danger inherent in men of their own association, which was linked to the invincible tendency to pride and domination and which led the multitudes to join together in "bands and companies" under different "leaders"; (3) by the invention of letters, which transformed into laws the first uncertain and ambiguous rules of living in common that had been trusted to oral tradition.[14]

The first phase of history is to some extent documented by those accounts of battles that men had with wild animals, present from the earliest ages among a variety of peoples. Profane history in the "infancy of all nations" is full of such accounts. Combats with wild animals and the killing of dragons are characteristic of the labors of the heroes of the earliest antiquity. Sphinxes, basilisks, winged dragons, and flame-spouting bulls are made of the fantastic projections of the sudden striking of serpents, the enormous size of crocodiles, and the irregular form of certain fish: those images were born of terror and of nightmares that reflected real and daily adventures.[15] The real situation of which those monstrous images were projections was made of terror and misery, of invincible and mutual hostility among men: "The danger men are in from one another," Horatio says,

> would naturally divide multitudes into bands and companies that would all have their different leaders, and of which the strongest and most valiant would always swallow up the weakest and most fearful.
>
> CLEOMENES: What you say agrees exactly with the accounts we have of the uncivilized nations that are still subsisting in the world, and thus men may live miserably many ages.
>
> HORATIO: The very first generation that was brought up under the tuition of parents would be governable, and would not every succeeding generation grow wiser than the foregoing?
>
> CLEOMENES: Without doubt they would increase in knowledge and cunning: time and experience would have the same effect upon them as it has upon others, and in the particular things to which they applied themselves they would become as expert and ingenious as the most civilized nations. But their unruly passions and the discords occasioned by contentions would be continually spoiling their improvements, destroying their inventions, and frustrating their designs.[16]

Religion and the fear of an invisible power were not enough to eliminate quarrels or to transform a situation of internecine strife into one of civilization: a "human power" was necessary to enforce the obligation of oaths and punish perjury[17] and to introduce an artificial equilibrium into a natural situation of disorder. This precivilized age saw the birth of language as expression by words. At a time when man's knowledge was restricted to limited contexts and he needed to obey only "the simplest Dictates of Nature," dumb signs and gestures substituted for discourse. For "untaught" men "it is more natural . . . to express themselves by Gestures, than by Sounds" but we are all born "with a Capacity of making ourselves understood . . . without Speech." There are signals to express pain, joy, love, wonder, and fear which are common to the species and we can communicate without speaking by crying, laughing, smiling, frowning, sighing, shouting, and using "the language of the Eyes." By this sort of language a wild couple "would at their first meeting intelligibly say more to one another without guile, than any civiliz'd pair would dare to name without blushing."

As it happened for all manifestations of culture—from agriculture to physics, from astronomy to architecture—the art of communication by words drew its life, as time passed, from a slow and gradual series of attempts ("by slow degrees . . . and length of time").[18] These attempts led to modifications in the structure of the vocal organs themselves. Organic evolution and cultural evolution were inextricably welded together as generations passed:

> From what we see in children that are backward with their tongues, we have reason to think that a wild pair would make themselves intelligible to each other by signs and gestures before they would attempt it by sounds; but when they lived together for many years, it is very probable that for the things they were most conversant with they would find out sounds to stir up in each other the ideas of such things when they were out of sight. These sounds they would communicate to their young ones, and the longer they lived together the greater variety of sounds they would invent, as well for actions as the things themselves; they would find that the volubility of tongue and flexibility of voice were much greater in their young ones than they could remember it ever to have been in themselves. It is impossible but some of these young ones would either by accident or design make use of this superior aptitude of the organs at one time or other, which every generation would still improve upon; and this must have been the origin of all languages, and speech itself, that were not taught by inspiration. I believe, moreover, that after language (I mean such as is of human invention) was come to a great degree of perfection, and even when people had distinct words for every action in life as well as everything they meddled or conversed with, signs and gestures still continued to be made for a great while to accompany speech, because both are intended for the same purpose.[19]

Mandeville was well aware that Bruno and Vanini "were both executed for openly professing and teaching of atheism." He saw this as no reason to transform them into two martyrs in the cause of freedom, however: for Mandeville, their sacrifice operated on the level of the passions, and their self-destructive decision should be interpreted as a manifestation of pride. "There is no pitch of self-denial that a Man of pride and constitution cannot reach, nor any passion so violent but he'll sacrifice it to another which is superior to it." An open profession of atheism is poles apart from the cynicism and the corrosive irony so often found in the pages of this "honest and clearsighted man."[20] References to the Bible and to revelation, the recurrent distinction between the language that God revealed to Adam and the language slowly and laboriously constructed by the first savage inhabitants of the earth,[21] seemed to his contemporaries—and still seem—completely external to Mandeville's argument.

Mandeville's reflections on language and on the stage of barbarism are set in an extremely broad context concerning vices and virtues, the construction of civilization and social equilibrium, how human nature is structured, and the function of religion and political power in society. The solutions that Mandeville offered to these problems led to the destruction of the presumed universality of a moral sense that enables man to distinguish good from evil in the same way that he distinguishes the beautiful from the ugly. Shaftesbury's ideas on "the goodness and excellency of our nature" seemed to Mandeville "as romantick and chimerical as they are beautiful and amiable." For Mandeville there are no Neoplatonic triads of Truth, Beauty, and Goodness: moral values are products of custom and society. The thesis of the structural uniformity of the passional nature of man made hypocrisy, exploitation, repression, and suffering elements necessary to early societies and to all societies. Following the lead of Bayle, Mandeville theorized a radical incompatibility between any form of social life and the Christian ideal of living together in love and virtue.[22]

The reactions to Mandeville's declarations, as is known, were many and impassioned. His theses had a decisive influence—think, for example, of Rousseau and Adam Smith. What sprang from the pages of the *Dialogues*, in any event, was a picture of the primitive barbarous state of mankind, of the "historicity" of man's experience, that was much more fully articulated and rich in detail than the picture Hobbes had traced. The world of primitive man appeared as an age of weakness and misery in which destruction, sickness, and death arrived with no explanation and no warning. From the presence of wild animals and of hurricanes, from an existence dominated by elementary needs, a terror was born that permeated the whole of life and that at first could not be fixed in any particular object. Only later the idea arose of an invisible cause that was an object of both fear and veneration, just as, for the savages' young, the father was an object of both fear and of love.[23] In these ages obsessive nightmares came to be experienced as daily reality, and life became populated with images of incredible monsters. The mentality of primitive adults was in every way like that of the infantile psyche that attributes life and senses to material objects and that projects its thoughts, passions, and sentiments onto external reality.

As in Lucretius and as in Hobbes, Mandeville projects the invention of language—which was preceded by gestural and symbolic forms of communication—against the vast historical and anthropological background of the evolution of the human species. This evolution was both biological and cultural, and it took place on an earth that during the course of time had undergone profound mutations. It had occupied long ages and was guided by a Providence that transformed the passions innate in each individual into a mechanism that works for the good of the species. Passions and the will to dominate guarantee the continuity of the species and the persistence of culture.

> The desire of dominion is a never-failing consequence of the pride that is common to all men; and which the brat of a savage is as much born with as the son of an emperor. This good opinion we have of ourselves makes men not only claim a right to their children, but likewise imagine that they have a great share of jurisdiction over their grandchildren. The young ones of other animals, as soon as they can help themselves, are free; but the authority which parents pretend to have over their children never ceases. How general and unreasonable this eternal claim is naturally in the heart of man we may learn from the laws which, to prevent the usurpation of parents and rescue children from their dominion, every civil society is forced to make, limiting paternal authority to a certain term of years. . . . [This] visible desire after government . . . is an undeniable instance of Divine Wisdom. For if all had not been born with this desire, all must have been destitute of it; and multitudes could never have been formed into societies if some of them had not been possessed of this thirst of dominion. Creatures may commit force upon themselves, they may learn to warp their natural appetites and divert them from their proper objects, but peculiar instincts that belong to a whole species are never to be acquired by art or discipline; and those that are born without them must remain destitute of them forever. Ducks run to the water as soon as they are hatched; but you can never make a chicken swim any more than you can teach it to suck.[24]

There is nothing to guarantee a historical process supported by this sort of Providence and in which divine wisdom is manifest in this way, nor is it necessarily progressive. Mandeville was decidedly and ferociously against any concession to primitivism: he considered romantic and "poetic"—even if refined and noble—the least nostalgia for the primitive world or the least inclination to regress toward an unattainable past. That past was made up not of sublime joys and innocent pastimes, but of suffering, terror, and barbarism. Mandeville's "disenchanted optimism" (Scribano, xxiii), his recognition of a sort of "invisible hand" operating in history arise from a double observation: first, "the seeds of every passion are innate to us and nobody comes into the world without them"; second, "what people most pretend to is the least and what they utterly deny their greatest motive." The Providence that operates in history is manifest in this "negation."[25] It is realized through this negation. And it is this negation that opens up the space needed for the introduction of "rational principles" and "virtues" in the life of both individuals and the collectivity. But man is in no way an infinitely perfectible being, and the progress of human history has nothing inebriating about it: "Mankind will always be liable to be reduced to savages."[26]

33·The Divine Legation of Moses

IT was the lot of William Warburton's reflections on the origin of language and writing to be isolated very early on from the difficult and complicated cultural context out of which they arose and within which they were so laboriously worked out. Book 4, section 4 of *The Divine Legation of Moses* was translated into French by Léonard de Malpeines. The two volumes published in Paris in 1744 were really a selection from the writings of Warburton and Fréret, in which Warburton's original text (*Legation* 4:2–6) was differently divided and appeared accompanied by additional observations and comments. Isolated from their theological context and from the discussions of chronology, myths, and Hebrew and Egyptian wisdom, these passages were open to interpretation as a typical expression of an Enlightenment position. Charles de Brosses referred to them, as did the articles on "Langue"and "Ecriture" in the *Encyclopédie*. Condillac used them for his chapter on writing in the *Essai sur l'origine des connoissances humaines*, and Rousseau made large use of them for his *Essai sur l'origine des langues*.

Warburton reached his conclusions on language and writing starting from an initial intention to draw up a point-by-point refutation of Spinoza and Toland. In the third chapter of his *Tractatus theologicus-politicus*, 'De Hebraeorum vocatione," Spinoza had denied the existence, among the Jews, of the ideas of the immortality of the soul and of a future state of punishment or reward. "The only respects in which the Hebrews surpassed other nations," Spinoza declares, "are in their successful conduct of matters relating to government, and in their surmounting great perils solely by God's eternal aid; in other ways they were on a par with their fellows, and God was equally gracious to all." As far as their intellectual life was concerned, "they held very ordinary ideas about God and nature." In the exercise of virtue and in the practice of the true life they were no different from other peoples:

> Their choice and vocation consisted only in the temporal happiness and advantages of independent rule. In fact, we do not see that God promised anything beyond this to the patriarchs or their successors. . . . Thus, the only reward which could be promised to the Hebrews for continued obedience to the law was security and its attendant advantages, while no surer punishment could be threatened

> for disobedience than the ruin of the state and the evils which generally follow therefrom. . . .It is enough for my purpose to have shown that the election of the Jews had regard to nothing but temporal physical happiness and freedom, in other words, autonomous government, and to the manner and means by which they obtained it; consequently to the laws insofar as they were necessary to the preservation of that special government; and, lastly, to the manner in which they were revealed. In regard to other matters, wherein man's true happiness consists, they were on a par with the rest of the nations.[1]

These theses of Spinoza reappear in the *Dialogues* of La Mothe le Vayer and in the *Theophrastus redivivus;* they come up in the *De legibus Hebraeorum* of Spencer; they are stated with particular force in the second of the *Letters to Serena* (1704) and in the *Origines judaicae* (1709) of John Toland.[2] By affirming that the belief in the immortality of the soul was born in ancient Egypt and that no trace of it could be found in the books of the Old Testament, Toland reached the point where he rejected the idea of a continuity in the development of religions—an idea that depended on an interpretation of non-Hebraic religions as degenerations or deviations of the original revelation to the Jews. Hebraic history, in Toland's interpretation, takes on a completely worldly and "political" dimension, and the traditional relationship of sacred and profane history is reversed.[3] Furthermore, while he rejects any sort of superiority in Hebraic history, Toland interprets the religion of Moses as a form of Spinozism and of pantheism, he cites Bruno, he takes over the central ideas of Renaissance Hermeticism, and he includes Mosaic wisdom in the tradition of the *prisca theologia*.[4] He also picks up a question that interested Bruno: the attempt to revive Egyptian magical religion in the modern world. In this view, Egyptian religion is the universal source of mankind's wisdom. Hebraic wisdom derived from it, as did the philosophies of the Greeks and the Persians. Divinity is equated with an infinite universe, eternal and material. This truth had been transmitted by mythical and allegorical paths to the learned of the ancient and the modern world, had been glimpsed by the sages of every age, and is present in every civilization. The initiate hid it in images, veiled it in fables, and kept it carefully concealed from the common people. Thales, Pythagoras, Plato, Orpheus, and Homer had direct or indirect contact (through Indian and Persian magicians) with the culture of the ancient inhabitants of Egypt. The Egyptians are the genuine source of all the wisdom and the religion of the pagan world:

> The Egyptians, who were the wisest of mortals, had a twofold doctrine: the one secret and in that very respect sacred; the other popular and consequently vulgar. Who is there, that is ignorant of their sacred letters, hieroglyphics, forms, symbols, enigmas, and fables? Far and near was spread the fame of the Egyptian philosophy, concealing things under the appearance of fables (says Plutarch) and in speeches that contain'd obscure indications and arguments of the truth.[5]

Bishop Warburton was faced with a series of ideas and affirmations that raised a complex knot of problems. He saw several fundamental theses emerging from them:

1. Atheists and libertines had denied that Hebraic history was superior to the history of other nations and, simultaneously, had denied the existence, among the Jews, of the doctrines of the immortality of the soul and a future state of punishments and rewards.
2. Atheists and libertines had asserted that these doctrines had appeared in history independently of the revelation God had conceded to the Hebrew people. Those doctrines, in their view, derived from ancient Egyptian wisdom, which was universal and primordial.
3. The followers of the Hermetic tradition had linked another thesis to this: Egyptian hieroglyphics contained a primordial and recondite wisdom, and they were an invention of the sages to hide the truth from the common people.

Warburton works out a different response to each of these three theses:

1. He counters the first thesis by affirming the divine, and not the human, origin of Mosaic law. Consequently, he reaffirms the superiority of Hebraic history over other histories and insists that, unlike the histories of other peoples, an "extraordinary providence" was a work within it.[6]

2. He accepts the "Spinozist" thesis that the idea of immortality and of a future state of punishments and rewards was absent in Hebrew thought, and he illustrates the origin and presence of these ideas among the earliest peoples.

3. He rejects the assertion of the "Hermetics" that hieroglyphic writing contained hidden significance. Hieroglyphics were a primitive form of writing, made of the pictures of things, and they were *used* for religious ends. It was the political astuteness of the Egyptian priests that led them to make the ignorant masses believe that behind that unsophisticated writing—which was tied to a primitive astral religion and at one time had been comprehensible to all—there lay concealed instead a sacred wisdom.

Voltaire judged the *Divine Legation of Moses* to be a "rhapsody in four thick volumes to show that God never taught the immortality of the soul during nearly four thousand years." To tell the truth, Warburton was not only a prolix writer, but a tortuous thinker. His aim was to refute the theses of the libertines and the freethinkers by using some of them for different ends.[7] It was, in fact, a lost cause, Warburton states, to try to demonstrate—against Hobbes, against Spinoza, against Toland—the presence of the doctrine of the immortality of the soul in the Old Testament. We must not push the texts too far, he insists, nor do outrage to the factual truth. We must instead recognize that Moses' thoughts did not contain a doctrine that all human legislators have considered indispensable proof of the historical unity of Mosaic teaching and of its nonhuman origin:

> In this Demonstration, therefore, which we suppose very little short of mathematical certainty, and to which nothing but a mere physical

possibility of the contrary can be opposed, we demand only this single *Postulatum*, that hath all the clearness of self-evidence; namely

'That a skilful Lawgiver, establishing a Religion, and civil Policy, acts with certain views, and for certain ends; and not capriciously, or without purpose or design.'

This being granted, we erect our Demonstration on these three very clear and simple propositions:

1. 'THAT TO INCULCATE THE DOCTRINE OF A FUTURE STATE OF REWARDS AND PUNISHMENTS, IS NECESSARY TO THE WELL-BEING OF CIVIL SOCIETY.
2. THAT ALL MANKIND, ESPECIALLY THE MOST WISE AND LEARNED NATIONS OF ANTIQUITY, HAVE CONCURRED IN BELIEVING AND TEACHING, THAT THIS DOCTRINE WAS OF SUCH USE TO CIVIL SOCIETY.
3. THAT THE DOCTRINE OF A FUTURE STATE OF REWARDS AND PUNISHMENTS IS NOT TO BE FOUND IN, NOR DID MAKE PART OF, THE MOSAIC DISPENSATION.'

Propositions so clear and evident, that,—one would think, we might directly proceed to our Conclusion, THAT THEREFORE THE LAW OF MOSES IS OF DIVINE ORIGINAL. Which, one or both of the two following SYLLOGISMS will evince.

I. Whatsoever Religion and Society have no future state for their support, must be supported by an extraordinary Providence.

The *Jewish* Religion and Society had no future state for their support:

Therefore, the *Jewish* Religion and Society were supported by an extraordinary Providence.

And Again,

II. The ancient Lawgivers universally believed that such a Religion could be supported only by an extraordinary Providence.

MOSES, an ancient Lawgiver, versed in all the wisdom of *Egypt*, purposely instituted such a Religion.

Therefore *Moses* believed his Religion was supported by an extraordinary Providence.[8]

Men's passions, Warburton concludes, are so capricious—sometimes through their fondness for paradoxes and sometimes through their love of systems—that these propositions, although self-evident, need an explicit defense. Libertines and unbelievers, in fact, deny the *major propositions* of both of these syllogisms, and many "Bigots amongst Believers" denied instead the minor premise of the first syllogism. The underlying purpose of the *Divine Legation of Moses* was to elucidate fully the premises of both syllogisms. Clarifying their meaning required a rigorous investigation of the politics, religion, and philosophical schools of the earliest times, as well as an examination of the first legislators' actions and of the first social institutions. Clarifying the meaning of the minor premise of the first syllogism, finally, required a detailed summary of the nature and the characteristics of Mosaic law, which would prove that it was exceptional and that, consequently, the history of the Hebrew people could not be reduced to the level of the history of the other nations.[9]

Warburton, like Vico, saw Bayle—that "indulgent foster-father of infidelity," who wrote an "apology for atheism"[10]—as the principal enemy to overcome. Bayle had declared in his *Pensées diverses sur la comète* that an atheistic society was possible, and thought he could detach society and civil life from religion. For Warburton, on the contrary (as before, like Vico), a society of atheists was impossible. The doctrine of the immortality of the soul and of a future state of rewards and punishments was the foundation of religion. The teaching of this doctrine was necessary to the life and to the very existence of every society. All peoples who have lived in society have founded their religion on this doctrine—with the exception of the Hebrew people.

It was from this angle of vision and to demonstrate these theses that Warburton took up Pomponazzi's and Cardano's theses—those deniers of the faith who, like Shaftesbury in his *Enquiry concerning Virtue and Merit*, had imagined a virtue independent of religion or who, like Mandeville in the *Fable of the Bees*, had imagined a religion independent of morality and virtue.[11]

> As the great foundation of our proposition, *that the doctrine of a future state of rewards and punishments is necessary to civil society*, is this, that *religion is necessary to civil society;* so the foundation of this latter proposition is, *that* VIRTUE *is so*. Now, to the lasting opprobrium of our age and country, we have seen a writer publicly maintain, in a book so intituled, that PRIVATE VICES WERE PUBLIC BENEFITS. An unheard-of impiety, wickedly advanced, and impudently avowed, against the universal voice of nature: in which *moral virtue* is represented as the invention of Knaves; and *christian virtue* as the imposition of fools; in which (*that his insult on common sense might equal what he puts on common honesty*) he assures his reader, that his book is a system of most exalted morals and religion, and that the *justice of his country*, which publicly accused him, was pure calumny.[12]

Like many of his contemporaries, Warburton saw clearly the connections between Hermeticism and the impious doctrines of Spinozism, materialism, the One-All, and the universal world soul. The Gnostics, the Manichaeans, the followers of Priscillian, he explains, had elaborated a doctrine of the soul as part of the divine substance, as something destined to be resolved into that substance. This doctrine was later transmitted to the Arabs, from whom the atheists and the Spinozists of the modern age had received it.[13] Strabo had stated that Moses rejected the Egyptian animal cults and Greek anthropomorphism because God is the One that contains all humanity, the sea, and the earth; that One that we call Heaven, World, and the Nature of all things. Strabo's statement was for Warburton "rankest Spinozism."[14] Why should a writer like Strabo attribute such a false and impious opinion to Moses? Warburton's response is typically tortuous: the absence in Mosaic law of any mention of the doctrine of the future state led Strabo to hold that Moses was a follower of the doctrine of the One. Strabo knew of Moses' contacts with Egyptian culture and his opinion was reinforced by the views of the Greek philosophers of his time, who saw the doctrine of the One as being of Egyptian origin. It was the followers of Hermeticism who had erroneously traced to ancient Egypt that "malignant notion," which had infected all of ancient Greek philosophy and later generated the "impious phantasm of Spinozism."

> The books which go under the name of TRISMEGISTUS and pretend to contain a body of the ancient Egyptian wisdom, being very full and explicit in favour of the doctrine of the TÒ ÈN, have very much confirmed this opinion: Now though that imposture hath been sufficiently exposed [by Isaac Casaubon], yet [those books] preserve, I do not know how, a certain authority amongst the learned, by no means due unto them.[15]

The Hermetic texts, the falsity of which Casaubon had demonstrated, were thus also responsible for the errors concerning the "birthplace" of Spinozism and of atheism, which had originated in Greece, not in Egypt.[16] The thesis that hieroglyphic writing was sacred and wisdom-based was indissolubly tied to the erroneous affirmations of the followers of Hermeticism. One first consequence of Warburton's stand against the authenticity of the Hermetic texts was the abandonment of the Neoplatonics' "dreams" and an explicit denial of the sacred character of hieroglyphics:

> It is pleasant to see [Kircher] labouring through half a dozen Folios with the Writings of late *Greek Platonists*, and the forged Books of *Hermes*, which contain a *Philosophy, not Egyptian*, to explain and illustrate Old *Monuments not Philosophical*. Here then we leave him to course his *Shadows of a Dream* thro' all the fantastic Regions of *Pythagorian Platonism*.[17]

In Warburton's pages, which are based in ideas that had already been expressed by Bacon and Wilkins, the hieroglyphics thus become a primitive form of writing by images.[18]

The belief that hieroglyphics were invented by the priests of ancient Egypt to veil their wisdom and conceal it from the vulgar, Warburton says, was a fatal error that had cast obscurity and confusion on our knowledge of the ancient world. Hieroglyphic writing consists of "images of things" (of the same sort as Mexican "pictures") and it arose, quite to the contrary, "by necessity for popular use." It was the expression of a universal custom characteristic of all primitive societies. In tracing out the "images of things," those early people made use of the "first and most natural way of communicating our thoughts."[19] Before the invention or the introduction of letters, the characters denoted "things" and not "sounds" and they emerged into history "by representation," as it happened, principally, with the Mexicans, by "analogy or symbols," as it happened with the Egyptians, or "by arbitrary institution," as it happened, through the centuries, to the writing of the Chinese. "All the *barbarous nations* upon earth" thus made use of hieroglyphics or "signs for things." The transition from simple representation, to analogy, to convention, coincided with the transition from rougher, more backward social forms to a more refined civilization.

Since hieroglyphics, as the "recording of meaning," appeared in close connection with "speaking by action" or gestural language, there is a precise correspondence between the history of writing and the history of language: the first passes "by a gradual and easy descent, from a *picture* to a *letter*" and the second from rudimentary language made up of sounds and of gestures to articulated languages:

> *Language*, as appears from the nature of things, from the records of history, and from the remains of the most ancient languages yet remaining, was at first extremely *rude*, *narrow* and *equivocal*. . . . Accordingly in the first ages of the world, mutual converse was upheld by a mixed discourse of words and actions. . . . Now this way of expressing the thoughts by action perfectly coincided with that, of recording them by picture. There is a remarkable case in ancient story, which shews the relation between *speaking by action* and *writing by picture*, so strong, that we shall need no other proof of the similar nature of those two forms. It is told by Clemens Alexandrinus (*Stromata* 5).[20]

In the age of pictures and speaking by actions—the age of the barbarous nations—men lived closed within and immersed in the senses. Incapable of conceptual thought and far from enjoying the rational and abstract knowledge that is proper to historical times, they made wide use of sense images: since visual or aural images were the simplest and most universal means of communication available to unlettered men, unable to think in abstractions, we can conclude that their use sprang naturally of "hard necessity."[21]

In the ages during which men were still incapable of abstract reasoning, "this rude manner of speaking by action" was slowly transformed into allegorical tales or fables. The "apologue or fable" is "a kind of speech which corresponds, in all respects, to writing by hieroglyphics." When language became an art—that is, a deliberate human construction, the fable "was contracted into a simile" and, with the advent of alphabetic writing, these similes were slowly transformed into metaphors. The more primitive forms of expression, both in language and in writing, did not fall into disuse after the advent of the more advanced forms: "The way of speaking by action was still used after the introduction of the apologue, and the apologue after that of simile and metaphor."[22]

When art took the place of nature, at the end of a long historical process, men sensed a great and decisive change, and they attributed the invention of the alphabet to the gods. In the course of this process, language and writing conditioned each other: a writing that drew on images of the fantasy paralleled the force of bodily, material images present in the first languages. The persistence of the primitive, fantastic, and sense-oriented figuration of the primitive world was responsible for the persistence, even among civilized nations, of the "poetic habit of personalizing every thing" that attributed sentiments and passions to natural objects. This habit—like imagination that "bodied forth the forms of things unknown" (Shakespeare)—populated the physical world with imaginary entities and was reflected in the arts and in the religious and civil life of the Greeks and the Romans:

> The influence *Language* would have on the first kind of writing, which was *hieroglyphical*, is easy to conceive. Language, we have shewn, was out of mere necessity, highly figurative, and full of material images; so that when men first thought of recording their conceptions, the writing would be, of course, that very picture which was before painted in the fancy, and from thence, delineated in words: Even long after, when figurative speech was continued out of choice, and adorned with all the invention of wit, as amongst the Greeks and Romans, and that the genius of the simpler *hieroglyphic* writing was again revived for ornament, in Emblems and Devices, the poetic habit of personalizing every thing, filled their coins, their arches, their altars, etc. with all kinds of imaginary Beings. All the qualities of the mind, all the affections of the body, all the properties of countries, cities, rivers, mountains, became the seeds of living things.[23]

Warburton finds his chief argument in his defense of the unity of Hebraic history and his quarrel with Hermeticism and Spinozism in the destruction of the hieroglyphic religion and of the myth of the wisdom of the Trismegistus. He declares that the *Divine Legation of Moses* definitively refutes the to then "unquestioned proposition" according to which hieroglyphics had been invented for reasons of secrecy.[24] And with the collapse of that false hypothesis fell the fantasies of those who, like Athanasius Kircher, had unduly extended the patrimony of antiquity by mixing together Hermeticism, magic, classical tradition, and mystery cults:

> But what must we think of KIRCHER, who mistook these Superstitions for the ancient *Egyptian* Wisdom; and setting up with this magic, and that other of the *Mysteries*, which the later *Platonists* and *Pythagoreans* had jumbled together in the Production of their *Fanatic-Philosophy*, at once ingrossed, in imagination, all the Treasures of Antiquity? However, to be just, it must be owned that he was misled by the Ancients themselves; some of whom imagined that the very first *hieroglyphics* were tainted with this magical pollution, just as some Moderns would have the first *Mysteries* to be corrupted by debauched practices.[25]

Warburton, as we have seen, was fond of less obvious solutions and more difficult theses. Unlike Stillingfleet, Vico, and Newton, he did not consider that recognizing the precedence of Egyption culture over the Hebraic constituted a danger for the truth of Christianity or for the exceptionality of Hebraic history. He was well aware that the freethinkers thought the praise of ancient Egypt to be a strong argument in support of their cause.[26] He held that advantage absolutely chimerical, however, and believed the opposite notion of a "low antiquity" of Egypt was merely a "fashionable doctrine" by which even Newton had been ensnared. Newton's prodigious discoveries in natural philosophy had induced many to consider him a sort of new King Midas capable of transforming any discourse into a demonstration. This explained why his *Chronology of the Ancient Kingdoms Amended* had wrongly met with a positive reception:

> But the sublimest understanding has its bounds, and, what is more to be lamented, the strongest mind has its foible. And this *miracle of science*, who disclosed all nature to our view, when he came to correct old Time, in the chronology of Egypt, suffered himself to be seduced, by little lying Greek mythologists and storytellers, from the *Goshen* of Moses, *into the thickest of the Egyptian darkness.*[27]

Newton had presented the reigns of ancient Egypt as both uncouth and relatively recent. This "contradicts every thing which Moses and the Prophets have delivered concerning these ancient people." Newton's admirers, Warburton concludes somewhat maliciously, may think that these denials of the truth of sacred history are no more harmful to religion that Newton's overturning of biblical astronomy: "I am of a different opinion; because, though the end of the sacred history was certainly not to instruct us in Astronomy, yet it was, without question, written to inform us of the various fortunes of the People of God; with whom, the history of Egypt was closely connected."[28]

Recognizing that the greater part of the Hebraic rites had been created in opposition to Egyptian superstitions did not in any way imply a denial of the divine origin of those rites. Warburton's position on this question was, once again, directly antithetical to Vico's: in full agreement with John Spencer and in open polemic with Hermann Wits, Warburton thought that the recognition of Moses' Egyptian training did not in any way constitute a hindrance to recognizing the divine character of his mission.[29] This permitted him to reject the thesis that Moses, directly inspired by God, invented writing:

> But if God was indeed the revealer of the artifice, how happened it that the history of so important a circumstance was not recorded? . . . Though I think it next to certain that Moses brought letters, with the rest of his learning, from Egypt, yet I could be easily persuaded to believe that he both enlarged the alphabet and altered the shapes of the letters.[30]

Warburton's "modern" interpretation of writing and hieroglyphics was not only tied to his thesis of the antiquity of Egypt. It was also connected to another thesis that he shared with many other authors of his time: that of the "double philosophy" or a double truth—one truth for the uninitiate and the ignorant masses, the other reserved to the learned and the followers of the mysteries.[31] Warburton interpreted hieroglyphics as a rough and primitive form of writing, but he also asserted that the myth of the hieroglyphics was born thanks to the political astuteness of the priests of ancient Egypt. The doctrine of a future state of punishments and rewards sprang, in antiquity, from this terrain:

> None of the ancient philosophers believed the doctrine of a future state of rewards and punishments, though on account of it's confessed necessity to the support of religion and consequently of civil society, all the theistical philosophers sedulously taught it to the people. . . . It will be proper to set together the public professions and the private sentiments of the ancient theistical philosophers who, notwithstanding they were for ever discoursing on the doctrine of the future state to the people, yet were all the while speculating in private on other different principles.[32]

The thesis of the libertines and those who theorized that religions were an imposture could be considered valid *in relation to the pagan world*. Religion was used by the legislators to invest their dictates with a sacred character:

> The next step the Legislator took, was to support and affirm the general doctrine of a *Providence*, which he had delivered in his laws, by a very circumstantial and popular method of inculcating the belief of a *future state of rewards and punishments*.
>
> This was by the institution of the *Mysteries*, the most sacred part of pagan Religion; and artfully framed to strike deeply and forcibly into the minds and imaginations of the people.[33]

The ancient opinion according to which hieroglyphics were invented by the Egyptians to hide their wisdom from the common people and veil it in mystery was thus without foundation. *After* alphabetical writing (which is an artificial product of the mind) had taken the place of the "natural" writing by images and signs for things, the priests used hieroglyphic writing for politico-religious purposes and *attributed to it* both divine origin and mysterious significance. In a long tradition that reaches from Lucan down to Kircher, the *invention* of hieroglyphic writing had been confused with the *use* that had been made of it for the purposes of political domination and the construction of a popular religion.[34] The mask of an imaginary wisdom had been placed on the still unformed face of a primitive humanity "immersed in the senses," which made use of sense images because it was incapable of formulating abstract ideas or using the reflective capacities of the intellect.

34·Impious Races and Language

WARBURTON'S writings on hieroglyphics, writing, and language have often been compared, as is known, with Giambattista Vico's treatment of the same subjects in the *Scienza Nuova*. Fausto Nicolini qualified Warburton's ideas as Viconian ideas "even without mention of the name of their first author."[1] Overly facile comparisons are a temptation that makes it worthwhile to insist on the profound difference between the tortuous arguments of a prolix theologian and the pages of a great philosopher. This is particularly true since parallels that might be set up between Vico's positions and Warburton's concerning the origin of hieroglyphics and the letters of the alphabet—as we shall see—occupy a historical terrain that is notably broader than the one Croce and Nicolini indicate.

It is perhaps appropriate, before we venture further into these obscure and barbarous ages, to attempt a brief summary and outline some essential points.

1. Vico considered any sign of endorsement of the thesis of an age-old, recondite wisdom in mankind to be an explicit denial of Providence.[2]

2. Vico, like many other authors of the second half of the seventeenth and the first part of the eighteenth century, saw the idea of a *lengthening* of the ages of human history and the hypothesis of the immense age of the pagan nations as affirmations that contradicted the truth of sacred history and that were dangerously close to materialist, Epicurean, and libertine theses concerning the eternity of the physical world.

3. By accentuating irrationality, barbarousness, and instinct as the dominant elements in the history of the earliest times—which was the direction that Vico would take—one could arrive at: (a) refuting those who supported a "boundless [*sterminata*] antiquity" and who based their arguments on the vainglory of the Egyptians and the Chinese to reject biblical chronology and reduce the Bible to the level of the chronicle of one particular people; (b) reinforcing—in opposition to the "chance" of the Epicureans—the idea of a meaning and a direction operating in *all* of the history of mankind, of a teleology present in it, and of a "provident God" without whom, given man's barbarous beginnings, "the world would know no other state than error, bestiality, brutality, violence, prideful arrogance, filth, and blood."

4. Only by accepting the hypothesis of the *bestioni* or bestial men and by insisting on the work of Providence could one explain why the earth is not today a "great, horrid, mute wilderness" without human population. These reasons alone make it possible "to prove false" Polybius, who states that "if the world had philosophers, there would be no need of religions," and Bayle, who claimed that "peoples can live in justice without the light of God."[3]

5. In his *De veritate religionis christianae* (1627), Hugo Grotius had attacked the impious thesis of the eternity of the physical world and of mankind with evidence of a ferine period of humanity taken from classical authors.[4] Vico repeats this sort of argument in the *Scienza Nuova*—and *precisely in opposition to the thesis of the eternity of the world*. When history began, men were either "sages" or "bestial men." The nature of beginnings, which are in all things simple and crude, makes the first of these impossible. Thus the beginnings of gentile humanity were simple and crude, and scholars should have meditated on this if "they did not, as they should not, wish to posit the eternity of the world."[5]

6. Grotius's hypothesis of the "simpletons" and Pufendorf's notion of the "destitutes" seemed to Vico inacceptable. Both could, in fact, be traced to Hobbes's notion of "violent, dissolute men," which was "identical in this matter" with the hypothesis of Epicurus. "Humanity could never possibly have begun" from the men that Hobbes, Grotius, and Pufendorf envisioned. Pufendorf's position was erroneous and "runs counter to the fact of sacred history," Grotius's was "Socinian," and Selden's posited common beginnings of the gentile nations and the Jews and failed to distinguish between God's people and the lost nations. None of these three authors "considered Providence," so none was able to trace the origins of natural law.[6]

7. For Vico, then, the hypothesis of the bestial men must be accepted, but, at the same time, the myth of the creation of Adam must not be compromised. Vico therefore had to avoid making the wandering in the wilderness coincide—as Lucretius and the Epicureans would have it—with man's appearance on the earth. To safeguard the truth of the biblical account, (a) the antediluvian ages must be kept outside of history;[7] (b) the six thousand years of Christian chronology must be respected; (c) the history of the Hebrew people must be kept rigorously separate from that of the Gentile nations; finally, (d) the period of *ferinità*, or wandering in the wilderness, must be presented as a divine punishment following the Deluge. For a Christian, those bestial men must be nothing but the degenerate sons of the men who once populated a particular region of the earth.

8. The "history" of which Vico speaks is therefore that of all "Gentile humanity" (*l'umanità gentilesca*), which takes its beginnings in the universal Deluge and in the descent into barbarism that followed the Flood. Vico's barbarians (as I have already pointed out) had "humanity" in their past. Vico is very clear on this point: men *arrived*, and for their impiety, at the state of solitude, weakness, and need that Grotius describes, at the state of total license that Hobbes describes, and at the state of abandon, without the help and care of God that Pufendorf describes. Man is unjust, "not by nature in the absolute sense, but by nature fallen and weak."[8]

What Vico has to say, then, about the barbarism of a mankind refers to "those first men, from whom the gentile nations later arose," to "the founders of gentile humanity," and to the men of the "impious races of the three children of Noah." After the Deluge, religion alone could keep men in society, but to free themselves of the servitude of the religion of God, creator of the world, the lost and impious races of Ham, Shem, and Japheth abandoned Asia (where they would have remained if, like the Jews, they had "persisted in humanity") and their impious ways "led them to dissipate themselves in a ferine migration, in which they penetrated the great forests of the earth." Little by little, after the dispersal of families, because of their "ferine upbringing," in which mothers abandoned their children and the children "must have come to grow up without ever hearing a human voice, much less learning any human custom," men reached the point where they forgot the language of Adam.[9]

Without language, guided only by the idea of satisfying their hunger, thirst, and sexual impulses, the men of the impious races "reached a state in which all human sense was stifled" and were "cast into a state of impiety" and of "bestial liberty." The Hebrew people did not undergo these wanderings, but inhabited Mesopotamia, both before and after the Flood. They enjoyed "extraordinary help from the true God" and they preserved their memories from the beginning of the world. The history of the Hebrew people, unlike that of all the others, was continuous and untainted by the fabulous. The bestial men living in a ferine state, on the other hand, succeeded in reemerging into humanity when they "imagined Jove." When they first subjected themselves to a higher power, that is, when religion was born, this was "the counsel employed by divine providence to halt them in their bestial wandering through the great forest of the earth, in order to introduce among them the order of human civil things."[10]

It was not true, as Bayle had sustained, that nations could stand without religion and men joined in nations "in a common belief in some sort of divinity." Without religions "not even languages would have been born among men." Only when they had "received religions" could the giants found "their native languages." Thus languages necessarily began among all the nations "from a divine sort." Thus, the linguistic sign originated in a divine name that passed down to the poets, who gathered the most ancient traditions of human history. The things that are necessary and useful to life came to be identified with animate and divine substances: "The things that were necessary or useful to mankind . . . they believed to be substances, and animate and divine substances, so that the later poets took on Jove for the thundering heavens, Saturn for the seeded earth, and Ceres for grain."[11]

The origin of languages among the Gentiles was tied to idolatry and to the cult of imaginary deities "falsely thought of as beings of supernatural power." The language of Adam was totally different, as it expressed the truth and the nature of things. The men of the impious races who had "wandered about in the wild state of dumb beasts" saw divine attributes in natural things. Adam, who knew the true God, had attributed names to things from their nature and based on a knowledge of their natural properties:

> Of greater truth than this false hypothesis or belief which identifies the useful things with divine substances may be the tradition even the

> philologists usually cite: that the first languages signified by nature. And from this we draw another proof of the truth of the Christian religion: that Adam, enlightened by the true God, imposed names on things from their nature: however it was not divine substances that enabled him to do so (since he understood true divinity), but natural properties.[12]

Although he "lacked words" (was *nudo di parlari*) Adam continued to be "inspired by the true God" and soon arrived at "a heroic, articulated language." Cain, however, who had to gather together the scattered giants in order to found his city, needed the idea of some sort of provident deity, and "to begin with them, had to begin [by using] a divine, mute language." The "sacred language" of the Hebrew people was also out of phase with Gentile language, as it began with the heroic phase instead of the divine phase. The "founders" of the Hebrew people, placed in a situation of "poverty of speech," remained "enlightened by knowledge of the true God, creator of Adam." "Hebrew began and remained a language of one God alone; gentile languages, although they arose necessarily from a god, later came to be multiplied so monstrously that Varro manages to enumerate a good thirty thousand [gods] among the peoples of Latium."[13]

Hebrew began and remained the language of one God, but languages among the Gentiles all began "of a divine sort" ("d'una spezie divina") and were tied to the cult of gods created by men's fantasy. In the language of the Jews, "one finds not a single mention of polytheism." Instead, "they must therefore have directed everything conducive to their continued existence . . . towards a single, provident and eternal god."[14] The beginnings of Hebrew, unlike all the other languages, had no initial connection with idolatry or with the attribution of animate or divine characteristics to things: sacred language thus seemed connected to the language of Adam.

In several passages in the *Diritto Universale* and the *Scienza Nuova Prima*, however, Vico came dangerously close to the irreligious theses, even while he reaffirms the difference between true religion and the religion of the Gentiles. He states in the *Diritto Universale* that the Hebraic language "is certainly all poetic, full of parables and similes" ("ferme omnis poetica est, parabolis et similitudinibus reperta").[15] In the *Scienza Nuova Prima* he went so far as to state that poetry "is shown to have been the first common language of all ancient nations, including the Hebrews" and that the people of God had spoken a "wholly poetic" language that "surpasses in sublimity that of Homer himself." Even the Jews, then, had known that same "poverty of speech" that characterized the history of all of the other peoples. In another passage of the same work Vico even says that the difficulty that children encounter in learning to speak is little in comparison to the "difficulties in pronunciation" experienced by the first men "of the dehumanized races of Cain," and even by Adam, "who gave things names," all of whom were "of rigid vocal organs" because of the robustness of their bodies.[16]

Here Vico comes close indeed to Hobbes, and he ran the risk of having Adam appear like a sort of crude *bestione*. These were dangerous paths, and they led to the wholly human origin of all language. They led necessarily to rejecting all suggestion of divine onomathesia, to stripping the problem of the relationship between the language of Adam and the Hebrew language of its meaning, and refusing any difference between the "sacred language" and the languages of the heathen nations.

The statement that "the origins of all things must by nature have been crude"[17] would have become valid not only for the descendants of the impious races, but for all mankind. The biblical prophets would have become comparable to Homer as "the poetic characters" or imaginative genera of the Hebrew people: rather than real persons, they would have become only the "names" in which—as it happened for Jove, Saturn, or Ceres—the "prelogical experiences" of a people who had not yet arrived at reason and civilization had been deposited. Would not the Hebrew prophets have thus become, as Spinoza would have it, men "endowed with unusually vivid imaginations, and not with unusually perfect minds"? And would not the Jews have appeared, as they did to Spinoza, men who "in respect to intellect . . . held very ordinary ideas about God and nature"?[18] Finally, would not sacred history have become one of the many "mythologies" that mankind had created?

As we have seen for the falling into bestiality of Cain's descendants (chapter 24), the definitive edition of the *Scienza Nuova* shows no trace of these statements about the poetic nature of the Hebraic language and Adam's difficulties in learning to speak. The definitive edition reaffirms the profound difference between divine onomathesia and the fantastic speech of the first languages of the Gentile peoples:

> For that first language, spoken by the theological poets, was not a language in accord with the nature of the things it dealt with (as must have been the sacred language invented by Adam, to whom God granted divine onomathesia, the giving of names to things according to the nature of each), but was a fantastic speech making use of physical substances endowed with life and most of them imagined to be divine.[19]

35·The *Ferini* between Heresy and Orthodoxy

STILLINGFLEET'S *Origines sacrae* was published in 1662; Woodward's principal works were written between 1695 and 1726; the *Sacred and Profane History* of Shuckford appeared in 1728; and Vico's *Scienza Nuova* was published, then laboriously enlarged and republished between 1725 and 1744. These authors, whose writings were published during the course of little more than a half-century had, as we have seen, a series of common adversaries: Epicurus and Lucretius, Lapeyrère and Spinoza, Bayle and Hobbes, the followers of the Hermetic tradition and the *prisca theologia,* the exponents of libertinism and deism, the champions of the eternity of the physical world and the theoreticians of atomistic and materialist philosophies. Beyond the differences—and they were often enormous—that existed among them, they all moved within the same group of problems, were interested in the same questions, operated on the same difficult and slippery terrain. They all aimed at defending the truth of the Christian faith and at refuting the great "heresies" ancient and modern. Furthermore, they all used—cautiously, and in defense of the biblical account, of the divine inspiration of Holy Writ, and of the recognition of Providence—some of the texts that were linked to the ideas of Lucretius and of Hobbes.

Stillingfleet was prodigiously active in combating the Socinians, he attacked the passages of Locke's *Essay on Human Understanding* that seemed to him in opposition to the Trinity, he attempted to demonstrate that no atheist ancient peoples had existed, and he strenuously rejected the theses that made religion derive from fear or the political astuteness of the first legislators. Dozens of the pages of his monumental work are devoted to polemics against Epicurus, against the materialists, against Cartesian mechanistic physics, against the attempts to lengthen the history of the world beyond the six thousand years permitted by the Bible, or against affirmations of the eternity of the world. Nevertheless, he insisted at length—precisely in explicit polemic with these heretical positions—on the "gradual increase of ignorance and barbarism," on the "stupidity," the lack of culture, and the existence of only "domestick employments," on the elemental struggle for existence, on the absolute absence of commerce and exchange that marked the history of mankind after the many "dispersons and plantations" of the sons of Noah.[1]

Woodward countered the atomists' view of the physical universe with a vision of its elegance, order, and admirable fashioning, and he aimed at proving the "Truth and Certainty of every individual Article throughout the whole *Mosaic* Narrative of the Deluge." He qualified Epicurus's work as "enormously absurd and senseless" and took his stand—as Vico was to do—against the sacrilegious notions of John Marsham and John Spencer, who saw ancient Egypt as the source of Hebraic wisdom.[2] Nevertheless, Woodward cites the famous verses of Horace's *Ars poetica*. Beginning from the view of the ancient Britons as "wild men" with neither laws nor government, who clothed themselves in animal skins, were incapable of cultivating their fields, fed on grasses and roots, and practiced human sacrifice, he arrives at a calm acceptance of the thesis of beastlike men who once populated the entire globe:

> Nor will it be thought strange that our progenitors should be, in those early times, thus rude and unciviliz'd, when 'tis known that several other great nations were likewise so 'till lately; nay that *all mankind quite round the globe were once so*, I mean at their first original in the ages that ensu'd *next after* the Deluge.[3]

Samuel Shuckford, as we have seen, declared that he had brought all of ancient profane history into perfect harmony with the Mosaic account and he attacked the impiety of Burnet, all the while insisting on the complicated process that led man, by a succession of gradual advances, to civilization. To keep faith with his vision of progress, Shuckford assumed the risk—which many other authors preferred to avoid—of giving Adam the characteristics of a "stupid" savage.

Vico and Warburton both aimed at safeguarding sacred history from all attempts to merge it with profane history, and they strove to defend the unity and truth of sacred history and reaffirm that special working of Providence within it that set it apart from profane history. Warburton speaks, in this connection, of an "extraordinary Providence," Vico of "extraordinary help from the true God."[4] Warburton and Vico both reacted to the questioning of the patrimony of Christian truth and the threat of its total destruction, to the "open war on all religions" declared by Spinoza, to the attempts of Spinoza, Hobbes, and Bayle to "take it on themselves to destroy all of human society,"[5] and to the thesis of a society of atheists not based on religion, the existence of which would signal the end of every possible form of civil commerce. They reacted to the "lasting opprobrium of our age and country" and to the "rankest Spinozism" of a materialistic and pantheistic view of the world, to the "Libertines and Unbelievers" and to the "scandal" of a reborn Epicureanism.[6] And their reaction was to take a stance that differed notably from the attitudes that had long been typical of the works of the more obtuse, intransigent, and fanatic among the champions of tradition, those whom Warburton called the "Bigots amongst Believers."[7] Warburton and Vico entered into their adversaries' terrain—which was, moreover, the terrain on which all of European culture had been in movement for more than a century.

It should not be necessary to point out that this was a rather treacherous terrain. The least affirmation concerning the relation of sacred history to profane history, the interpretation of the Mosaic narrative, the earliest ages of history, the origins of language and myth, or chronology lent itself to contrasting judgments, to different interpretations, and, above all, to ambiguous uses. Warburton intended to emphasize the separation of sacred from profane history, but he accepted the Spinozist thesis that the Jews had no doctrine of the immortality of the soul or of a future state of punishments and rewards. He meant to keep faithfully to the Mosaic account, but he made broad use of the typically libertine theses of a "double philosophy" and of religion as a means of control over the ignorant masses. He inveighs against Hobbes but at the same times insists on the idea of a primitive humanity, incapable of abstract thought, and who thought in corporeal images and populated the world with figures, symbols, and the "personalizations" of ideas.

All we need to do is simply read the texts of Marsham and Spencer, of Huet and Warburton, of Lapeyrère and Spinoza, and it will keep us from taking seriously the parallels—which have enjoyed some degree of success—that have been drawn between the theses that Vico actually did advance and the truly radical and libertine theses of which the so-called Neapolitan atheists were accused.

The Neapolitan atheists, according to Giambattista Menuzzi's accusation, upheld certain specific doctrines: the existence of men in the world before Adam, the view of man as merely a grouping of atoms, man as identical to the other animals, the origin of religions from the imposture of "cunning men" who founded cities and kingdoms and who passed themselves off as sons of gods "to be esteemed and venerated by the peoples, but who in reality believed they were not so," and the imposture of Jesus Christ, who was condemned for that reason by the Jews.[8]

The quarrel between those who support a "secular" Vico and those who defend a "devout" Vico has lost all interest. To reproach someone for the use of the term *ambiguity* when he attempted to transfer his researches onto a terrain less sterile than the traditional one makes no sense whatsoever. The "ambiguous" positions that Woodward, Shuckford, Warburton, and Vico took during the first forty years of the eighteenth century, and their attempt to *change the sense* of the antireligious theses of Lucretius and Hobbes, were in fact to be the object of *both* the praise of certain up-to-date and unprejudiced theologians *and* the bitter criticism of the "bigots" and of the more obtuse and determined defenders of tradition.

Thomas Baker, as we have seen, judged Woodward's *Essay* difficult to reconcile with Scripture, which does not speak of "total" transformations of the terraqueous globe. In Baker's judgment the work, beyond its stated pious intentions, ended up objectively offering a grave danger to religion.[9] In a letter to Woodward dated 16 May 1699, John Edwards points out all of the passages in the *Essay* that might lend themselves to being interpreted as heterodox. Woodward's defense (which takes up thirty-seven closely written manuscript pages) is a characteristic document. It is full of citations of the *De rerum natura*, it refers to Lucretius as "a mighty master of the learning of the still earlier ages," it declares the state of "primitive integrity" of mankind to be less brief than his correspondent thought, since it occupied "all that time yt pass from the creation down to the end of Saturnus reign, which was sometime after the Deluge" while at the same time it energetically combats the accusation of heterodoxy:

> I was really surprized to find a Gentleman of your humanity and upon whose friendship I had so great relyance, representing my opinions to the world as repugnant to the sacred writings and disrespectfull to Moses, at the same time I had reason to believe you were fully satisfyed there was no room for any such suspicions, and your scruples of it [?] kind were removed by my letter of Febr. 16th. 96.[10]

Doubts of the type that Baker expressed in a letter not written for publication were expressed publicly by John Arbuthnot, a friend of Swift's and collaborator of Pope's, in his *Examination of Dr. Woodward's Account of the Deluge* of 1697: "I wish the compilers of theories would have more regard to Moses's Relation, which surpasses all the accounts of philosophers as much in wisdom as it doth in authority. Now, I believe it will be hard to reconcile the Doctor's history of the Deluge with that of Moses."[11]

In that same year, John Harris greeted Woodward's *Essay* enthusiastically as a book that confirmed the Mosaic account as perfectly condordant with the phenomena of nature.[12]

Shuckford, in the *Creation and Fall of Man*, does not spare his criticisms of Newton's *Chronology*, and he protests against the attempt to explain some parts of the Mosaic account by interpreting them as allegorical tales or fables:

> The Shadow of *Allegory* seems to give us some Appearance of knowing, what we do not plainly understand; and an unexamined Hearsay of *eastern Sages*, their *Mythology* and *Literature*, amuses us with a Colour of being very learned, whilst perhaps we really mistake the Rise and Design of the very *Literature* we have Recourse to, in endeavouring to resolve into it *Moses's* Narration, which most evidently sets before us Particulars absolutely incapable of admitting any *allegorical* interpretation whatsoever.[13]

Shuckford's declarations of fidelity to the Mosaic account were not enough, as we have seen, to avoid accusations of having denied the wisdom of Adam, of having outraged divine justice, of having made Adam into a figure of ignorance, a "poor simpleton," and an idiot.[14]

When Vico appealed to the ferine origins of mankind, he did not intend either to compromise the myth of the creation of Adam or weaken the distinction and lessen the gap between sacred history and profane history. In Vico, the thesis of the *ferini*—which arose from a reiteration of Lucretian ideas and had been reinforced by observations of "savages"—was subject, as we have seen, to two preliminary reservations. First, it did not include the Jews, who lived in isolation, had preserved their records, unlike other peoples, "from the very beginning of the world," and, also unlike the others, had never been giants of "disproportionate strength and stature." Second, the period of man's origins did not include antediluvian times, which were excluded from the historical picture that Vico traced:

those origins followed the Deluge as a result of the punishment God inflicted on the impious builders of the Tower of Babel.[15]

Emanuele Duni, who was a professor at the Sapienza in Rome and whom it would be difficult to number among the Spinozists and supporters of atheism, began from the same reservations and supported the full legitimacy of the thesis that the first men, "from whom the first familial societies drew their origin, led a wandering and ferine life, dispersed in the wilderness and among the wild animals, without the use of language and without fixed mates."[16]

The Dominican Bonifazio Finetti (*in saeculo* Germano Federico, and not Gian Francesco, which was the name of his brother), on the contrary, held these theses to be impious and dangerous. In 1756 Finetti published a *Trattato sulla lingua ebraica e sue affini* in which he reaffirms the thesis of the Hebraic language as "of wholly celestial origins . . . having God himself for inventor and first teacher." Nevertheless, he limited himself to judging it "unlikely" that man, although gifted with reason and therefore capable of fashioning his own language, had created the first language unaided, as others had sustained. Finetti solved the problem—as Shuckford had done and many others had tried to do—by a compromise, recognizing the divine origin of language but also its further development by the efforts of human industry:

> I do not deny that man, being gifted with reason, could not form a language for himself and by himself; but that the first was formed in that manner, whatever others might say, I esteem improbable; because for that much time is needed. We know from Scripture that Adam and Eve, immediately after their creation, spoke with God and to each other. . . . It is quite credible, therefore, that the Lord impressed on them the idea of just a few of the most necessary words, that is, only the primordial ones, leaving the rest to their industry. For this reason, not only did he not infuse in Adam the name of the animals, but not even the name by which a woman should be called.[17]

In his *De principiis iuris et gentium adversus Hobbesium, Pufendorfium, Thomasium, Wolfium et alios* of 1764, Finetti took his stand against Machiavelli, Bayle, Spinoza, Collins, and Rousseau. The last chapter of volume 12 is devoted to a refutation of the thesis that mankind went through a ferine state before it reached civil society. Finetti accepts Vico's hypothesis, which Duni had repeated in the *Saggio sulla giurisprudenza universale* of 1760, on the basis of a set of basic assumptions: the solitary life, impermanent, unstable mating, the lack of a formed language, the absence of all rules of conduct. Each one of these assertions seemed to Finetti connected to the Lucretian doctrine of men born of the earth like mushrooms and to a vision of the world as a blind concourse of atoms, a position irreconcilable with divine revelation and even with reason. Why on earth should the descendants of the three sons of Noah, who lived in domestic society, ever have abandoned the example of their progenitors to embrace a highly miserable and brutal life? How, in the space of about two centuries, when the memory of the Deluge was still fresh, could men have been reduced to a condition of doltish barbarousness? How could they have forgotten life organized in families and the use of reason? How could we think that man lost his natural faculties, those faculties that make up his very essence: the capacity to reflect on his own actions and the capacity to unite, separate, and abstract, and thus to form new ideas?[18]

The thesis of a ferine state and a reduction of men to the level of beasts, Finetti reiterated in his famous *Apologia* of 1768, was born of Vico's inability to curb "his dear and most fertile inventiveness." The ferine state was "indecorous to divine Providence, pernicious to religion, and favorable to the libertines." The very idea of it "foments, at least indirectly, the errors of the libertines." In the hands of a libertine, that idea became "a sewer black with filth"; it "can lend arms to impugn religion"; it "can give rise to most harmful abuses"; it will seem "more than a little convenient to anyone who would use it to impugn or cast doubt on Divine Revelation." There was a particular reason for the strong tone of Finetti's argument: Nicolas Boulanger "had seen that he could deny reality to many sacred personages using the same reason by which Vico denies it to so many profane personages." Furthermore, Rousseau had garbed the same thesis "in ingenious and seductive eloquence" in the second *Discourse,* even though, unlike Vico and Duni, Rousseau limited himself to presenting his views of the state of nature and original barbarism as only a fiction and a hypothesis.[19]

The tortuous, difficult path that had led humanity out of barbarism to civil order was, for Vico, decisive proof of the presence in history of a Providence that "disposes to a universal end" the particular ends of human actions and that uses those narrow ends, beyond the conscious purposes of individuals and of peoples "to preserve the human race upon this earth."[20] To acknowledge that brutish state of barbarism and to assert a generalized dehumanizing of men seemed to Finetti, on the contrary, an impious denial of the work of Providence:

> What could be harder to accept, more difficult to harmonize with the idea that we have of Providence, than to imagine that the Lord, after having made the supreme gesture of preserving mankind from total destruction through the Ark, should then abandon it so completely that men reached the point where they became like beasts, all religion lost, the use of reason lost, all language lost, all companionship lost, in few words, humanity lost, falling into a bestial state, rather, worse that that of the beasts themselves?[21]

Did not the "capricious origins" that Vico had established support the libertine thesis that man's soul was identical to that of the beasts? From one point of view, the men of Vico's prehistory were men, because they had the *faculty* of thinking and living humanly, but from another point of view they were not men, because they were deprived, consistently and for centuries, of all acts of thought and all human life. If it were true that men lived for a long time like speechless beasts, only to return subsequently to a human mode of living, why cannot the libertines declare that there will come a time when the animals, which we call brutes and beasts, "will finally choose reason" and will become reasoning creatures just like us? And how can we claim there is an essential and substantial difference between a human soul and that of the beasts if in the orangutan "can be seen traces of reason greater than those that Vico attributes to his savage and ferine men"? If men lived for centuries "with only material, corporeal, and concrete ideas," where, in man, is the essence that distinguishes him from the beasts? What is the use, in light of the materialists' theses, of Duni's affirmation that this was a provisory state in which man kept his spiritual faculties and out of which men emerged when they began to live in a human manner?

> The impious materialist will render thanks to him for the materialism he admits during the ferine state, and then he will laugh at all the rest, and will tell him either than even today all ideas are material and became refined only with education and with practice, or that materialism became more subtle with the passing of the years, so that pure matter is the source from which all human ideas spring.[22]

Emanuele Duni, on the other hand, considered that the horror with which Finetti viewed an "outlaw and ferine state" at the roots of human history was completely unjustified. Only by an appeal to that idea—which obviously must be limited to the history of the Gentile nations—was it possible, Duni said, to account for the development of history, for change in popular customs, for the origins and the progress of institutions, for inventions and for governments, all of which "we observe in the great theater of this world of men." That statement agreed with the testimony of the ancients—not only of Horace, but of Plato, Aristotle, and Cicero. It was supported and proven true by works "that fill the bookstores" concerning the customs and the brutal life "of so many men discovered in our day by Europeans, who lead a savage, lawless, and vagabond life perhaps worse than the one to which the ancient writers attest."[23]

To counter Finetti's outmoded positions, Duni cites long passages from the *Origine des lois, des arts et des sciences* of Antoine Yves Goguet, and he refers his adversary, for fuller information, to the notes to that work, which list the works of the modern historians "that are in everybody's hands."

> Tell me then, if you do not believe in the tradition of the ancient writers, will you at least lend faith to living witnesses who attest to the same facts? Will you at least lend faith to those who, with their own eyes, have seen them in Europe? If it should ever happen that these facts prove true—as indeed they are—what would you do with all of your arguments on their *impossibility?* Dear signor Finetti, there is no half-measure here: either you must resolve to deny any human faith even in your own senses, or you must strike out of your work all of your reflections on the impossibility of such facts.[24]

Duni certainly did not intend to oppose the power of facts to the truth of the Christian faith. He held it possible to adhere to those truths and, at the same time, to recognize the existence of those facts. To do this, he did not need to take a position in radical opposition to tradition. In fact, only the existence of an original ferine state could, in his eyes, explain the difference between the history of the chosen people and that of other peoples and explain the origins of the idolatry, barbarism, and materialism of pagan religions:

> Do you not realize that without such origins of human things standing behind the Gentile nations, the way is completely closed to understanding and explaining the surest monuments of history and the customs of the earliest peoples, as well as of many other peoples discovered in our days and that even now are being discovered by us Europeans? I have already said in my essay and I repeat here that I do not intend to reason on the origin and creation of the world, much less of the Hebrew nation, but only on the origins of the Gentile nations. You might like it if the first founders of the Gentile nations had been gifted with innocence of customs, but, my dear signor censor, how can you explain the origins of idolatry, barbarism, and inhumanity in the practices of their horrid religions, full of uncompromising materialism? . . . Such as the inhumanity of their laws and customs, the vestiges of which have long been conserved? . . . Such as, finally, the progress of such nations? . . . How can you deny the various developments of such customs, which, gradually freeing themselves of materialism, we find, in fact, purer in the later age than in the infancy of all the nations?[25]

Finetti's *Apologia*, as is known, was to remain unanswered. But in his *Scienza del costume* of 1786 Duni reiterated the full orthodoxy of his "Viconian" thesis and its continuity with a tradition that went back to Cicero and to the Church Fathers. Had Cicero not asserted that there was once a time when men lived dispersed and without any form of natural or civil law? In the *De inventione* (1:1) had he not explicitly referred to a remote time when men wandered through the countryside *more bestiarum*, foraged for their food like the wild animals, and lived without reason and without religion? Had not John Chrysostom (*Ad Stagirium*, 2) deplored the ferine state of men after the Deluge? Had he not stated that in the age of Noah, men were "more brutal than the wild animals themselves"? Had he not written in the fifth of the *Epistles to Titus* that before the advent of Christ, hate reigned among men, parents killed their children, crime and treachery flourished everywhere, and thievery was considered a virtue? And had not Saint Basil the Great, in his epistle 325, written of the Magusans that it was impossible to communicate with them by language, and that they had no writing and lived according to customs that were without reason? And had not Saint Cyril taught (*Contra Iulianum*, 5) that among the Chaldeans and the Persians it was permissible to mate with one's own mother or sisters and to do the most opprobrious things without incurring the slightest reproach? And had he not said that the Scythians considered anger a virtue and homicide an honorable act? Had not Athanasius spoken, in his *Discourse on the Incarnation*, of men's falling from virtue and their reduction to the irrationality of brutes ("adeo ut iam amplius rationales non viderentur, sed bruti ex moribus, irrationalesque censerentur")? Had not Saint Jerome seen men eating human flesh? And had he not stated that the peoples of Scotland *pecudum more lasciviunt* and that Persians, Medes, Hindus, and Ethiopians coupled with their mothers, grandmothers, daughters, and granddaughters? Finally, did not the *De praeparatione evangelica* (2:5, 7), the *De demonstratione evangelica* (6:10; 7:2; 8, preface) and the *Storia ecclesiastica* (1:2) of Eusebius of Caesarea contain ample evidence of the existence of the "lawless and ferine" state of mankind?

> The extravagance of [Finetti's] inferences excited my curiosity to read the best interpreters concerning this question, Fathers and Doctors of the Church, and I was surprised, thinking that my theologian [Finetti] had ventured, against all rules of theology, to interpret sacred history completely contrary to the written legacy of those interpreters and approved Doctors of the Church, who not only adopted the tradition of the barbarous and ferine state of the earliest Gentile peoples, but even made express use of that tradition as one of the weightest arguments to confute gentilism.[26]

The polemic between Duni and Finetti was brought back to light in our own century in the context of a quarrel in the best Risorgimento tradition over the orthodoxy or the impiety of Vico. When Finetti's *Apologia* was published, however, no one thought to note that Duni's *Risposta* in defense of Vico had been judged, in 1766, "highly adapted to morality [*buon costume*] and quite in conformity with the rules of the Catholic religion" by Agostino Giorgi, public professor of Holy Scripture at the Sapienza in Rome.

The dispute between *ferini* and *antiferini*—as an article in the *Magazzino Italiano* of the following year emphasizes—had been enough to divide the Sapienza of Rome "into two parties" and to set off a sort of war there. The anonymous reviewer of Finetti's work did not interpret the dispute on the level of opposition of heresy to orthodoxy: he saw a "peaceful accommodation" between the two "illustrious adversaries" as both appropriate and necessary. In man in the ferine state, the author of that article states:

> I do not imagine a bear or other more horrid beast, but I see only a beast deprived of the noble exercise of reason, unable to use other faculties than those of the senses, . . . who feels no other passions than those that directly regard his own preservation, . . . who is moved by and believes pertinent to himself only what is present. Now, in that dispersal of the peoples, . . . in that first solitude of families and confusion of languages and change of country and extreme scarcity of everything, there must necessarily have been produced . . . a total bewilderment, a loss of all morality, a true stupor, a drowsiness of

all the faculties. Except for that family that was particularly helped and chosen by God.

This exception respected, the theses of Vico and Duni seemed to the reviewer completely acceptable, even if Vico's system did not persuade him and Vico's overly enigmatic style was not much to his liking. Within the covers of a review that had published a bitter attack on Rousseau's "savage," Vico's positions seemed reconcilable with revelation: "If what is meant by ferine state is an uncultivated reason, still accompanied by an ungovernable pride, then I could come to accept it; better, I believe it an opinion reconcilable with both revelation and proper philosophy, and with the constant persuasion of men."[27]

Even the thesis of the "barbarism" of ancient peoples had been used, and was still to be used, within "orthodox" positions. Seven years after the publication of the definitive edition of the *Scienza Nuova,* Laurent François published his *Les preuves de la religion de Jésus-Christ contre les spinozistes et les déistes* (1751). This work, written to furnish arms to combat the materialists and the incredulity of the libertines, contrasts the destiny of the sons of Noah with that of a humanity fallen into barbarism. The sons of Noah, "who established themselves in the place, or not far from the place where they had lived," built cities, founded kingdoms, preserved the laws, religions, and arts. As for the other peoples, "the farther they were from those regions, the more barbarous, rough, and ignorant they were." The "families" were involved in protecting themselves from the climate, struggling with their environment, and warding off ferocious wild animals: "It is natural that, distracted by so many worries, fatigues, anxieties, and miseries, they paid little attention to the arts and they fell into ignorance and barbarism."[28] François contrasts the civilization of the first monarchies, which flourished near the land that Moses had designated as "at the center of mankind" with the uncivilized and rough culture of the Greeks and the Nordic and Italic peoples. Fable, myth, and poetry became the essential characteristics of the world that found expression in Homer:

> For how long were the Greeks, the peoples of the North, the Gauls, and the Italians without regulated forms of government, without manners, without the use of writing and the fine arts? . . . Only formless fables, based in some weak trace of a confused and uncertain tradition, are spoken of. . . . Josephus states that Homer did not write his poem, but sang it from memory, that he taught it to singers, who also sang it and taught it to others, until it was written down in later times. . . . According to Strabo, fable and poetry were in honor in that land long before history and philosophy.[29]

A great tradition of thought, a revolutionary vision of man and of the world appear in François's pages as if emptied of force, inserted into another tradition and accommodated to old demands and old problems. An almost innumerable series of texts—allowance made for differences in the stature of both men and texts—give a similar impression. One example is *Of the Origins and Progress of*

Language (1773—76) of James Burnet, better known as Lord Monboddo, which appeared nearly twenty years later. Monboddo was undeniably a "lunatic" and eccentric thinker, but he was not about to assume a revolutionary position. He clearly opposes Hobbes, Spinoza, and Hume, criticizes Cartesian mechanics and Newton, and declares himself a bitter enemy of the new experimental philosophy and of the "materialism" that he saw rampant among his contemporaries. He poses as the defender of the ancient theism against its modern degenerations and supports the ancient wisdom of the Egyptians, yet says that he sees "the true system of human nature" in the famous verses of Horace (1:3):

> When living beasts first crawled on earth's surface, dumb brute beasts, they fought for their acorns and their lair with nails and fists, then with clubs, and so from stage to stage with the weapons which need thereafter fashioned for them, until they discovered language by which to make sounds express feelings. From that moment they began to give up war, to build cities, and to frame laws that none should thieve or rob or commit adultery. [Wickam trans.]

A history of man, Lord Monboddo adds in a note, "would be nothing else than a commentary upon these few lines." This conception of man and of his history, he declares, has nothing revolutionary or degrading about it: it refers to the human condition after the Fall and after the Flood, and it increases rather than diminishes man's appreciation of the wisdom and the provident works of God.[30]

Vico had appeared to Finetti as a "man of great genius, of even greater fantasy and inventiveness, but of little judgment."[31] Warburton, exactly thirty years earlier, had faced similar reproaches: he had been charged with a fundamental ambiguity and with being incapable of carrying out an intransigent defense of orthodoxy. It had seemed possible, furthermore, that many of the theses he had adopted in the *Divine Legation of Moses* could furnish new arguments to the libertines and could be used by the champions of atheism. Warburton was a tenacious adversary of Hobbes and Spinoza, of Bayle and Shaftesbury, of Toland and Tindal, of Mandeville and of Bolingbroke. Nevertheless, his theses were subjected to ferocious attacks. One critic said:

> If a *Clergyman* writes, under pretence of *defending* Revelation, in the very same Manner that an *artful Infidel* might naturally be supposed to use in writing *against* it, such a Writer must excuse me if I suspect his Faith, and condemn his Book; and if he should be in *Holy Orders*, my Indignation would rise the higher, and my Endeavours to lessen his Credit would be the warmer. I hear the Infidels have already sounded the Praises of this Book, and I will do them the Justice to own, they generally know *what* they commend. This I am very sure of; the Author must be a *subtle Enemy* to Revelation, or a very indiscreet Friend. . . . He ought to be *discountenanced* by all the Clergy who are sincere Friends to the *Established Religion*, and *hindered*, as far as

it can be done by *Christian* and *Legal* Methods from any further Advancement in the Church.[32]

These words were written by William Webster on 14 February 1738. On the other hand, no more than four months before (18 October 1737), Bishop Hare had heaped Warburton with praise for having taken a position in opposition to Bayle's ("the best defence of Atheism that was even made") and against the author of the *Fable of the Bees* (whom he had "thoroughly confuted in a very few pages").[33] In 1741 Bishop Thomas Sherlock, known for his positions of enlightened tolerance, wrote to Warburton: "I think these two volumes must silence the suspicions which some seemed willing to propagate, of your being no friend to Revelation."[34]

Reducing the suspicious to silence was not an easy task in those circles.[35] The Reverend Arthur Ashley Sykes wrote in 1744 that Bishop Warburton charged Newton with falsifications, errors, confusions, and contradictions. But had he not claimed that the Egyptian priests had extended their annals back an unreasonably long time? And had not Newton, to the contrary, defended biblical truth against the false doctrines that put the earliest civilizations back tens of thousands of years before the creation?[36]

Vico and Warburton were not the only thinkers to assume a notably cautious position concerning the denial of the traditional thesis of the divine origin of language, which had been brought to a crisis by Hobbes and by Richard Simon. Some of the most illustrious and best-known figures in Enlightenment thought did so as well, starting with the authors of the two fundamental essays on the question, Condillac and Rousseau. Condillac, in 1746, resorts to the expedient of the double genesis of language in an attempt to reconcile the idea of an onomathesia of divine inspiration with a more naturalistic explanation. Furthermore, he begins his essay be refusing the dangerous thesis, advanced by Hobbes and reiterated by Shuckford, that Adam and Eve gradually learned language "by the effect of experiences." "Immediately after their creation they were rendered capable, by the extraordinary assistance of the Deity, of reflecting and of communicating their thoughts to each other."[37]

For Condillac, although Bishop Warburton had rejected the positions of Gregory of Nyssa and Richard Simon and had reaffirmed the truth of the biblical account (according to which God taught Adam language together with religion), he had nevertheless in some ways come too close to the theses of Hobbes and Shuckford and had accentuated the "very poor and narrow" character of Adam's language, which "we cannot reasonably suppose . . . any other than what served his present occasions." For poor and narrow needs, poor and narrow language. Condillac prefers not to venture onto this mined terrain. He cites a significant passage of the French translation of Warburton in a note and, in good Cartesian manner, he presents Warburton's thesis as "very judicious" and a "romance" of undeniable "probability," based on the supposition that "some time after the deluge two children, one male, and the other female, wandered about in the deserts, before they understood use of any sign. . . . And who knows but some nation or other owes its original to an event of this kind? Let me then be permitted to make the supposition."[38]

Rousseau, elusive as always, while summarizing the reasons for his disagreement with Condillac's theses in the second *Discourse*, "On the Origin and Foundations of Inequality among Men," declares himself "frightened by the multiplying difficulties, and convinced of the almost demonstrated impossibility that languages could have arisen and been established by purely human means."[39]

The expression *presque démontrée* serves in fact to belie the authenticity of that conviction. Rousseau returns to the problem of this reconciliation of theory and biblical truth, more directly and with closer reference to the biblical text, in the *Essai sur l'origine des langues:*

> Adam spoke, Noah spoke; but it is known that Adam was taught by God himself. In scattering, the children of Noah abandoned agriculture, and the first common tongue perished with the first society. That had happened before there was any Tower of Babel. Solitaries, isolated on desert islands, have been known to forget their own tongue. . . . Scattered over the vast wilderness of the world, men would relapse into the stupid barbarism in which they would be if they were born of the earth. In pursuing such natural ideas, it is easy to reconcile the authority of Scripture with the ancient monuments; and one is not reduced to treating as fables traditions as ancient as the peoples who have transmitted them to us.[40]

The "primitive times" of which Rousseau speaks are those of "the period of time from the dispersion of men to any period of the human race that might be taken as determining an epoch." In those times "the sparse human population had no more social structure than the family, no laws but those of nature, no language but that of gesture and some inarticulate sounds." The first men were hunters and shepherds; they did not practice agriculture. The objection, which springs from the Bible, that Cain was a farmer and Noah planted grapes

> does not conflict with my thesis. I have said what I understand by primitive times. In becoming a fugitive, Cain was compelled to give up agriculture, and the wandering life of Noah's descendants forced them to give it up, too. . . . During the first dispersal of the human race, until the family was instituted, and man had stable habitation, there was no more agriculture.[41]

If we did not have the testimony of the Scriptures, *then* we could believe in the existence of purely spontaneous, natural processes. Describing those natural processes becomes legitimate and possible, and this description includes the biblical account within it, making it *one* account among other possible accounts of the same process. Rousseau argues here on a level that is very different from the one characteristic of Vico's philological linguistics. He takes a step that Vico did not want to take, or never had the courage to take.

The point is an important one. The barbarous stupor into which, according to the biblical account, the sons of Noah fell is for Rousseau similar in all points to the state in which men "born of the earth"—according to Lucretian doctrine—were supposed to have been found.[42] The process of the differentiation of languages is "natural" and would have taken place in any event, even without the historical existence of the Tower of Babel. "The ages should not be confused," Rousseau says:

> The patriarchal period that we know is very remote from primitive times. Scripture lists ten intervening generations at a time when men were very long-lived. What did they do during these ten generations? We know nothing about it. Living almost without society, widely scattered, hardly speaking at all: how could they write? And given the uniformity of their isolated life, what events would they have transmitted to us?[43]

Beyond his concessions to orthodoxy, Rousseau's ultimate positions tended toward recognizing the indissoluble tie between language and society and recognizing a time scale enormously longer than the traditional one, appropriate to a process that led man from a state of mute bestiality to the civil life of historical times. Rousseau recognizes the validity, finally, of the thesis that the earth now differs substantially from its prehistoric state:

> The original condition of the earth was quite different from what it is today, when we see it as embellished or marred by the hand of men. The chaos that poets attribute to the elements actually reigns in their own productions. In those remote times, when revolutions were frequent, when a thousand accidents changed the nature of the soil and the face of the earth, everything grew confusedly. Trees, vegetables, shrubs, pasture: no species had had time to appropriate land better suited to itself, on which it could suppress others. They would separate slowly, little by little, until suddenly a revolution would occur, which would confuse everything.[44]

When he adopted an enormously long time scale, Rousseau (once again, unlike Vico) took a "modern" position: between the "pure state of nature" and when there was need for languages there must have been "an immense distance"; the first invention of languages cost men "inconceivable difficulties and infinite time"; "thousands of centuries" were needed to develop, one by one, the operations of which the human mind is capable.[45]

Language and society, for Rousseau, are interdependent phenomena. Because the transition from insociability to sociability and from the state of nature to the civilized state necessarily implies the transition from an original mutism to the use of language, it is impossible to decide whether society has to be already established when language is invented or the contrary. In Rousseau's eyes, Condillac made the precise error of making this impossible choice and of supposing exactly what Rousseau puts up to question, "a kind of society already established among the inventors of language." The same error—which is also that of transferring ideas that arise out of society to arguments on the state of nature—is committed by those who ascribed the birth of language to the commerce of fathers, mothers, and their children. Man in the primitive state, in fact, knew "neither houses, nor huts, nor property of any kind," and males and females mated "fortuituously, depending on encounter, occasion, and desire, without speech being a very necessary interpreter of the things they had to say to each other." The errant life did not leave time for any idiom to take form. To say that the mothers taught words to their children "shows well how one teaches already formed languages, but it does not teach us how they are formed." Furthermore, if we then think we have the better of this first difficulty, a worse one comes up: if men needed speech in order to learn to think, they needed even more to be able to think in order to discover the art of speech: "One can hardly form tenable conjectures about the birth of this art of communicating thoughts and establishing intercourse between minds."[46]

Many of the ideas found in late seventeenth-century thought and in the work of Condillac return in the article in the *Encyclopédie* under "Langage": the "limited" nature of the language of the first peoples; the mixture of words, gestures, images, and actions "in the conversation in the first centuries of the world"; the presence, among the biblical prophets, of a language of actions that reflects "a perpetual representation by sense images" necessary because the prophets were speaking to an uncultured people who could understand no other terms; the affirmation that differences of climate, temperament, and customs were the natural causes of linguistic differentiation. A first language of mankind is once more connected, as an operative part of the process of nature, with an original barbarism. And once more, Warburton's works are cited:

> To judge things by their nature, says Mr. Warburton, we would not hesitate to adopt the opinion of Diodorus of Sicily and of other ancient philosophers, who thought that the first men lived for a time in woods and caverns like beasts, articulating, like them, only confused and indeterminate sounds, until, having joined together for their reciprocal needs, they arrived by degrees and after some time at forming more distinct and varied sounds by means of signs or arbitrary marks on which they agreed, so that one who spoke could express the ideas that he wanted to communicate to others.
>
> This origin of *language* is so natural that a father of the Church, Gregory of Nyssa, and Richard Simon, a priest of the Oratory, have both striven to confirm it; but revelation should have instructed them that God himself taught *language* to men.

Condillac, the author concludes, had written of the natural process of the formation of languages *en qualité de philosophe*. Therefore, even though God taught language,

> it would not be reasonable to suppose that this *language* had extended beyond the immediate necessities of man, and that man had not himself had the capability of understanding it, enriching it, and perfecting it. Daily experience teaches us the contrary. Thus, the first *language* of the peoples, as the monuments of antiquity prove, was necessarily very sterile and very limited.[47]

If we did not have the indisputable testimony of the Scriptures, and *if* revelation did not certify the contrary, we would have to believe in purely terrestrial and spontaneous processes, in a purely natural genesis of language and writing, in a slow passage from the original barbarism to civilization. Different positions, ranging from a reiteration and defense of Epicureanism and libertine ideas to cautious attempts to insert new ideas into traditional views took cover behind this sort of reasoning. The question, in any event, had certainly not been invented by the authors of the late seventeenth century or by the philosophes. It had been used as early as the end of the cinquecento, precisely in the context of the origin of writing, by Michele Mercati:

> From which we see that if we did not have the above-mentioned information by which it is proven that the manner of writing our letters is most ancient and was held by the first fathers, we would have to agree, according to nature, that the first manner of writing which was introduced among men had been this: the one that pictures things themselves that we are attempting to explicate.[48]

In 1768 Herder asserted that the thesis of a human origin of languages permitted us "not to untie the knot of this investigation, but simply to cut through it, following the idea that occurred to Alexander in the temple of Gordium." He also declared quite explicitly that his guides in this matter were "two blind pagans, Diodorus Siculus and Vitruvius, and two Catholic Christians, Saint Gregory and, for me even more sainted, Richard Simon.[49]

36·The Death of Adam

THE problem of the monogenesis or the polygenesis of languages, as Sebastiano Timpano had seen clearly,[1] was subordinate, during the course of its history, to the problem of the origin of language in general. The latter, in turn, was part of the even broader question of the origin of humanity and of civilization. Discussions of the origins of civil life often intertwined with those of the history of the earth and changes in nature. Many of the pages that were devoted to the origin of language and writing penned between the middle of the seventeenth and the middle of the eighteenth centuries need to be rescued, as far as their relation to Vico is concerned, from the persistent temptations of a historiography based on "divinitory intuitions" and evaluated against the background of the discussions on geology and theology, sacred history and profane history, and theories on civilization and language. Exactly as it has occurred, for more than a half-century now, for the history of astronomy, physics, and biology, which have benefited remarkably from no longer being considered the exclusive territory of astronomers, physicists, and biologists.

Only the way in which the discussions and the analyses of these great problems were resolved; only the results that arose from this process; only the slow history of the dissolution of the myth of Adam in European culture gave different directions—and more or less modern meanings—to the discussions on language and on the origins of language and writing that took place during those centuries.

What Schlegel had to say (1808) about Sanskrit as an intact, spiritual, and "divine" language, born whole and with a perfect structure among a people privileged by God, free from the "rough" and material origins that are proper to other languages, serves to demonstrate the persistence—even among the so-called founders of scientific linguistics—of the connections between the general views of the history of the physical world and the specific problems of the science of language.

The so-called Viconian theses of a historical development for language and of its slow emergence into the course of history are present in various ways in European culture from the age of Bacon to the middle of the eighteenth century and beyond. These theses accompanied both the destruction of the myth of Adam and the affirmation of "rough" and "stupid" beginnings for mankind. They joined with the ideas of the slow growth of civilization, a progressive rationalizing of the instincts, and a transition from the age of sense and symbolic images to that of reason and conceptual abstractions. The notions of man's crude beginnings and of civilization's gradual development were undeniably tied to the Epicurean tradition and to atheistic and libertine currents, and they had been taken up again, in the modern age, by Hobbes, Richard Simon, Gassendi, Toland, and Mandeville. But they had made their way, laboriously, even in circles much more closely tied to tradition and among scholars and philosophers who were far from willing (just as Vico was unwilling) to espouse the more extreme theses of Lapeyrère or Spinoza—theses that seemed destructive of the Judeo-Christian myth of the divine origin of man and his institutions. If we proceed on the basis of radical oppositions between "progressives" and "reactionaries," we will end up barring the way to any true understanding. New and truly revolutionary ideas were indeed coming forward during that century and a half, and even as part and parcel of solidly consolidated positions. Bacon and Wilkins, Dalgarno and Stillingfleet, Woodward and Shuckford, Warburton, Vico, and Lord Monboddo certainly had no intention of taking a position outside of the Christian community, nor of espousing the theses of an Epicurean and libertine atheism. Many of these authors were, quite to the contrary, strenuous and implacable adversaries of these theses.

Like many of their contemporaries, these writers tried to hold firm to the truth of the biblical account without rejecting—as not a few of the more fanatical defenders of tradition were doing—the ideas on man and on the physical world that could be found in a number of classical authors and that had been and were still receiving critical examination in the modern world. As Frank Manuel has noted in connection with Newton, a Protestant who wrote on the history of the world around 1700 and who did not want to assume decidedly impious or heretical positions had to combine an unconditional acceptance of the biblical account, a historicization of an euhemeristic sort of the ancient pagan myths, and the achievements of later Greek and Roman historians. If such a person had the necessary prerequisites, he could attempt an even more difficult operation: to try to integrate into this whole a fourth element: the results of the new physical and astronomical sciences.[2]

Arthur Sykes, in his impassioned eulogy of Newton (which was also a polemical stand against Warburton) envisioned this harmonious combination of elements as having already been realized:

> Matters of Chronology, in those early Times, are certainly very intricate: and Sir *Isaac* has shewn so very great Sagacity in adjusting and reconciling the fabulous stories of the Antients, and in making them accord, that if he has not hit upon the Truth, he has made the whole so very *probable*, so consistent with the course of Nature, and with itself, and above all with sacred History, that it is not easy to shake so compact and well united a Building.[3]

That unity and that compactness were in reality only imaginary, and the harmony that appeared to Sykes already effected was in reality only an impossible ideal. Different elements were becoming intermixed with the wide-ranging and centuries-old discussion of chronology and of the origins of civilization, and the most disparate combinations were coming to be made. Many currents met on that terrain: the heritage of the Church Fathers, the new philosophies of Bacon and Descartes, the work of the theorists of natural law and the observations of travelers and philosophers on the "savages," the considerations of humanists and jurists, the work of students of myths and of chronology, biblical studies, a new evaluation of pagan religions, the tradition of Platonism and the *prisca theologia*, the heritage of Epicureanism and of Lucretius, the debates on languages, the estimates of the age of Egypt and China, and the impressive results of the new scholarship and the new science of nature. There is an idea, as Frank Manuel has rightly emphasized,[4] that ran through all of eighteenth-century culture: the idea that the first inhabitants of the earth were *fundamentally* different from civilized men; that their mentality was in some measure comparable to that of children, ignorant peasants, or the mentally ill; the idea that there was a sort of common ground among them—that of the primitive mentality—which is far from rational thought and from the scientific dimension, operates according to different categories, and is linked to the concreteness of its images and to the sense-oriented nature of its symbols. But not every discourse that emphasized the "beast-men" or depicted the rough and primitive character of the earliest ages of history or that accentuated the irrational, the senses, and the passions that ruled the first inhabitants of the earth in the age after the Flood was intended to express acceptance of the impious consequences that Lucretius in the ancient world and Hobbes in the modern had drawn from those affirmations.

Combating the libertines and the deists, and furnishing arguments to the libertines and the deists: at first glance contradictory, these two undertakings more than once took on the dimensions of a *single undertaking*. Vico and Warburton—and many other authors with them—could be attacked bitterly as destroyers of the faith and, during the same years, warmly praised as its shrewdest champions. Epicurus and Lucretius, Lapeyrère and Hobbes, Spinoza and Bayle, however, provided a background to every treatise and every discussion, sometimes (as it has been said of Vico regarding Bayle) in an almost obsessive manner.

The discussion of the barbarism of man's origins and of man's ties with nature, of the animality in man, and of magico-mythical forms of thought was to have a fairly long and complicated road before it. If we followed along this road, we would have to refer frequently, and with different (and occasionally even opposed) purposes to Hobbes, Mandeville, Vico, and Rousseau: to the works that had given new life and breath to the Lucretian vision of man as emerging out of animality and out of nature.

Man had for many centuries conceived of himself as at the center of a universe limited in space and time and created for his benefit. He had imagined himself as inhabiting, from the moment of creation, an earth immutable in time. He had built for himself a history of a few thousand years that equated humanity and civilization with the nations of the Near East, then with Greece and Rome. He had thought himself different, by his essence, from the animals, lord of the world and lord and master of his own thoughts. He was soon to find himself, in the new century, called to account, adjusting to the destruction of all of these certitudes and contending with a different, less narcissistic, but certainly more dramatic image of man.

The death of Adam was a slow death. But in the history of ideas, as in the history of individuals, the resistance put forth and the defense mechanisms are no less important, and certainly no less interesting to analyze, than the achievements and the discoveries of the truth.

Notes

CHAPTER 1

1. Plot 1705, 112. See Adams 1954, 258–59.

2. Lister, in *Philosophical Transactions* 6, 76:2281–84. See Adams 1954, 259; Rudwick 1976, 61–64.

3. Hooke 1665, *Obs*. 17; Ray 1692. See Rudwick 1976, 65.

4. Moro 1740, 9–14.

5. Rudwick 1976, 44. See also Beringer 1954; Toulmin 1965; Cannon 1960; Burke 1966.

CHAPTER 2

1. Schweigger 1665, in Adams 1954, 88–89.

2. Campanella 1:13; 3:6, 13. See Adams 1954, 255–56; Rudwick 1976, 56–58.

3. Kircher 1678, 2:6.

4. Ibid., 2:6, 347.

5. Debus 1977, 55–56, 84–85, 93–95, 75–76, 318.

6. See Gliozzi 1977, 286–367.

7. Robinson 1696, dedication to William Nicholson (unnumbered pages); 33.

8. Ibid., additional remarks (unnumbered pages); preface (unnumbered pages); 96. On fossils, see p. 105.

9. Robinson 1694, 1. See also p. 21.

10. Ibid., 3.

11. Ibid., 4.

12. Dickinson 1702, 221–25, 239–45, 140–54.

CHAPTER 3

1. Bacon 1887–1892, 3:330. See also ibid., 4:293–95; Rossi 1968, 11–12; 214–19; Rossi 1970, 118–21; Rossi 1971a, 118–25.

2. Hooke 1705, 1–4. See Davies 1964; Davies 1968; Keynes 1960; Gunther 1930; Hall 1966; Patterson 1949, 1950.

3. Hooke 1705, 334, 435, 411. See Haber 1966, 56–57.

4. Hooke 1705, 289.

5. Ibid., 408, 412, 439.

6. Ibid., 290.

7. Ibid., 290, 293.

8. Ibid., 290, 298, 325, 327.

9. Ibid., 327–28.

10. Ibid., 335. For the preceding, see 435, 291.

11. Ibid., 328. For the preceding, see ibid., 411, 389. See also Haber 1966, 56–57.

12. See Haber 1966, 54–57; Toulmin-Goodfield 1965, 90; Rudwick 1976, 73.

13. Hooke 1705, 328.

14. Ibid., 396. For the preceding, see ibid., 394, 396.

15. Ibid., 397, 374.

16. Hooke 1935, 205.

CHAPTER 4

1. Adams 1954, 359–64; Rudwick 1976, 66.
2. Steno 1910, 2:223.
3. Ibid.
4. Ibid., 2:186. See also Scherz 1956, 1958, 1969.
5. Steno 1910, 2:222, 221–24.
6. Ibid., 224.
7. Scilla 1670, 15, 74–75.
8. Ibid., 153–54.
9. Ibid., 33, 21, 26, 22, 29.
10. Ibid., 38–39.
11. Ibid., 119, 86–87, 56, 58–59.
12. Ibid., 100.
13. Ibid., 107–8. For the preceding see ibid., 106–7.
14. Ibid., 99, 85, 86–87, 158.
15. Ibid., 129.
16. Ibid., 154; *L'autore a chi legge* (unnumbered pages); 130, 9, 10–11, 53–54.
17. Ibid., 10.
18. Ibid., 18. For the preceding, see 105, 13.
19. Rudwick 1976, 58. See also Scilla 1670, 14.
20. Scilla 1670, *L'autore a chi legge* (unnumbered pages).
21. Ibid., 52, 105. The citation to Ciampoli, 106.
22. Leibniz 1749, 45; Maillet 1748, 2:22–26.
23. Wotton, in Arbuthnot 1697, 65–84.
24. Ibid., 66–68, 83.
25. Harris 1697, 226.

CHAPTER 5

1. Stillingfleet 1710, 2:266.
2. Ibid., 265–66.
3. Ibid., 271, 274, 275–76.
4. Ibid., 279–80.
5. Ibid., 292. For the preceding, see 282, 291, 293.
6. Ibid., 294.
7. Ibid., 294–95.
8. Ibid., 272–73.
9. Ibid., 273.
10. Descartes, *Correspondance* (ed. Adam-Milhaud), 1:421–22. See also Descartes 1969, 12 (E. Garin, introduction). On time, see Puech 1952.

CHAPTER 6

1. Hale 1805, 36–37. Burnet's biographical sketch, 7–84.
2. Hale 1677, "To the Reader" (unnumbered pages).
3. Ibid., 182, 183–85.
4. Ibid., 193. For the preceding, see 195–96, 190–92.
5. Ibid., 200. For the preceding, see 193, 198–99.
6. Ibid., 201.
7. Ibid., 201, 286, 339, 211, 213, 239.

CHAPTER 7

1. See Burnet 1680, 1684, 1689, 1690. See also Burnet 1965. There were many editions during the eighteenth century, some of which will be referred to in the text.

2. Burnet 1680, 60 (1965, 93).

3. Ibid., 17, 10–15 (1965, 35). On the Flood, see also Bligny 1973.

4. Burnet 1680, 52, 30, 45; Burnet 1690, 1:74–75 (1965, 54, 51, 31).

5. Burnet 1680, 53; Burnet 1684, 107 (1965, 89).

6. Burnet, *Sacred Theory of the Earth,* introduction to chapter 1, edition of 1722, 1–2.

7. Burnet 1733, 299.

8. Burnet 1680, 83 (1965, 110); Nicolson 1959, 206.

9. Burnet 1684, 109 (1965, 91). See also Nicolson 1959, 212.

10. See Haydn 1950, 529–31.

11. Burnet 1684, 139. See also Tuveson 1951. Engel 1950 is unaware of either the works of Tuveson or those of Nicolson.

12. See Nicolson 1959 and Aubin 1934; Boch 1959, 445–47; Haller 1940; Hodgen 1964; Fleckenstein 1949; Raimondi 1966.

13. Burnet 1684, 109.

14. Burnet 1689, dedicatory epistle (unnumbered pages), 240–41. See Jacob 1972, 274–75. On millenarianism, see Hill 1971, Jacob 1976a, 1976b. See Rossi 1977 for fuller bibliography.

15. Burnet 1733, 260–61.

16. Ibid., 268.

17. Burnet 1727, 8.

18. Burnet 1733, 400 (1693, 47–48). For the preceding, Burnet 1733, 286–87, 392–93, 398–400.

19. Ibid., 404–5 (1693, 53–54); see also 403–4.

20. Ibid., 405 (1693, 54).

21. Ibid., 432–34 (1693, 74).

CHAPTER 8

1. *The Correspondence of Newton* (ed. Turnbull) 2:329–34. See also Manuel 1963, 8, 295, for a discussion of the theses in Westfall 1958, 200.

2. Manuel 1963, 164.

3. See Milhaud 1911, 325; Manuel 1963, 165; Casini 1973, 42–43.

4. Newton 1721, 377–78.

5. Newton 1779, 3:171–72 (1934, 544).

6. Boyle 1772, 4:68–69. See also Jacob 1969; Campodonico 1978.

7. Turgot, *Oeuvres* 1808, 2:96. See also Vartanian 1953, 126.

8. Condillac 1749, 406, 368.

9. Nicolson 1959, 182. For the preceding, see Descartes, *Correspondence* 1:421; *Principia* 3:44, 4:1; *Discourse,* 5; *Le monde/The World,* vols. 6–7. See also Descartes 1970, 1:274, 279, 280; 1983, 181.

10. A Mersenne, 15/4/1630, in *Correspondence* 1:129. See also ibid., 252–53.

11. *Principia* 3:45 (1983, 105–6). For the preceding, *Discourse,* 5 (Descartes 1970, 1:109); *Principia* 3:44, 3:1. On cosmology, see Daubrée 1880.

12. Fontenelle 1749, 1:139, 157–58, 1 (1688, 155–56, preface).

13. Fontenelle 1749, 158. See also Meyer 1958. On the spread of Cartesianism in England, Pacchi 1973 is important.

14. Bayle 1734, 4:438b.

15. Ibid. 4:440a, 439b.

16. Ibid. 4:439a.

17. Ibid. 4:439b.

18. Kant 1953, 183–84, 191 (*Werke: Gesammelte Schriften* [Berlin, 1905], 2:137, 144).

CHAPTER 9

1. Leibniz 1875–90, 4:283, 341.

2. Ibid., 283. See also Descartes 1904, 9:125–26 (1983, 108) (*Principia* 3:47).

3. Leibniz 1875–90, 4:341, 344, 299, 283. The best work on the subject of good is Heinekamp 1969.

4. See Hintikka 1972, 156–57; Lucretius 5:v. 420–30; Hobbes, *De corpore* 2:4, 10 (1966, 1:129).

5. Leibniz, 1875–90, 6:216–18 (1951, 234–36).

6. Descartes (Adam and Tannery, eds.) 6:45; Descartes 1980, 23–24; 1970, 108. For the preceding: *Correspondence* (Adam and Milhaud, eds.) 1:183; Descartes 1969, 59, 61; Descartes (Adam and Tannery, eds.) 6:42–44 (1980, 23, 24; 1970, 1:107–9).

7. Descartes 1969, 61. See also Descartes (Adam and Tannery, eds.) 11:37.

8. Leibniz 1875–90, 4:431; 6:106–07.

9. Ibid., 6:107; 7:406.

10. Belaval 1960, 454–56.

11. Leibniz 1875–90, 2:12; 7:20, 23, 37, 40. See also Mugnai 1976, 179–82.

12. Leibniz 1875–90, 2:40.

13. Ibid., 51.

14. Ibid., 3:565.

15. Ibid., 5:43.

16. Ibid., 3:572–73. See also ibid., 288: Hintikka 1972, 161.

17. Lovejoy 1936, 261; Hintikka 1972, 157ff.

18. Leibniz 1875–90, 3:221; 6:263.

19. Ibid., 6:263.

20. Ibid., 4:445; 7:19, 23.

21. Ibid., 6:262.

22. See Solinas 1973, the best global study of the *Protogaea*, p. 21.

23. Leibniz 1749, 86 (See also French translation, ed. Bertrand de Saint-Germain, Paris, 1859, 133).

24. Ibid., 2 (tr. Saint-Germain 1859, 1, 3).

25. Ibid., 3 (tr. Saint-Germain 1859, 4).

26. Ibid., 4 (tr. Saint-Germain 1859, 5, 11).

27. Ibid., 5–7.

28. Ibid., 7 (tr. Saint-Germain 1859, 9). For the preceding see ibid., 8–9 (tr. Saint-Germain 1859, 16, 11).

29. Solinas 1973, 44–45.

30. Leibniz 1749, 12, 41.

31. Ibid., 15 (tr. Saint-Germain 1859, 23).

32. Ibid., 35 (tr. Saint-Germain, 1859, 55–56).

33. Ibid., 29–30 (tr. Saint-Germain 1859, 46–48).
34. Ibid., 46 (tr. Saint-Germain 1859, 75).
35. Ibid., 33–34 (tr. Saint-Germain 1859, 53–54).
36. Ibid., 44–46 (tr. Saint-Germain 1859, 75). On the genesis of metals, 19–20.
37. Leibniz 1875–90, 7:69. See also Rossi 1971a, 130–35.
38. Leibniz 1875–90, 7:69.
39. Leibniz 1749, 18 (tr. Saint-Germain 1859, 27).
40. Ibid., 18–19, 21 (tr. Saint-Germain 1859, 28, 31). See also Rossi 1971a, 139–47; Rossi 1977, 153–57.
41. Leibniz 1749, 31 (tr. Saint-Germain 1859, 49). For the preceding, ibid., 19.
42. Ibid., 10 (tr. Saint-Germain 1859, 14).
43. Ibid., 41 (tr. Saint-Germain 1859, 66).
44. Ibid., 62–63, 66 (tr. Saint-Germain 1859, 97f).
45. Leibniz 1875–90, 3:579. See also Lovejoy 1936, 279.
46. Leibniz 1875–90, 5:296. See also Leibniz, *Monadology*, 73–74.
47. Leibniz 1710, 112.
48. Leibniz 1875–90, 3:221, 250.
49. Ibid., 221.
50. Buffon, tr. Smellie 1791, 1:126–27, 125.
51. Leibniz 1710, 111–12.
52. Leibniz 1875–90, 3:566.
53. Ibid., 6:262–63.
54. Ibid., 3:221.
55. Leibniz 1967, 2:118. See also Brucker 1743, 4:621–22.

CHAPTER 10

1. Collier 1934, 114, 120–21.
2. Whiston 1737, 78.
3. Ibid., 81, 66.
4. Ibid., 71. For the preceding, see ibid., 68, 69, 87, 72.
5. *The Correspondence of Newton* (ed. Turnbull), 2:329–34. See also Nicolson 1963, 235; Casini 1969, 92–96.
6. Whiston 1737, 79. For the preceding, see ibid., 78.
7. Ibid., 82.
8. Ibid., 83.
9. Ibid., 137, 140, 139, 413.
10. Whiston 1753, 261.

CHAPTER 11

1. Jacob 1969, 313-14, 308. See also Jacob-Guerlac 1969; Jacob 1972; Jacob 1974.
2. Bentley 1693, 4, 7, 29; Bentley 1836, 3:75.
3. Warren 1690, "To the Reader" (unnumbered pages); Graverol 1694, 64–65, 197; Lovell 1696.
4. Keill 1698, 14, 21.
5. Ibid., 5, 10, 6–8, 11–14.
6. Ibid., 14–15, 16, 19–20.
7. Ibid., 32, 21.

8. Ibid., 170. For the preceding see ibid., 37, 171–72.

9. Ibid., 52–54, 55–57, 176. See also Buffon 1749, 1:180–82; Whiston 1753, 35.

10. King 1776, 1:221–22. On the controversy about Burnet's work, see Willey 1962, 32ff.; Nicolson 1963, 224–70; Tuveson 1950, 54–74; Manuel 1959, 135–36; Casini 1969.

11. Aulisio 1733, 132. For the preceding, see Bentley 1699, 32; Tuveson 1964, 119.

12. See Bonanate 1972, 47–52.

13. Derham 1754, 82; Arbuthnot 1697, 28–29; Arbuthnot 1710–12.

14. Hume 1935, 202, 239, 215.

15. Ibid., 187.

CHAPTER 12

1. Vallisnieri 1721, 15.

2. Ibid., 30.

3. Ibid., 31.

4. Ibid., 56–57. For the preceding, see ibid., 55.

5. Ibid., 73–74.

6. Ibid., 75, 105.

7. Ibid., 107.

8. Vallisnieri 1726, 8–9. The first edition was 1715.

9. Ibid., 2. For the preceding, see ibid., 7–8; Vallisnieri 1721, 110; Vallisnieri 1733, 1:313.

10. Vallisnieri 1733, 1:33; 2:145, 285. On Vallisnieri's method, see L. Geymonat et al. 1962; Baldini 1976; Savelli 1961; Spallanzani 1977 (although the *Lettere critiche* are incorrectly dated as 1726 instead of 1721). On the more general related problems, see Badaloni 1968b; Roger 1963; Solinas 1967; Tega 1971.

11. Vallisnieri 1721, 108.

12. See Adams 1954, 365.

13. Moro 1740, 231.

14. Ibid., 1, 2, 5, 7.

15. Ibid., 15–25; 82, 83.

16. Ibid., 31, 34, 35.

17. Ibid., 36–57, 82, 85.

18. Ibid., 238. See also ibid., 38, 39, 40, 211, 212, 239, 264, 391.

19. Ibid., 86, 95, 101–2. See also ibid., 106, 112, 144–15, 117.

20. Ibid., 194, 195–202 (for the ancients); 202–10 (for the moderns). See in particular ibid., 202, 205, 206–07.

21. Ibid., 207–08.

22. Ibid., 214–15. For the preceding, see ibid., 213.

23. Ibid., 240, 238, 218–20, 243. On Delos, ibid., 250ff.

24. Ibid., 246. On large islands, see ibid., 254.

25. See Adams 1954, 368. On continents, see ibid., 295. On primary and secondary mountains, see ibid., 262–63, 271.

26. Moro 1740, 271–72. On plains strata, see ibid., 288–94.

27. Ibid., 300. For the preceding, see ibid., 288–94.

28. Ibid., 424–25. For his summary, see ibid., 426–32.

29. Ibid., 430.

30. Ibid., 431–32.

CHAPTER 13

1. Hampton 1955. See also Roger 1953.

2. Maillet 1748, 2:61–62 (1968, 361, 161). For the preceding and what follows immediately, see 1:ix, xii, xiv, xxx, xxxii; 2:67, 85, 96, 98. (See also Maillet 1797, vii, 190–91.)

3. *La religion vengée* 1757–62, 17:300–306. For the eternity of matter and the origin of marine life see ibid., 18:13ff., 18ff.

4. François 1764, 114–15. For the preceding see ibid., 104, 106.

5. Nonotte 1774, 2:23–27.

6. Holbach 1794, 4:10–12.

7. Ibid., 12.

8. *Encyclopédie*, "Tremblements de Terre." Naville 1967 pays little attention to geology.

9. Holbach 1794, 1:174–75. For the preceding see ibid., 173–74.

10. Ibid., 1:182–83, 181.

11. Ibid., 1:149-50.

12. Ibid., 4:249–50.

13. Ibid., 4:170–71, 242.

14. Voltaire 1876–92, 10:2–3.

15. Besterman, 1976, 287. It is difficult, however, to agree that Voltaire's position on the problem of fossils could be qualified as a "pardonable mistake" or that he suffered "little mishaps."

16. Voltaire 1877–83, 22:400–437.

17. Voltaire 1876–92, 24:129–37; 20:234–371; 18:20–25; 27:400–415.

18. Ibid., 24:129, 133.

19. Ibid., 24:130, 131; 27:407.

20. Ibid., 18:23; 27:400. For the preceding, see 24:130.

21. Ibid., 20:235–37; 24:131, 133, 135.

22. Ibid., 24:133, 135.

23. Ibid., 24:134; 27:401.

24. Ibid., 27:402, For the preceding, see ibid., 27:401, 403.

25. Ibid., 24:135. See Haber, "Fossils and the idea of a Process of Time in Natural History," in Glass et al., *Forerunners of Darwin* 1968, 228.

26. Voltaire 1876–92, 27:415, 404.

27. Maupertuis 1768, 1:xii–xiii.

28. Voltaire 1876–92, 27:406, 415.

29. See Lovejoy 1936, 282–83; 274–77.

30. Robinet 1766, 244.

31. Ibid., 1:193, 194, 197.

32. Ibid., 1:206–7, 205.

33. Ibid., 1:207–10, 213–17, 224–25; 4:177, 187.

34. Ibid., 4:212. For the preceding, see ibid., 4:198ff., 206, 209–10, 211.

35. Ibid., 4:213, 243.

36. Robinet 1768, 1, 2, 4, 36, 37.

37. Buffon, tr. Smellie 1791, 1:97, 108, 112, 113, 118–19, 129, 132.

38. Buffon, 1981, 148–49.

39. Buffon, tr. Smellie 1791, 1:57–58.

40. Ibid., 513–14.

41. Buffon 1981, 153. For the preceding citations from Newton, see notes 4 and 5 to chapter 8, above.

42. Renzoni, in Buffon 1959, 522. See also Roger 1962, xlv–xlvi.

43. Diderot 1956, 243.

CHAPTER 14

1. Venturi 1958, 232.
2. As cited in Hampton 1955.
3. Rousseau 1959, 1:373.
4. Boulanger 1768, 1:8–10.
5. Venturi 1947.
6. Boulanger 1768, 1:1.
7. Diderot, "Précis sur la vie et les oeuvres de Boulanger," in Boulanger 1794, 1:9.
8. *SNP* 104; *SN* 369; *SNP* 101. See also *SN* 369; *SNP* 96–97; *SN* 369, 372.
9. *SNP* 40; *SN* 349, 1132, 1178. See also Rossi 1969, 49.
10. *SN* 2.
11. *SN* 331.
12. Adams 1935. See also Berlin 1976, 120; Piovani 1969, 260, 262, 265. For the preceding, *SN* 331, 147.
13. Badaloni 1961, 389, 172–76. See also Badaloni 1968. On the state of the dead, see Walker 1964.

CHAPTER 15

1. Buffon, in Roger 1962, lxvi.
2. Ibid., 3–4 (see also Haber 1966, 136).
3. Ibid., 205–6 (see also Smellie translation 1791, 3:172–73).
4. Diderot 1981, 9:95, 94.
5. Jaki 1978, 78–79. See also Jaki 1974.
6. Wright 1750, 49–50.
7. Ibid., 74.
8. Ibid., 75.
9. Ibid., 76, 79.
10. Munitz 1957, 23–24, 247; Jaki 1977.
11. Kant 1910, 313–14 (tr. Hastie, University of Michigan Press, Ann Arbor, 1969, 144–45).
12. Buffon, in Roger 1962, 40–41. On Bruno and Kepler, see Rossi 1971, 238–44; Pascal 1954, 1105–6 (tr. Stewart 1965, no. 43). See also Roger 1962, lvii.

CHAPTER 16

1. Herschel 1912, 1:158. See also Haber 1966, 227; Greene 1959.
2. Gillispie 1959, 41. On Hutton, see Bailey 1967; Davies 1968. On Herder see Verra 1966.
3. Hutton 1795, 1:200.
4. Ibid., 3. See also ibid., 5–6.
5. Ibid., 16–17.
6. Ibid., 2:545–47.
7. Ibid., 1:183, 562.

8. Ibid., 4:221–22.
9. Diderot 1981, 9:94.
10. Cited in Lovejoy 1968, 366.
11. See Rudwick 1976, 110.
12. Hutton 1795, 1:20.
13. Ibid., 1:15.
14. Ibid., 1:19, 20, 21.
15. Ibid., 1:18–19.
16. Ibid., 1:187.
17. Ibid., 1:182.
18. Ibid., 2:180ff., 409; 3:49, 100, 654ff.
19. See Toulmin-Goodfield 1965, 154; Blum 1962.
20. Kelvin 1894, 2:10ff. See Haber 1966, 289.
21. Darwin 1963, 425, 274.
22. Kelvin 1854, 61; 1871, cv, cii; Tait 1876, 167–68. See also Bellone 1978, 28, 200, 212.
23. Toulmin-Goodfield 1965, 223–24. See also Haber 1966, 288–90; Eiseley 1975, 205–22; Toulmin-Goodfield 1965, 222–24; Rudwick 1976, 257–60; Burchfield 1975.
24. See Lerner 1972, 78–81; Brouwer 1975. On Hutton, see also Dott, in Schneer 1969, 122–41.

CHAPTER 17

1. See Walker 1972; Iverson 1961; Yates 1964 on these themes.
2. Scott 1924–36, 1:39. See also Yates 1964, 182; Rossi 1977, 109–47.
3. Tindal 1732, 4.
4. Dickinson 1702, 203.
5. Marsham 1696, 155.
6. Marsham 1723, 146, 225.
7. Spencer 1705, 777–78.
8. Spencer 1686, 670.
9. Corsano 1968, 13. For comment on the same erroneous conclusions in Nicolini, see Rossi 1969, 124.
10. *Acta Eruditorum* 1684, 213–18. For comment on Spencer's book, see ibid., 113–20.
11. Morhof 1732, 2:563. For the preceding, see Simon 1680, "Avertissement" (unnumbered pages).
12. Leydekker 1704, 7; Simon 1708, 511; Woodward 1726, 107.
13. Toland 1704, 20.
14. Ibid., 39, 40.
15. Toland 1709a, 146.
16. Ibid., 118–19. For the preceding, see ibid., 162.
17. Toland 1704, 130. For the preceding, see 1709a, 195, 1704, 71.
18. Toland 1709b, 20.
19. See Middleton 1752: *De medicorum apud veteres Romanos conditione*, 4:177–245; *Germana quaedam antiquitatis eruditae monumenta*, 4:1–176; *The History of the Life of M. T. Cicero*, London, 1741; *A Treatise of the Roman Senate*, 3:379–471; *De latinarum literarum*

pronunciatione, 441–55. The first volume of the latter contains Middleton's response to pamphlets attacking Middleton 1748. Some mention in Stephen 1962, 1:213ff.

20. Middleton 1733 (which is in answer to Williams 1733), in Middleton 1752, 2:305.
21. Middleton 1752, 2:74–75. For the comparison with Hume, see Stephen 1962, 1:227.
22. Middleton 1752, 2:131, 190–92 226–27; 3:193.
23. Ibid., 2:131, 190–95, 226–27; 3:193.
24. Ibid., 2:225. For the preceding, see ibid., 2:224–26 (which contains a strong attack on Newtonian ideas).
25. Rotheram 1753, 52, 65, 67–68, 79–81.
26. Williams 1733, 36, 39.
27. Middleton 1752, 2:316. On Fontenelle, see ibid., 3:187.
28. Tindal 1732, 4, 76–78.

CHAPTER 18

1. Lapeyrère 1655, 26 (1656, 18–19). For the preceding, see ibid., 7–8, 13.
2. Ibid., 27–28. A listing of writings attacking Lapeyrère can be found in Engelcken 1707, 4–10. On the subject, see also Mahudeau 1915; McKee 1944; Pastine 1971.
3. Lapeyrère 1655, 29 (1656, 22).
4. Ibid., 118 (1656, 132). On Kepler, see Rossi 1971, 241–42.
5. Lapeyrère 1655, 108, 113, 185, 198, 216.
6. Ibid., 113, 149.
7. Ibid., 154, 196 (1656, 215–16).
8. Ibid., 183 (1656, 202).
9. Ibid., 160, 167, 69–70, 143–44, 160.
10. Ibid., 148 (1656, 164).

CHAPTER 19

1. Morhof 1732, 1:45–46.
2. Ursinus 1656, 7–8.
3. Schoock 1662, 274–76; Le Prieur 1656, *Censura* (unnumbered pages); *Lectori* (unnumbered pages); Simon 1730, 2:27.
4. Pezron 1687, 11.
5. Shuckford 1731–37, 1:18, 24–45.
6. Vossius, I., 1659a, iii.
7. Horn 1659a, 70–71.
8. *SN* 50. See also Martini 1658.
9. Webb 1669, 48. For the preceding, see ibid., 47, 50, 95, 109. On Webb, see Ch'en Shou-Yi 1935; Formigari 1970, 55. On the general subject, see Dawson 1967; Garin 1971, 1975; Guy 1963; Van Kley 1971; Zoli 1972, 1973, 1974.
10. *Memorie Istoriche* 1700, 69–72
11. Etiemble 1966, 56.
12. See Landucci 1972, 89.
13. See Pinot 1932, 239.
14. Shuckford 1731–37, 1:29.
15. Dortous de Mairan 1759, 104.
16. Renaudot 1718, xxxviii–xxxix, 372–73.
17. *Recueil des lettres* 1713–76, 26:19–22.

18. Dortous de Mairan, 1759, 203–5.
19. Voltaire 1817 (*Oeuvres complètes*), 6:2:878.

CHAPTER 20

1. Horn 1655, 7, 305, 312.
2. Vossius, I., 1659a, dedication (unnumbered pages); i–iii.
3. Ibid., v–vi.
4. Ibid., vii, ix, xxxiv–xlviii, lv, xliv–xlvii.
5. Ibid., xlviii–xlix.
6. Ibid., xlix.
7. Steinmann 1960, 180–84.
8. Vossius, I., 1659a, liv.
9. Horn 1659a, 2^{r}.
10. Ibid., 2^{v}.
11. Ibid., 1–3, 3^{v}–4^{v}.
12. Ibid., "Ad lectorem" (unnumbered pages); 2^{r}, 4^{v}.
13. Ibid., 36.
14. Vossius, I., 1659b, 27–28, 38, 30.
15. Horn 1659b, 50, 54.
16. Vossius, I., 1659c, 40–41.
17. Horn 1668, dedication (unnumbered pages).
18. Pezron 1687, 10 (emphasis added). See also ibid., 9.

CHAPTER 21

1. Bochart 1646a, 1.
2. Ibid., 11.
3. Vossius, G., 1668, 58, 62, 74.
4. Toland 1709, 109–13. On Huet, see Dupront 1930.
5. Fayus 1709, 174–75, 176.
6. Lafitau 1724, 1:36.
7. Perrault 1702, 303.
8. Tindal 1732, 2–4.
9. Arnauld 1775–83, 3:400–401.
10. See Busson 1948, 367.
11. Toland 1709, 192–93.
12. Fayus 1709, praefatio (unnumbered pages); 2, 35, 37, 23.
13. This is the *parere* (opinion) of Giannone on Costantino's work: see Ricuperati 1965, 624.

CHAPTER 22

1. Webb 1669, 51, 52, 55.
2. Horn 1659b, 64; Horn 1659a, 49.
3. Le Prieur 1656, 101.
4. Ursin 1661, 80–81. See also ibid., 90–95, 117.
5. Le Prieur 1656, 103, 111.
6. Schoock 1662, 133.
7. Horn 1659a, 3, 33–34, 46–47, 48.

8. Vossius, I., 1659, xxxii, xxxv, xxxviii, xlv–xlviii.
9. Vossius, G., 1659, 2, 7.
10. Hale 1677, 148.
11. Ibid., 185.
12. Leydekker 1704, 115. For the preceding, see ibid., 27, and the title.
13. Martianay 1693, 355–56, where he sees in the Horn-Voss dispute something more than a simple question of chronology; Du Pin 1708, 18; Renaudot 1718 and Pinot 1932, 219, 235; Tyssot de Patot, *Lettres* 8:1. See also Pinot 1932, 241.
14. Tyssot de Patot 1710, 178–79 (1733, 110).
15. Newton 1779–85, 5:142–93.
16. *The Original of Monarchies*, in Manuel 1963, 211.
17. Calmet 1741, 36, 152, 162; 1723.
18. Giannone 1940, 1:34–35.
19. Goguet 1820, 3:16, 247 (1976, 275, 270).
20. *Lettres philosophiques* 17; *Essai sur les moeurs*, chaps. 1–4; Nonotte 1772, 87.
21. *Free Thoughts* 1734, 136, 138, 139, 140.
22. François 1764, 284, 395, 396.
23. Chateaubriand, préface, 73, 78, 79 (1976, 124–36).
24. Maillet 1748, 2:61–62 (1968, 362, n. 9).
25. Ibid., 2:63–65 (1968, 363, n. 9).
26. Ibid., 2:67 (1968, 364, n. 9).
27. Ibid., 2:88 (1968, 171).
28. Lamennais 1823, 4:182.

CHAPTER 23

1. Troiani 1977, 274 (Josephus, trans. Thackeray, Loeb Classical Library, 1926, 2:15, lines 152ff).
2. *SN* 125, 126.
3. *DU* 313–14. See also ibid., 312, 313. The same problem, with a more general reference to "great difficulties," can be found in *SNP* 297 and *SN* 807.
4. *DU* 316.
5. Ibid., 314–15.
6. Ibid., 341.
7. Ibid., 342–43.
8. Ibid., 348.
9. Ibid., 348–50.
10. Ibid., 350.
11. Ibid., 378, 327–28.
12. Vico 1947, 218.
13. *SN* 126.
14. *SNP* 274.
15. *SN* 126–27, 121–23.
16. *SN* 166.
17. *SN* 54.
18. *SNP* 27.
19. *SNP* 30.
20. *SN* 44.

21. *SNP* 97; SN 48–50, In *SN* 58 Vico again takes a stand against Marsham.
22. *SNP* 211.
23. *SN* 50.
24. Nicolini 1949, 93–99.
25. *SN* 51.
26. *SN* 123.

CHAPTER 24

1. Momigliano 1965, 788.
2. Reale 1970, 97 and Sina 1978, 104 insist on the third point. The fourth point is in Sina 1978, 101.
3. The expression is Momigliano's, 1965, 781.
4. Reale 1970, 101.
5. Ibid., 93, which restates theses of Badaloni 1961, 387. For my reservations concerning Badaloni's Vico, see Rossi 1969, 154, 174–78.
6. *SN* 51, 53–54, 125, 126, 166, 94, 95, 172, 371–72, 195, 301, 313.
7. *SN* 53–54, 165.
8. See Reale 1970, 92. For the preceding, see *DU* 2:340, 312, 313, 341.
9. Sina 1978, 104, 100.
10. *SNP* 25.
11. *SNP* 25; *SN* 165.
12. *SNP* 103.
13. *SN* 51; *SNP* 103.
14. *SNP* 212, 88; *SN* 52, 103, 118, 524.
15. *SN* 378; *SNP* 293, 408.
16. *SNP* 88; *SN* 118; *SNP* 87.
17. *SNP* 213–16; *SN* 735.
18. *SNP* 256, 293; *SN* 53–54.
19. *SNP* 406.
20. *SNP* 406, 407; *SN* 1308; *SN* 373, 377, 734, 736.
21. Landucci 1972, 315. See also ibid., 320, 325–26.
22. *SNP* 21, 211, 408, 409, 211; *SN* 1091. See also Kubler 1977.

CHAPTER 25

1. *SN* 9, 53, 54, 59, 93, 166, 172.
2. *SNP* 30, 95, 293; *SN* 94, 95, 1154.
3. *SNP* 13, 127; *SN* 1384.
4. *SN* 395, 1384. See also note 7 below.
5. *SN* 68, 207, 208, 209; *SNP* 87.
6. *SN* 179, 334, 1043, 1110. See also *SNP* 78, 269, 476.
7. *SNP* 24; *SN* 334. See also *SNP* 476; *SN* 1110.
8. *SN* 396; *SNP* 30, 47.
9. *SNP* 27.
10. *SNP* 32.
11. *SNP* 88.
12. Le Prieur 1656, 13.
13. *SNP* 18, 26, 544; *SN* 397, 1109.

14. See *DU* 53; *SN* 135. See also on this point the accute observations in Landucci 1972, 292–95. See also Nicolini 1949; Bulferetti 1952.

CHAPTER 26

1. *SN* 95.
2. *SN* 50. For the preceding, see *SN* 45, 47, 48, 121.
3. "La religione dei geroglifici e le origini della scrittura," in Rossi 1969, 81–131.
4. Nicolini 1949, 151; Nicolini 1949–50, 1:171.
5. *SNP* 97; *SN* 45, 83, 99.
6. Giannone 1969, 1:169.
7. Ibid., 163, 169. See also Mannarino 1976.
8. Fréret 1796, 1:69. See also Cantelli 1964, 29–30.
9. Fréret 1796, 13:300–301.
10. Fréret 1796, 6:609–12.
11. Bougainville, "Eloge de M. Fréret" in Fréret 1796, 1:18–19.
12. Ibid., 1:19.
13. *SN* 48; Fréret 1758, preface, xxxvii.
14. Fréret 1796, 13:116–20.
15. Bougainville, in *Mémoires* 23:333. See also Cantelli 1974, 28, 394.
16. See Bastian 1959; Daniel 1968.

CHAPTER 27

1. See Iversen 1961.
2. Calmet 1741, 16–17. See also Duret 1619, 1–13. On the origin of language in Epicurean thought, see Chilton 1962.
3. Ricuperati 1970, 330–31.
4. Calmet 1741, 19–20. See also Gregory of Nyssa, *Contra Eunomium* 2:216ff.; Leclerc 1685, *Lettre,* 19; Simon 1685, 85.
5. Calmet 1741, 35–36. For the preceding, see ibid., 29, 32.
6. De Mauro 1970, 59–60.

CHAPTER 28

1. Lapeyrère 1655, 116–17 (1656, 143, 200, 130). On Adam's compilation of a dictionary, subsequently passed on until the age of Noah, see ibid., 121.
2. See Formigari 1970, 47.
3. De Mauro 1970, 56–67.
4. See the bibliographical information in De Mauro 1970, 57.
5. Hobbes 1966a, 3:11, 16, 48 (*Leviathan*); 4:28–29.
6. Hobbes 1966a, 3:16.
7. Hobbes 1966a, 2:xvii (*Philosophical Rudiments*); 3:35, 16.
8. Hobbes 1966a, 3:16, 34–35; 2:xvi–xvii.
9. Ibid., 3:35.
10. Ibid., 3:16, 35–36, 28.
11. Hobbes 1966b, 2:88–89 (*De homine*). See also Hobbes 1978.
12. Ibid., 2:90–92; Hobbes 1966a, 1:36 (*Elements of Philosophy. The First Section Concerning Body*); ibid., 4:25; ibid., 3:24–25.
13. Hobbes 1966b, 2:1–13 (*De homine*).

14. Hobbes 1966a, 2:2–4 (*Philosophical Rudiments*); Hobbes 1966b, 2:90–92, 94 (*De homine*); Hobbes 1966a, 3:18. See also Hobbes 1978, 40.

15. Hobbes 1966a, 2:12 (*Philosophical Rudiments*); ibid., 3:113; ibid., 4:84–85.

16. Hobbes 1966a, 3:18–19; Hobbes 1966b, 2:89. See also Hobbes 1978, 38.

17. Hobbes 1966a, 1:21, 31–33; 4:21. On the difference between concrete and abstract names, see Gargani 1966, 272–73.

18. Hobbes 1966a, 3:22–23; Hobbes 1966b, 2:90, 89 (*De homine*). See also Hobbes 1978, 39, 38.

19. Spinoza, *Opera* 3:56–57; Spinoza 1959b, 101–2 (1901, 330). See also Brunelli 1977; Parkinson 1969; Savan 1958.

20. Spinoza, *Opera* 3:135; Spinoza 1959b, 185–86 (1901, 366).

21. Spinoza 1981, 69, 74–75; Spinoza 1959b, 185.

22. Spinoza 1981, 76, 63, 62.

23. Spinoza, *Opera* 1:8 n.2 (*De intellectus emandatione*) (1901, 9, n.2).

24. Bacon 1887–92, 3:599. The Latin text of *Cogitata et visa* has been translated by B. Farrington, *The Philosophy of Francis Bacon*, Liverpool, 1964, 81.

25. Spinoza, *Opera* 1:27–28 (1901, 33).

26. Ibid., 4:199 (1963, 122).

27. Spinoza 1981, 153–54 (*Opera* 1:182–83).

28. Spinoza, *Opera* 2:179. See also Spinoza 1908, 1:107.

29. Spinoza 1981, 85–86.

30. Spinoza 1981, 7, 28 (*Opera* 1:41, 127). See also Spinoza 1959b, 81, 176; *Opera* 1:72.

31. Spinoza 1908, 189.

32. Ibid., 175.

33. Spinoza, *Opera* 1:10 (*De intellectus emendatione*) (1901, 14–16).

34. Spinoza 1908, 38, 39, 40, 175, 9.

35. Spinoza, *Opera* 1:10 (*De intellectus emandatione*) (1901, 9–10).

36. Cited in Vernière 1954, 1:128. On the fortune of Spinoza, see also Colie 1959, 1963.

CHAPTER 29

1. Stillingfleet 1662, "To the Reader" (unnumbered pages). See Tulloch 1874, 1:144–54.

2. Stillingfleet 1662, 16.

3. Stillingfleet 1710, 2:12.

4. Ibid., 2:365.

5. Ibid., 2:367. For the preceding, see ibid., 2:357.

6. Ibid., 2:12–13.

7. Ibid., 2:335, 362.

CHAPTER 30

1. Woodward 1695, 245, 82, 89.

2. Ibid., 85. For the preceding, see ibid., 90–91.

3. Ibid., 86, 58, 165.

4. Ibid., 55, 56. See Allen 1963, 105.

5. Woodward 1695, 57–58.

6. Woodward 1723, 25–26, 32.
7. Woodward 1695, 166.
8. Woodward 1726, 105–6.
9. Woodward 1695, preface (unnumbered pages).
10. Cambridge University Library MS. 7647 (Woodward Correspondence), fol. 24, letter to John Edwards of 26/6/1699.
11. Ibid., fol. 37.
12. Ibid., fol. 9, letter to John Edwards of 23/5/1699.
13. Ibid., letter 58.
14. Woodward 1726, 44–45.
15. MS. 7647, cited above, letter 28.
16. Ibid., letter 35.

CHAPTER 31

1. Shuckford 1731, 1:107, 108, 109.
2. Ibid., 111.
3. Ibid., 111, 113.
4. Ibid., 112–13.
5. Ibid., 114.
6. Ibid., 115.
7. Ibid., 238, 239–40.
8. Ibid., 242.
9. Ibid., 240–43.
10. Shuckford 1753, 2:125–28.
11. Ibid., lxxxi–lxxxii.
12. *Spicilegium shuckfordianum*, 26.
13. Hobbes 1966a, 3:18–19 (*Leviathan* 1:4).

CHAPTER 32

1. Mandeville 1962, 171, 169; The same references apply to what follows immediately.
2. Mandeville 1732, 31.
3. Mandeville 1962, 219.
4. Ibid., 172.
5. Ibid., 161; Mandeville 1924, 1:281.
6. Mandeville 1962, 173.
7. Ibid.
8. Ibid., 262.
9. Ibid., 172–73.
10. Ibid., 214.
11. Ibid., 214–15.
12. Ibid., 215, 217, 221.
13. Ibid., 218.
14. Ibid., 254, 259, 261.
15. Ibid., 232.
16. Ibid., 259–60.
17. Ibid., 260.
18. Ibid., 261–62. On this question, see also Mandeville 1924, 1:138–41; 2:286–87.
19. Mandeville 1962, 261–62.

20. Mandeville 1924, 1:214. See I. Primer's preface to Mandeville 1962, 4; Marx, *Capital*, trans. Eden and Cedar Paul, Everyman's (1930, 1974), 678.

21. Mandeville 1962, 222, 234.

22. Mandeville 1924, 2:357. See Scribano 1978a, 1978b. On Mandeville, see Garin 1934, 1958 (in *Giornale critico della filosofia italiana* 37:500–509); Goretti 1958; Talluri 1951; Young 1959.

23. Mandeville 1962, 210.

24. Ibid., 211–12.

25. Mandeville 1924, 1:281. Mandeville's statement concerning the governors of charity schools doubtless would apply to men in general. See also Mandeville 1974, and M. E. Scribano, preface, xxiii.

26. Mandeville 1962, 257.

CHAPTER 33

1. Spinoza 1908, 46–48.
2. La Mothe le Vayer 1716, 326–27. See also Gregory 1972, 214; Spencer 1705, 11ff.
3. Toland 1704, 19–20, 21–44.
4. Toland 1709a, 117. See also Yates 1964; Walker 1972.
5. Toland 1720, 70.
6. Warburton 1788, 1:50 (*Divine Legation*).
7. Ibid., 2:353. See Evans 1932, 53, 281.
8. Warburton 1788, 1:50.
9. Ibid., 1:51–52, 127.
10. Ibid., 1:126, 128.
11. On Pomponazzi and Cardano, see ibid., 1:66–74. On Bayle, see ibid., 741ff. On Mandeville and Shaftesbury, see ibid., 80–81, 112–14.
12. Ibid., 1:112.
13. Ibid., 2:116. See also Yates 1964; Walker 1972, Rossi 1969, 103, 107, 121, 135.
14. Warburton 1788, 2:117.
15. Ibid., 2:117–18.
16. Ibid., 2:127. See also ibid., 2:433ff.
17. Ibid., 2:110.
18. See Rossi 1960, 142–53, 216–31.
19. Warburton 1788, 2:387–88.
20. Ibid., 2:405–9.
21. Ibid., 4:139–40.
22. Warburton 1788, 2:413, 409. For the following, see ibid., 2:429.
23. Ibid., 2:448.
24. Ibid., 2:452.
25. Ibid., 2:451–52.
26. Ibid., 2:439.
27. Ibid., 2:492. There is brief mention of Warburton's polemic with Newton in Manuel 1963, 180–81.
28. Warburton 1788, 2:493.
29. Ibid., 2:590, 641–42, 648, 653. For his evaluation of Spencer and Wits, see ibid., 584–85.
30. Ibid., 2:437.

31. See Manuel 1959, 65–69.
32. Warburton 1741, 2:8.
33. Warburton 1788, 1:166–67.
34. Ibid., 2:413, 429, 452.

CHAPTER 34

1. Nicolini 1949, 163. See also ibid., 127. On the "fabulous" history traced by Nicolini, see Rossi 1969, 82–85, 86–131.
2. *SNP* 13, 127; *SN* 1384.
3. *SNP* 476; *SN* 179, 334, 1100.
4. Grotius 1640, 16 and note. See Landucci 1972, 322.
5. *SNP* 27.
6. *SNP* 18, 19, 47, 59.
7. *SN* 13, 54, 166, 172, 371, 372.
8. *SN* 310, 369.
9. *SNP* 40; *SN* 195, 301, 369.
10. *SN* 1097. For the preceding, see *SNP* 241, 293; *SN* 53, 54, 241, 313.
11. *SNP* 306. For the preceding, see *SNP* 241, 303, 476. See also Pagliaro 1969, 420; Apel 1975, 409ff.
12. *SNP* 306. For the preceding, see *SNP* 9; *SN* 230.
13. *SNP* 303. For the preceding, see *SNP* 253, 264, 406.
14. *SNP* 264, 303.
15. *DU* 380.
16. *SNP* 253, 264, 192, 367.
17. *SN* 367.
18. Spinoza 1908, 27, 46–47.
19. *SN* 401.

CHAPTER 35

1. Stillingfleet 1710, 2:11–12.
2. Woodward 1726, 107.
3. Woodward 1723, 32. See also ibid., 35.
4. Warburton 1788, 1:50; *SN* 329, 313.
5. *SN* 1214; *Carteggio*, 188.
6. Warburton 1788, 1:111; 2:17.
7. Ibid., 1:51.
8. Vico, 1947, 218.
9. MS. 7647, Cambridge University Library, Woodward Correspondence, letter 35.
10. Ibid., letter 116, fols. 9, 23, 24, 25.
11. Arbuthnot 1697, 28–29.
12. Harris 1697, 44.
13. Shuckford 1753, 2:v.
14. *Spicilegium shuckfordianum* 26.
15. *SN* 13, 54, 166, 172, 371, 372.
16. Duni 1775, 141. The thesis is also already present in Duni 1760, however.
17. Finetti 1756, 2.
18. Finetti 1764 and Duni 1845, 3:25–27.

19. Finetti 1936, 54, 59, 68, 84, 85, 21, 24, 86.
20. *SN* 1108. See also *SN* 132, 133, 312, 341.
21. Finetti 1936, 69.
22. Ibid., 84. For the preceding, see ibid., 69, 83–85.
23. Duni 1845, 3:46–47.
24. Ibid., 48.
25. Ibid., 40–41.
26. Ibid., 4:142–47.
27. *Magazzino Italiano* 1 (1767):192–95 in *Giornali Veneziani* 1962, 292–95. See also ibid., introduzione, xxiii–xxv.
28. François 1764, ix, 408.
29. Ibid., 408, 409.
30. Monboddo 1774–92, 1:410, note. See also Greene 1971, 250. On Monboddo, see Cloyd 1972; Knight 1900.
31. Finetti 1936, 26.
32. Webster, *The Weekly Miscellany*, 279, cited in Evans 1932, 68.
33. Warburton 1841, 105.
34. Ibid., 86.
35. There is a list in Evans 1932, 296–306, of the books and pamphlets connected with the "Warburtonian controversies."
36. Sykes 1744, 223.
37. Condillac 1947, 2:1:1 (1971, 169).
38. Ibid., 2:1:15 (1971, 169–71).
39. Rousseau 1972, 53 (1964, 126). See Derrida 1967; Starobinski 1967 on this question.
40. Rousseau, in Verri 1970, 203–5 (1967, 36).
41. Ibid., 195, 201 (1967, 34, 31).
42. Ibid., 201 (1967, 36).
43. Ibid., 203 (1967, 35–36).
44. Ibid., 215 (1967, 42).
45. Rousseau 1972, 50, 51 (1964, 121, 119).
46. Rousseau, in Verri 1970, 269; Rousseau 1972, 50–51 (1964, 120–21.)
47. Lucca edition, 1758–71, 9:195.
48. Mercati 1589, 98. See Rossi 1969, 99–102.
49. Herder-Monbaddo 1973, 90–91.

CHAPTER 36

1. Timpanaro 1972, 86. See also Timpanaro 1953.
2. Manuel 1963, 37.
3. Sykes 1744, 222–23.
4. Manuel 1959, 141.

Bibliography

TEXTS

Adams, John

1790 *Curious Thoughts on the History of Man Chiefly Abridged or Selected from the Celebrated Works of Lord Kaimes, Lord Monboddo . . . and Montesquieu*, Dublin.

Agricola, Georgius (Bauer, Georg)

1546 *De natura fossilium libri X*, Basileae.

1556 *De natura fossilium. De re metallica*, Basileae.

Aldrovandi, Ulisse

1648 *Musaeum Metallicum*, Bononiae.

Arbuthnot, John

1697 *Examination of Dr. Woodward's Account of the Deluge . . . with a Letter to the Author concerning an Abstract of Agostino Scilla's Book on the same Subject*, London.

1710–12 "An argument for Divine Providence taken from constant Regularity observ'd in the Birth of both Sexes." In *Philosophical Transactions* 27:186–90.

1751 *The Miscellaneous Works of the late Dr. Arbuthnot*, 2 vols., Glasgow.

Arnauld, Antoine

1775–85 *Oeuvres*. Paris.

Aulisio, Domenico d'

1733 *Delle scuole sacre libri due*, Naples.

Bacon, Francis

1887–92 *The Works*, London.

1960 *The New Organum and Related Writings*, edited by Fulton H. Anderson. New York: Liberal Arts Press

1964 *The Philosophy of Francis Bacon with New Translation of Some Fundamental Texts*, B. Farrington, Liverpool: Liverpool University Press.

1975 *Scritti filosofici*, ed., P. Rossi, Turin: Utet.

Bayle, Pierre

1734 *Dictionnaire historique et critique. Cinquième édition, revue, corrigée et augmentée de remarques critiques avec la vie de l'auteur par M. Des Maizeaux*, 5 vols. Amsterdam.

Beaumont, John

1693 *Considerations on a Book entitled the Theory of the Earth, published by the learned Dr. Burnet*, London.

1694 *A Postscript to a Book entitled Considerations . . .*, London.

Bentley, Richard

1693 *A Confutation of Atheism from the Origin and Frame of the World. A Sermon preached at St. Martin's in the Fields, November the 7th 1692, being the seventh of the Lectures founded by the Honourable R. Boyle*, London.

1699 *The Folly and Unreasonableness of Atheism demonstrated from the Advantage and Pleasure of a religious Life, the Faculties of Human Souls, the Structure of animate Bodies, and the Origin and Frame of the World: in Eight Sermons preached at the Lecture founded by the Honourable Robert Boyle esquire*, London.

1836–38 *The Works*, London.

Berewood, Edward

1622 *Enquiries touching the Diversity of Languages and Religions through the chief Parts of the World*, London.

Bianchini, Francesco

1925 *La storia universale provata con monumenti, e figurata con simboli degli antichi* (1697), Venice.

Bochart, Samuel

1646a *Geographiae sacrae pars prior Phaleg seu de dispersione gentium et terrarum divisione facta in aedificatione turris Babel*, Cadomi.

1646b *Geographia sacrae pars altera Chaanan seu de coliniis et sermone Phoenicum*, Cadomi.

Bonnet, Charles

1770 *La Palingénésie philosophique, ou, Idées sur l'état passé et sur l'état futur des êtres vivans*, Paris.

Boulanger, Nicolas Antoine

1768 *L'antiquité dévoilée par ses usages ou examen critique des principales opinions, cérémonies et institutions religieuses et politiques des différens peuples de la terre*, Amsterdam.

1794 *Oeuvres*, Amsterdam.

1955 *Extraits des Anecdotes de la nature*, in Hampton, 1955; see below, Studies.

Bourguet, Louis

1715 *Dissertation sur les pierres figurées*, Paris.

1729 *Lettres philosophiques sur la formation de sels et des crystaux . . . avec un mémoire sur la théorie de la terre*, Amsterdam.

1735 *Spiegazioni di alcuni monumenti degli antichi Pelasgi trasportata dal francese con alcune osservazioni sovra i medesimi*, Pesaro.

Boyle, Robert

1772 *The Works*, London.

Brucker, Jakob

1742–47 *Historia critica philosophiae a mundi incunabulis ad nostram usque aetatem deducta*, Lipsiae.

Buffon, George Louis Leclerc, Comte de

1749 *Histoire naturelle, générale et particulière*, Paris.

1791 *Natural History, General and Particular, by the Count de Buffon*, trans. William Smellie, 3rd ed., 9 vols., London.

1959 *Storia naturale. Primo discorso sulla maniera di studiare la storia naturale. Secondo discorso; storia e teoria della terra*, ed. M. Renzoni, Turin: Boringhieri.

1960 *Epoche della natura*, ed. M. Renzoni, Turin: Boringhieri.

1962 *Les époques de la nature*, critical edition, J. Roger, *Mémoires du Muséum National d'Histoire Naturelle*, Série C, Sciences de la Terre 10, Paris.

1981 *From Natural History to the History of Nature: Readings from Buffon and his Critics*, trans and ed. John Lyon and Phillip R. Sloan, Notre Dame: University of Notre Dame Press.

Bulwer, John

1644 *Chirologia or the naturall Language of the Hand*, London.

1975 *Chirologia . . .* (reprint London, 1644), New York: AMS Press.

Burnet, Thomas
1680 *Telluris theoria sacra: orbis nostri originem et mutationes generales, quas aut jam subiit aut olim subiturus est, complectens . . .* Londini.
1684 *The Sacred Theory of the Earth, containing an Account of the Original of the Earth and of all general Changes which it hath already undergone or is to undergo 'till the Consummation of all Things*, London.
1689 *Telluris theoria sacra. Libri duo posteriores de conflagratione mundi et de futuro statu rerum*, Londini.
1690 *The Sacred Theory of the Earth*, London.
1693 *Archeologiae philosophicae*, chapters 7 and 8, in Charles Blount et al., *The Oracles of Reason*, "The 7th and 8th Chapters of Dr. *Burnet's Archeologiae Philosophicae*, together with his Appendix to the same concerning the *Brachmin's* Religion, all written Originally in *Latin*, and now rendered into *English*, by Mr. *H. B.*" London.
1727a *De fide et oficiis Christianorum* liber, Londini.
1727b *De statu mortuorum et resurgentium tractatus*, Londini.
1733 *Archaeologiae philosophicae*, Londini.
1759 *The Sacred Theory of the Earth: Containing an Account of its Original Creation, and of the General Changes which it hath undergone, or is to undergo, until the Consummation of all Things*. 7th ed. 2 vols., London.
1965 *The Sacred Theory of the Earth, with an Introduction by Basil Willey*, London: Centaur Press.
Calmet, Augustin
1723 *Trésor d'antiquitez sacrées et prophanes*, Amsterdam.
1741 *Il tesoro delle antichità sacre e profane*, Venice.
Campanella, Tommaso
1620 *De sensu rerum et magia*, Francofurti.
Cardanus, Hieronimus (Cardano, Girolamo)
1550 *De subtilitate libri XX*, Nuremberg.
Chateaubriand, François-René
1802 *Le génie du Christianisme*, Paris: Flammarion.
1976 *The Genius of Christianity*, trans. Charles I. White, New York: H. Fertig.
Colonna, Fabio
1616 *Aquatilium et terrestrium aliquot animalium . . . observationes*, Romae.
1759 *De glossepetris dissertatio* (1616), in Scilla 1759.
Condillac, Etienne Bonnot de
1749 *Traité des sistèmes*, The Hague.
1947 *Essai sur l'origine des connoissances humaines*, in *Oeuvres philosophiques*, ed. G. Le Roy, Paris.
1971 *An Essay on the Origin of Human Knowledge*. Facsimile 1756, trans. Thomas Nugent, introduction by Robert G. Weyant, Gainsville, Florida: Scholars Facsimiles and Reprints.
Dannhaewerus, Johannes C. (Dannhauer, Johann Conrad)
1656 *Praeadamita Utis; sive, Fabula primorum hominum ante Adamum conditorum explosa*, Starssbing.
Darwin, Charles
1963 *On the Origin of Species*, New York: Heritage Press.

Derham, William
1721 *Astro-Theology; or, a Demonstration of the Being and Attributes of God, from a Survey of the Heavens* (1702), London.
1754 *Physico-Theology; or, a Demonstration of the Being and Attributes of God, from his Works of Creation . . .* (1712), London.
Descartes, René
1897–1957 *Oeuvres*, ed. C. Adam and P. Tannery, Paris: L. Cerf.
1936–63 *Correspondance*, ed. C. Adam, G. Milhaud, Paris.
1967 *Opere*, introduction by E. Garin, Bari: Laterza.
1969 *Il mondo. L'uomo*, ed. E. Garin, Bari: Laterza.
1970 *The Philosophical Works of Descartes*, trans. Elizabeth S. Haldane and G. T. Ross, 2 vols., Cambridge: Cambridge University Press.
1980 *Discourse on Method and Meditations on First Philosophy*, trans. Donald A. Cress. Indianapolis, Cambridge: Hackett.
1983 *Principles of Philosophy*, trans. Valentine Rodger Miller and Reese P. Miller, Boston, Dordrecht, London: D. Reidel.
Dezallier d'Argenville, Antoine Joseph
1757 *L'histoire naturelle éclaircie dans deux de ses parties principales . . . La Conchyliologie*, Paris.
Dickinson, Edmund
1655 *Delphi Phoenicizantes, sive, Tractatus, in quo Graecos, quicquid apud Delphos celebre erat . . . è Josuae historiâ . . . effinxisse . . . ostenditur . . . Appenditur Diatriba de Noae in Italiam adventu . . .*, Oxoniae.
1702 *Physica vetus et vera; sive Tractatus de naturali veritate hexaëmeri Mosaici*, Londini.
Diderot, Denis
1956 *Oeuvres philosophiques*, ed. Paul Verrière, Paris: Garnier.
1967 *Opere filosofiche*, ed. Paolo Rossi, Milan: Feltrinelli.
1975– *Oeuvres complètes*, publiées sous la direction de Herbert Dieckmann, Jean Fabre et Jacques Proust; avec les sions de Jean Varloot, Paris.
Dortous de Mairan, Jean Jacques
1759 *Lettres . . . contenent diverses questions sur la Chine*, Paris.
Duni, Emanuele
1760 *Saggio sulla giurisprudenza universale*, Rome.
1775 *Scienza del costume o sistema del diritto universale*, Naples.
1845 *Opere complete*, 5 vols., Rome.
Du Pin, Louis Ellies
1708 *Bibliothèque universelle des historiens; contenant leurs vies, l'abrégé, la chronologie, la géographie, et la critique de leurs histoires . .* .2 vols., Amsterdam.
Duret, Claude
1619 *Thresor de l'histoire des langues de cest univers . . .* Yverdon.
Edwards, John
1695 *Some Thoughts concerning the several Causes and Occasions of Atheism . . . with some brief Reflections on Socinianism and on a late Book entituled the Reasonableness of Christianity as deliver'd in the Scriptures*, London.
1697 *Brief Remarks on Mr. Whiston's New Theory of the Earth*, London.

Engelcken, Hermann Christoph
1707 *Dissertatio theologica praeadamitismi recens incrustati examen complectens*, Rostochii.
Fayus, Jacobus
1709 *Defensio religionis necnon Mosis et gentis judaicae contra duas dissertationes Joh. Tolandi*, Ultrajecti.
Finetti, Bonifacio
1756 *Trattato della lingua ebraica e sue affini*, Venice.
1764 *De principiis iuris naturae et gentium adversus Hobbesium, Pufendorfium, Thomasium, Wolfium et alios libri XII*, Venetiis (for the criticism of Vico and Duni, see book 3, chap. 19 and Duni 1845, 3:17–39).
1768 *Apologia del genere umano accusato di essere stato una volta una bestia. Parte prima in cui si dimostra la falsità dello stato ferino degli antichi uomini dimostrata colla S. Scrittura*, Venice.
1936 *Difesa dell'autorità della Sacra Scrittura contro G. B. Vico*, introduction by B. Croce, Bari: Laterza (reprint of Finetti 1768).
Fludd, Robert
1638 *Philosophia Moysaica*, Oppenheim.
Fontenelle, Bernard le Bovier de
1688 *A Discovery of New Worlds. Translated by Mrs. A. Behn*, London.
1719 *Entretiens sur la pluralité des Mondes*, Amsterdam.
1749 *Trattenimento sopra la pluralità dei mondi*, Paris and Venice.
1966 *Entretiens sur la pluralité des Mondes*, ed. Alexandre Calame, Paris: Didier.
François, Laurent
1751 *Preuves de la religion de Jésus-Christ contre les spinosistes et les déistes*, 4 vols., Paris.
1764 *Pruove della religione di Gesù Cristo contro gli spinosisti e i deisti*, Venice.
Free Thoughts
1734 *Free Thoughts concerning Souls in four Essays . . . to which is added an Essay on Creation*, London.
Fréret, Nicolas
1758 *Défense de la chronolcgie fondée sur les monumens de l'Histoire Ancienne, contre le systême chronologique de M. Newton*, Paris.
1796 *Oeuvres complètes*, Paris.
Gesner, Konrad
1656 *De rerum fossilium, lapidum et gemmarum maximè, figuris & similitudinibus* liber . . ., Tiguri.
Giannone, Pietro
1940 *Il Triregno*, ed. A. Parente, 3 vols., Bari: Laterza.
Gimma, Giacento
1730 *Della storia naturale delle gemme, delle pietre, e di tutti li minerali . . .*, Naples.
Giornali Veneziani
1962 *Giornali veneziani del Settecento*, ed. M. Berengo, Milan: Feltrinelli.

Goguet, Antoine Yves
1820 *De l'origine des loix, des arts, et des sciences; et de leur progrès chez les anciens peuples*, 3 vols., Paris.
1976 *The Origin of Laws, Arts, and Sciences, and their Progress among the most Ancient Nations*, (reprint Edinburgh 1761), 3 vols., New York: AMS Press.
Graverol, Jean
1694 *Moses vindicatus . . . adversus Thomas Burnetii Archeologias Philosophicas*, Amstelodami.
Grotius, Hugo
1640 *De veritate religionis christianae*, Parisiis.
Hale, Sir Matthew
1677 *The Primitive Origination of Mankind, considered and examined according to the Light of Nature*, London.
1805 *The Works Moral and Religious . . . to which are prefixed his Life and Death, by Bishop Burnet D. D. and an Appendix to the Life . . .*, 2 vols., London.
Halley, Edmund
1715 "A Short Account of the Cause of Saltness of the Ocean and of Several Lakes that emit no Rivers, with a Prosal by Help thereof, to discover the Age of the World," in *Philosophical Transactions of the Royal Society* 29, no. 344, 296–300.
Harris, John
1697 *Remarks on some Late Papers relating to the Universal Deluge and to the Natural History of the Earth*, London.
1784 *Ideen zur Philosophie der Geschichte der Menschheit*, Riga, Leipzig
Herder, Johann Gottfried von
1954 *Saggio sull'origine delle lingue*, ed. G. Necco, Rome.
1967 *On the Origin of Language*, trans. and ed. John H. Moran and Alexander Gode, New York: F. Ungar.
1968 *Reflections on the Philosophy of the History of Mankind*, trans. T. O. Churchill, abridged and with an introduction by Frank Manuel, Chicago: University of Chicago Press.
1971 *Idee per la filosofia della storia dell'umanità*, ed. V. Verra, Bologna: Zanichelli.
Herder, Johann Gottfried and Monboddo, James Burnet, Lord
1973 *Linguaggio e società*, ed. N. Merker and L. Formigari, Bari: Laterza.
Herschel, William
1912 *Scientific Papers*, London.
Hobbes, Thomas
1966a *The English Works of Thomas Hobbes of Malmesbury; Now first collected and Edited by Sir William Molesworth, Bart.*, London: John Bohn, 1839–1945, 11 vols. (reprint, Scientia Verlag Aalen).
1966b *Thomae Hobbes Malmesburiensis Opera Philosophica quae latine scriptsit omnia in unum corpus nunc primum collecta studio et labore Gulielmi Molesworth*, Londinii, apud Joannem Bohn, 1839–1845, 5 vols. (reprint, Scientia Verlag Aalen).
1978 *De Homine*, trans. Charles T. Wood, T. S. K. Scott-Craig, and Bernard Gert, and *De Cive*, trans. Thomas Hobbes (also known as *Philosophical Rudiments*

concerning Government and Society), ed. with an introduction, Bernard Gert, Gloucester, Mass.: Peter Smith, 1978.

Holbach, Paul Henri Thiry, Baron d'

1794 *Système de la nature ou des loix du monde physique et du monde moral*, 6 vols., Paris.

1889 *The System of Nature: or, Laws of the Moral and Physical World*, trans. H. D. Robinson, Boston.

Hooke, Robert

1665 *Micrographia*, London.

1705 *The Posthumous Works . . .*, ed. R. Waller, London.

1935 *Diary 1672–1680 Transcribed from the Original in the Possession of the Corporation of the City of London*, ed. H. W. Robinson and W. Adams, with a preface by Sir F. G. Hopkins, London.

1961 *Micrographia* (facsimile London 1665), preface R. T. Gunther, New York: Dover.

Horn, Georg (Hornius)

1655 *Historiae philosophicae libri septem*, Lugduni Batavorum.

1659a *Dissertatio de vera aetate mundi qua sententia illorum refellitur qui statuunt Natale Mundi tempus annis minimum 1440 vulgarem aeram anticipare*, Lugduni Batavorum.

1659b *Defensio dissertationis de vera aetate mundi contra castigationes I. Vossii*, Lugduni Batavorum.

1666 *Arca Noae sive historia imperiorum et regnorum a condito orbe ad nostra tempora*, Lugduni Batavorum.

1668 *Arca Mosis; sive, Historia mundi; quae complectitur primorida rerum naturalium omniumque artium ac scientiarum*, Lugduni Batavorum et Roterodami.

1670 *Historia naturalis et civilis, ad nostra tempora libri septem*, Lugduni Batavorum.

Huet, Pierre Daniel

1690 *Alnetanae quaestiones de concordia rationis et fidei*, Parisiis.

Hume, David

1970 *Dialogues concerning Natural Religion*, ed. Nelson Pike, Indianapolis and New York: Indiana University Press.

1971 *Opere*, ed. E. Lecaldano and E. Mistretta, Bari: Laterza.

Hutton, James

1788 "Theory of the Earth, or an Investigation of the Laws discernible in the Composition, Dissolution and Restoration of the Land upon the Globe," in *Transactions of the Royal Society of Edinburgh*," 1:2, pp. 209–304.

1792 *Dissertation on the Different Subjects in Natural Philosophy*, Edinburgh.

1794a *An Investigation of the Principles of Knowledge and of the Progress of Reason, from Sense to Science and Philosophy*, Edinburgh.

1794b *A Dissertation upon the Philosophy of Light, Heat and Fire*, Edinburgh.

1795 *Theory of the Earth, with Proofs and Illustrations*, 2 vols., Edinburgh.

1899 *Theory of the Earth . . .*, vol. 3, ed. Sir A. Geikie, London: Geological Society.

Hyde, Thomas

1700 *Historia religionis veterum Persarum*, Oxonii.

Josephus, Flavius
1925 *Works*, vol. 1, "Against Apion," trans. H. St. J. Thackeray, Loeb Classical Library.
1974 *The Works of Flavius Josephus*, Grand Rapids: Baker Book House.
Kant, Immanuel
1900 *Kant's Cosmogony*, trans. W. Hastie, Glasgow: J. Maclehose & Sons.
1910 *Allgemeine Naturgeschichte und Theorie des Himmels* (1755) in *Kant's Werke*, Band I, 1, Vorkritischer Schriften, 1, Berlin: Reimer.
1955 *Scritti precritici*, Bari: Laterza.
1969 *Universal Natural History and Theory of the Heavens*, introduction by Milton Karl Munitz, Ann Arbor: University of Michigan Press.
Keill, John
1698 *An Examination of Dr. Burnet's Theory of the Earth with some Remarks on Mr. Whiston's New Theory of the Earth*, Oxford.
Kelvin, William Thomson, Lord
1854, 1871 Report, British Association for the Advancement of Sciences.
1894 *Popular Lectures and Addresses*, London.
King, Charles
1705 *An Account of the Origin and Formation of Fossil Shells*, London.
King, William
1776 *Original Works*, London.
1972 *The Original Works*, ed. John Nichols (reprint London 1776), Westmead: Gregg International.
Kircher, Athanasius
1652–54 *Oedipus aegyptiacus*, Romae.
1675 *Arca Noë in tres libros digesta sive de rebus ante diluvium, de diluvio et de rebus post diluvium a Noemo gestis*, Amstelodami.
1678 *Mundus subterraneus in XII libros digestus quo divinum subterrestris mundi Opificium mira ergasteriorum naturae in eo distributio, Protei regnum, universae denique naturae maiestas et divitiae summa rerum varietate exponuntur*, Amstelodami.
1679 *Turri Babel sive Archontologia*, Amstelodami.
Lafitau, Joseph-François
1724 *Moeurs des sauvages ameriquains, comparées aux moeurs des premiers temps*, Paris.
Lambert, Johann Heinrich
1965–68 *Gesammelte philosophische Werke*, Hildesheim.
1976 *Cosmological Letters*, ed. S. L. Jaki, Edinburgh: Scottish Academic Press.
Lamennais, Félicité Robert de
1823 *Essai sur l'indifférence en matière de religion*, 4 vols., Paris.
La Mothe le Vayer, François de
1716 *Cincq dialogues faits à l'imitation des anciens, par Oratius Tubero*, Frankfurt.
Lapeyrère, Isaac de
1655 *Systema theologicum ex Praeadamitarum hypothesi. Praeadamitae. Sive exercitatio super versibus duodecimo decimotertio & decimoquarto, capitis quinti Epistolae D. Pauli ad Romanos. Quibus inducuntur primi homines ante Adamum conditi*, n. p. (Amstelodami).

1656 *Men Before Adam, or a Discourse upon the Twelfth, Thirteenth, and Fourteenth verses of the Fifth Chapter of the Epistle of the Apostle Paul to the Romans . . .*, London.

La Religion Vengée

1757–62 *La religion vengée ou réfutation des auteurs impies dédiée à Monseigneur le Dauphin par une société des gens de lettres*, 18 vols., Paris.

Leclerc, Jean

1685 *Sentimens de quelques théologiens de Hollande . . .* , Amsterdam.

Lehmann, Johann Gottlob

1756 *Versuch einer Geschichte von der Flötz-Gebürgen*, Berlin.

Leibniz, Gottfried Wilhelm

1710 "Epistola G. G. Leibnitii ad autorum dissertationis de figuris animalium quae in lapidibus observantur et lithozoorum nomine venire possent," in *Miscellanea Berolinensia* 1:111ff.

1749 *Protogaea; sive, De prima facie telluris et antiquissimae historiae vestigiis in ipsis naturae monumentis dissertatio ex schedis manuscriptis viri illustris in lucem edita a C.L. Scheidio*, Goettingae.

1875–90 *Die philosophischen Schriften*, Herausgegeben von C. I. Gerhardt, 7 vols. Berlin (reprint edition Hildesheim: G. Olms, 1965).

1951 *Theodicy. Essays on the Goodness of God the Freedom of Man and the Origin of Evil*, trans. E. M. Huggard, ed. Austin Farrer, London: Routledge & Kegan Paul.

1963 *Saggi filosofici e lettere*, ed. V. Mathieu, Bari: Laterza.

1967 *Scritti filosofici*, ed. D. O. Bianca, Turin: Utet.

Leland, John

1754 A View of the principal Deistical Writers that have appeared in England . . . , 3 vols., London.

Le Prieur, Philippe

1656 *Animadversiones in librum Praeadamitarum. In quibus confutatur nuperus scriptor, et primum omnium hominum fuisse Adamum, defenditur*, Parisiis.

Leydekker, Melchior

1704 *De republica Hebraeorum. Libri XII. Subijcitur archaeologia sacra, qua historia creationis et diluvii mosaica contra Burneti profanem telluris theoriam asseritur*, Amstelodami.

Lhuyd, Edward

1707 *Archaeologia britannica, giving some Accounts . . . of the Languages, Histories, and Customs of the original Inhabitants of Great Britain . . .* 1: *Glossography*, Oxford.

Lister, Martin

1685–92 *Historia Conchyliorum*, Londini.

Lovell, Archibald

1696 *A Summary of Material Heads which may be enlarged and improved into a complete Answer to Dr. Burnet's Theory of the Earth*, London.

Maillet, Benoît de

1748 *Telliamed; ou Entretiens d'un philosophe indien avec un missionaire françois sur la diminution de la Mer, la formation de la Terre . . .*,2 vols., Amsterdam.

1797 *Telliamed; or, The World Explain'd*, Baltimore.

1968 *Telliamed . . .*, trans. and ed. Albert V. Carozzi, Urbana: University of Illinois Press.

Mandeville, Bernard de

1732 *An Enquiry into the Origin of Honour, and the Usefulness of Christianity in War*, London.

1924 *The Fable of the Bees*, ed. F. B. Kaye, 2 vols., New York

1962 *The Fable of the Bees*, ed. I. Primer, New York.

1974 *Ricerca sulla natura della societa*, ed. M. Scribano, Bari: Laterza.

1978 *Dialoghi fra Cleomene e Orazio*, ed. G. Belgioioso, Lecce: Milella.

Marsham, John

1696 *Canon chronicus aegyptiacus, hebraicus, graecus*, Londini.

1723 *Canon chronicus aegyptiacus, hebraicus, graecus*, Londini.

Martianay, Jam

1693 *Continuation de la Défense du texte hébreu et de la Vulgate, par les véritables traditions des Eglises chrétiennes, et par toutes sortes d'anciens monumens . . . contre Isaac Vossius, et contre les livres du P. Pezron*, Paris.

Martini, Martino

1658 *Sinicae historiae decas prima . . .*, Monachii.

Maupertuis, Pierre Louis Moreau de

1768 *Oeuvres*, 3 vols., Lyon

Mémoires

1733 Mémoires de l'Académie des Inscriptions et Belles Lettres.

Memorie Istoriche

1700 *Memorie istoriche della controversia de' culti chinesi, lettere de' Signori Superiori e Direttori del Seminario delle Missioni Straniere di Parigi al Sommo Pontefice Innocenzio XII intorno all'idolatria e superstizioni della China. In italiano e in francese. Due pareri di centoventi dottori dell'Università di Parigi con una raccolta di varie principali scritture de' Padri della Compagnia de Gesù e de' Signori Missionari del Clero secolare di Francia sopra la medesima controversia, molte delle quali si danno in luce per la prima volta*, Cologne.

Mercati, Michele

1589 *De gli obelischi di Roma*, Rome

1719 *Metallotheca, opus postumum (auctoritate et munificentia Clementis undecimi . . . e tenebris in lucem edita)*, Romae.

Middleton, Conyers

1729 *A Letter from Rome, shewing an Exact Conformity between Popery and Paganism:* or, *The Religion of the Present Romans to be derived entirely from that of their Heathen ancestors . . .*, London.

1730 *A Letter to Dr. Waterland; containing some Remarks on his Vindication of Scripture; in answer to a Book, intituled Christianity as Old as the Creation. Together with a Sketch or Plan of another Answer to the said Book*, London (also in Middleton 1752, 2:135–37).

1733 *Remarks on some Observations adress'd to the Author of the Letter to Dr. Waterland, by the Author of the Letter*, London (also in Middleton 1752, 2:297–316).

1748 *A Free Inquiry into the Miraculous Powers, which are supposed to have subsisted in the Christian Church*, London (also in Middleton 1752, 1:1–88).
1752 *The Miscellaneous Works of the late Reverend and learned Conyers Middleton . . . Containing all his Writings except the Life of Cicero; many of which never before Published . . . in Four Volumes with a Complete Index for the Whole*, London.

Monboddo, James Burnett, Lord
1774–92 *Of the Origin and Progress of Language*, 6 vols., Edinburgh.
1779–99 *Antient Metaphysics: or the Science of Universals*, 6 vols., Edinburgh.
1974 *Of the Origin and Progress of Language*, 6 vols. (reprint Edinburgh 1772–89), Hildesheim, New York: G. Olms.

More, Henry
1646 *Democritus Platonissans, or an Essay upon the Infinity of the World out of Platonick Principles*, Cambridge.
1726 *Enchiridion Metaphysicum*, London
1968 *Democritus Platonissans . . .* (reprint 1646), Los Angeles: Andrews Clark Memorial Library.

Morhof, Daniel G.
1732 *Polyhistor, literarius, philosophicus et practicus*, Lubecae.

Morin, Etienne
1694 *Exercitationes de lingua primaeva ejusque appendicibus*, Amstelodami.

Moro, Anton-Lazzaro
1740 *De' crostacei e degli altri corpi marini che si trovano su' monti, libri due*, Venice.

Muratori, Ludovico Antonio
1758 *De paradiso regnique caelestis gloria . . . adversus Thomae Burneti Britanni librum De Statu mortuorum*, Veronae.

Newton, Isaac
1721 *Opticks, or, a Treatise of the Reflexions, Refractions, Inflexions and Colours of Light . . .*, London.
1728 *The Chronology of Ancient Kingdoms Amended . . .*, Dublin.
1757 *La cronologia degli antichi regni emendata*, trans. P. Rolli, Venice.
1779–85 *Opera quae extant omnia. Commentariis illustrabat Samuel Horsley*, 5 vols., Londini.
1934 *Sir Isaac Newton's Mathematical Principles of Natural Philosophy and his System of the World*, trans. Andrew Motte, 1929, translation revised by Florian Cajori, Berkeley: University of California Press.
1952 *Opticks, or A Treatise of the Reflections, Refractions, Inflections and Colours of Light*, New York: Dover.
1959– *The Correspondence of Isaac Newton*, ed. H. W. Turnbull, J. P. Scott, A. R. Hall, L. Tilling, 5 vols., continuing, Cambridge.

Nonotte, Claude François
1772 *Dictionnaire philosophique de la religion, où l'on établit tous les Points de la Religion, attaqués par les Incrédules*, Avignon.
1774 *Dizionario filosofico della religione, dove si stabiliscono tutti i punti della religione attaccati dagl'increduli e si risponde a tutte le loro obbiezioni, opera*

dell'autore degli Errori di Voltaire, prima edizione italiana dopo la seconda francese, 4 vols., Venice (anonymous).

Palissy, Bernard

1888 *Oeuvres de Maistre Bernard Palissy*, ed. B. Fillon, 2 vols., Niort.

Pascal, Blaise

1954 *Oeuvres complètes*, ed. J. Chevalier, Paris: Gallimard (Bibliothèque de la Pléiade).

1965 *Pascal's Pensées*, trans. and ed. H. F. Stewart, New York: Pantheon.

Perrault, Nicolas

1702 *Le morale des Jésuites extraite fidellement de leurs livres imprimez avec la permission et l'approbation des Supérieurs de leur Compagnie*, Mons.

Pezron, Paul

1687 *L'antiquité des tems rétablie et défenduë contre les Juifs et les nouveaux chronologistes*, Paris.

Plattes, Gabriel

1639 *A Discovery of an infinite Treasure hidden since the World's beginning*, London.

Plot, Robert

1705 *The natural History of Oxford-shire*, Oxford.

Pluche, Noël Antoine

1732 *Le spectacle de la nature, ou Entretiens sur les particularités de l'histoire naturelle*, 8 vols., Paris.

1740 *Histoire du ciel considéré selon les idées des poètes, des philosophes et de Moïse*, The Hague.

Ramazzini, Bernardino

1691 *De fontium mutiniensium admiranda scaturigine tractatus physio-hydrostaticus*, Modena.

Ray, John

1691 *The Widsom of God manifested in the Works of Creation*, London.

1692 *Miscellaneous Discourses Concerning the Dissolution and Changes of the World*, London.

1693 *Three physico-theological Discourses, concerning:* I *the Primitive Chaos, and Creation of the World.* II *the General Deluge, its Causes and Effects.* III *the Dissolution of the World, and Future Conflagration . . .* , London.

Recueil des Lettres

1713–76 *Recueil des lettres édifiantes et curieuses*, 32 vols., Paris.

Renaudot, Eusèbe

1718 *Anciennes relations des Indes et de la Chine, de deux voyageurs Mahométans qui y allèrent dans le neuvième siècle*, Paris.

1808–14 "An Account of the Travels of two Mohammedans through India and China . . . ," in John Pinkerton, ed., *A General Collection of the Best and most Interesting Voyages and Travels . . .* , London.

Robinet, Jean Baptiste René

1766 *De la nature*, 4 vols., Amsterdam.

1768 *Considérations philosophiques de la gradation naturelle des formes de l'être ou les essais de la nature qui apprend à faire l'homme*, Paris.

Robinson, Thomas

1694 *The Anatomy of the Earth*, London.

1696 *New Observations on the Natural History of this World of Matter and this World of Life, in Two Parts. Being a Philosophical Discourse grounded upon the Mosaick System of the Creation and the Flood. To which are added some Thoughts concerning Paradise, the Conflagration of the World and a Treatise on Meteorology, with Occasional Remarks upon some Late Theories, Conferences and Essays*, London.

1709 *An Essay towards a Natural History of Westmoreland and Cumberland . . . To which is annexed, a Vindication of the Philosophical and Theological Paraphrase of the Mosaick System of the Creation, &c.*, London.

Romano, Damiano

1753 *Dello stato naturale dopo la prevaricazione di Adamo, insufficiente per la sicurezza dell'uomo, Dissertazione apologetica*, Naples.

Rotheram, John

1755 *The Force of the Argument for the Truth of Christianity drawn from a collective View of Prophecy. In Three Parts A Brief State of the Argument; A Defence and further Illustration of the Argument; a Brief State of the Question whether Prophecies or Miracles afford a stronger Evidence for the Truth of Christianity. Occasioned by Dr. Middleton's Examination of the Lord Bishop of London's Discourses*, Oxford.

Rousseau, Jean-Jacques

1959–69 *Oeuvres complètes*, publiées sous la direction de B. Gagnebin et M. Raymond, Bibliothèque de la Pléiade, 4 vols., Paris.

1964 *The First and Second Discourses*, trans. Roger D. Masters and Judith R. Masters, ed. Roger D. Masters, New York: St. Martin's Press.

1967 *On the Origin of Language*, trans. and ed. John H. Moran and Alexander Gode, New York: Frederick Ungar.

1970 *Essai sur l'origine des langues*, in appendix to Verri 1970 (see below, Studies).

1972 *Opere*, ed. Paolo Rossi, Florence: Sansoni.

Scaliger, Joseph Juste

1583 *De emendatione temporum*, Frankfurt.

1666 *Scaligeriana sive excerpta ex ore Josephi Scaligeri. Per FF.PP.*, Hagae-Comitum.

Schoock, Marten

1643 *Admiranda Methodus novae philosphiae Renati des Cartes*, Ultrajecti.

1658 *Tractatus de Anima Belluarum variis disputationibus propositus in Academia Gronigae et Ommelandiae*, Gronigae.

1662 *Fabula Hamelensis sive disquisitio historica qua, praemissa generali dissertatione de historiae veritate, ostenditur commenti rationem habere quae vulgo circumferuntur de infausto exitu Puerorum Hamelensium . . . Accessit diatriba qua probatur Noachi deluvium toti terrarum orbi incubuisse. Adversus virum quendam celeberrimum*, Gronigae.

Schweigger, Johannes Christophorus

1665 *De ortu lapidum*, Wittenberg.

Scilla, Agostino

1670 *La vana speculazione disingannata dal senso. Lettera risponsiva circa i corpi marini, che petrificati si trovano in varij luoghi terrestri*, Naples.

1759 *De corporibus marinis lapidescentibus quae defossa reperiuntur . . . Addita dis-*

sertatione Fabii Columnae De glossopetris, Romae (first edition, Rome 1752. Translation in Latin of Scilla 1670; see also Colonna 1759).

Shirley, Thomas

1727 *A Philosophical Essay, declaring the Probable Causes, whence Stones are produced in the Greater World. From which Occasion is taken to Search into the Origin of all Bodies, discovering them to proceed from Water and Seeds. Being a Prodromus to a Medicinal Tract concerning the Causes, and Cure of the Stone in the Kidneys and Bladder of Men* (1672), London.

Shuckford, Samuel

1731–37 *The Sacred and Prophane History of the World Connected*, 3 vols., London.

1753 *The Creation and Fall of Man. A Supplemental Discourse to the Preface of the First Volume of the Sacred and Profane History of the World Connected*, London.

Sibiscota, George

1670 *The Deaf and Dumb Man's Discourse*, London.

Simon, Richard

1680 *Histoire critique du Vieux Testament*, Paris.

1685 *Histoire critique du Vieux Testament*, Rotterdam.

1708 *Bibliothèque critique*, Amsterdam.

1730 *Lettres choisies*, 3 vols., Amsterdam.

Smith, Adam

1880 *The Theory of Moral Sentiments*, ed. Dugald Stewart, London.

1963 "Of the Origin and Progress of Language" (1672), in *Lectures on Rhetoric and Belles Lettres*, ed. John M. Lothian, New York, London: T. Nelson (1971, Carbondale, Illinois: Southern Illinois University Press).

1976 *The Theory of Moral Sentiments*, ed. D. D. Raphael and A. L. Macfie, Oxford: Clarendon Press.

Spencer, John

1670 *Dissertatio de Urim et Tummim*, Cambridge.

1686 *De legibus Hebraeorum ritualibus et earum rationibus libri tres*, Hagae Comitum (also, The Hague 1705).

Spicilegium Shuckfordianum

1754 *Spicilegium Shuckfordianum: or, A Nosegay for the Critics. Being some Choice Flowers of Modern Theology and Criticism, gathered (this Spring Time) out of Dr. Shuckford's Supplemental Discourse on the Creation and Fall of Man*, London.

Spinoza, Benedict de

1882–84 *Opera*, ed. J. Van Vloten and J. N. P. Land, Hagae Comitum.

1901 *Improvement of the Understanding, Ethics and Correspondence of Benedict de Spinoza*, trans. R. H. M. Elwes, New York, London: M. Walter Dunne.

1908 *Tractatus Theologico-Politicus (A Theologico-Political Treatise)* in *The Chief Works of Benedict de Spinoza*, trans. R. H. M. Elwes, 2 vols., London: George Bell.

1959a *Etica*, ed. S. Giannetta, Turin: Boringhieri.

1959b *Epistolario*, ed. A. Droetto, Turin: Einaudi.

1963 *Earlier Philosophical Writings . . .*, trans. Frank A. Hayes, Indianapolis: University of Indiana Press.

1972 *Trattato teologico politico*, ed. A. Droetto and E. Giancotti Boscherini.

1981 *Ethics*, trans. George Eliot, ed. Thomas Deegan, Salzburg: Institut fur Anglistik und Amerikanistik, Universität Salzburg.

Steno, Nicolaus (Steensen, Niels, Stenone)

1669 *De solido intra solidum naturaliter contento dissertationis prodromus. Ad serenissimum Ferdinandum II Etruriae Ducem*, Florentiae.

1910 *Opera philosophica*, ed. V. Maar, 2 vols., Copenhagen.

1928 *Prodromo*, trans. G. Montalenti, Rome.

1944–47 *Opera theologica cum proemiis ac notis germanice scriptis ediderunt K. Larsen et G. Scherz*, 2 vols., Hafniae.

1952 *Epistolae et epistolae ad eum datae*, ed. G. Scherz., 2 vols., Hafniae.

1956 *Indice di cose naturali*, in Scherz 1956.

1968 *The Prodromus of Nicolaus Steno's Dissertation concerning a solid Body enclosed by a Process of Nature within a Solid*, ed. J. G. Winter, with a foreword by W. H. Hobbs, New York: Hafner.

1969 *Geological Papers*, ed. G. Scherz, Odense University Press.

Stillingfleet, Edward

1662 *Origines sacrae or A Rational Account of the Grounds of Christian Faith, as to the Truth and Divine Authority of the Scriptures, and the Matters therein contained*, London.

1709 *Origines Sacrae or a Rational Account of the Grounds of Natural and Reveal'd Religion. The Eighth Edition, to which is now added Part of another Book upon the Same Subject Written A.D. 1697 Publish'd from the Author's own Manuscript*, London (vol. 2 of Stillingfleet 1709–10).

1709–10 *Works*, 6 vols., London.

Sykes, Arthur Ashley

1744 *An Examination of Mr. Warburton's Account of the Conduct of the Antient Legislators, of the Double Doctrine of the Old Philosophers, of the Theocracy of the Jews, and of Sir Isaac Newton's Chronology*, London.

Tindal, Matthew

1732 *Christianity as Old as the Creation; or, The Gospel, a Republication of the Religion of Nature*, London.

Toland, John

1702 *Christianity not mysterious*, London.

1704 *Letters to Serena* (facsimile edition, Stuttgart 1964).

1709a *Origines Judaicae*, London.

1709b *Adeisidaemon; sive, Titus Livius a superstitione vindicatus . . . Annexae sunt ejus dem Origines Judaicae*, Hagae Comitis.

1720 *Tetradymus*, London.

Tyssot de Patot, Simon

1710 *Voyages et aventures de Jacques Massé*, Bordeaux.

1733 *The Travels and Adventures of James Massey*. London.

Ursin, Johann Heinrich

1656 *Novus Prometheus Prae-Adamitarum plastes ad Caucasum . . . relegatus et religatus, schediasma*, Francofurti.

1661 *De Zoroastre Bactriano, Hermete Trismegisto, Sanchoniathone phoenicio eorumque scriptis et aliis contra. Mosaicae Scripturae antiquitatem*, Norimbergae.

Usserius, Jacobus (Ussher, James)
1650 *Annales Veteris Testimenti a prima mundi origine deducti*, Londini.
Vallisnieri, Antonio
1721 *De' corpi marini, che su' Monti si trovano, della loro origine; e dello stato del Mondo avanti 'l Diluvio, nel Diluvio, e dopo il Diluvio: Lettere critiche*, Venice.
1726 *Lezione accademica intorno l'origine delle fontane*, Venice.
1733 *Opere fisico-mediche*, Venice.
Vico, Giambattista
1911 *L'autobiografia, il carteggio e le poesie varie*, ed. B. Croce, Bari: Laterza.
1913 *La Scienza Nuova*, ed. F. Nicolini, Bari: Laterza.
1931 *La Scienza Nuova Prima*, ed. F. Nicolini, Bari: Laterza.
1944 *The Autobiography of Giambattista Vico*, trans. Max Harold Fisch and Thomas Goddard Bergin, Ithaca: Cornell University Press.
1947 *Autobiografia di G. B. Vico*, ed. F. Nicolini, Milan: Bompiani.
1948 *The New Science of Giambattista Vico*, translated from the third edition (1744) by Thomas Goddard Bergin and Max Harold Fisch, Ithaca: Cornell University Press.
1953 *La Scienza Nuova Seconda*, ed. F. Nicolini, Bari: Laterza.
1968 *Il Diritto Universale*, ed. F. Nicolini, Bari: Laterza.
1982 *Vico: Selected Writings*, ed. and trans. Leon Pompa, Cambridge and New York: Cambridge University Press.
Voltaire, François Marie Arouet, called
1772 *Elémens de la philosophie de Newton . . .* , Neuchatel.
1876–92 *Oeuvres complètes*, 46 vols., Paris: Hachette.
1877–83 *Oeuvres*, ed. Moland, 52 vols., Paris.
1968 *La filosofia de Newton*, ed. P. Casini, Bari: Laterza.
Vossius, Gerardus Johannes
1642 *De theologia gentili et physiologia christiana; sive, De origine ac progressu idolatriae libri IV*, Amstelodami.
1654 *De veterum poetarum temporibus libri duo*, Amstelodami.
1659 *Chronologiae sacrae Isagoge, sive de ultimis mundi antiquitatibus ac imprimis de temporibus rerum hebraeorum dissertationes VIII*, Hagae Comitum.
1668 *De theologia gentili et physiologia christiana sive de origine et progressu idolatriae deque naturae mirandis quibus homo abducitur ad Deum libri IX . . . Posteriores quinque libri, ex auctoris autographo nunc primum prodeunt*, Amstelodami.
1695–1701 *Opera in sex tomos divisa*, Amstelodami.
1976 *De theologia gentili et physiologia Christiana* (reprint Amsterdam 1641), New York: Garland Pub.
Vossius, Isaac
1659a *Dissertatio de vera aetate mundi qua ostenditur Natale mundi tempus annis minimum 1400 vulgarem aeram anticipare*, Hagae Comitum.
1659b *Castigationes ad scriptum G. Hornii de aetate mundi*, Hagae Comitis.
1659c *Auctarium castigationum ad scriptum de aetate mundi*, Hagae Comitis.
1661 *De septuaginta Interpretibus, eorumque translatione et chronologiae dissertationes*, Hagae Comitum.

Warburton, William
1741 *The Divine Legation of Moses*, London.
1744 *Essai sur les hiéroglyphes des Egyptiens*, 4 vols., Paris.
1764–65 *The Divine Legation of Moses*, 5 vols., London.
1788 *Works*, 7 vols. plus A Supplental Volume, London.
1789 *Tracts, by Warburton and a Warburtonian; not admitted into the Collections of their Respective Works*, London.
1841 *A Selection from the Unpublished Papers of the Right Reverend William Warburton by the Rev. Francis Kilvert*, London.
1977 *Essai sur les Hiéroglyphes des Egyptiens*, ed. J. Derrida and P. Tort, Paris: Aubier Flammarion.

Warren, Erasmus
1690 *Geologia: or a Discourse concerning the Earth before the Deluge*, London.

Webb, John
1669 *An Historical Essay endeavouring a Probability that the Language of the Empire of China is the Primitive Language*, London.

Whiston, William
1698 *A Vindication of the New Theory of the Earth from the Exceptions of Mr. Keill and Others*, London.
1700 *A Second Defence of the New Theory of the Earth from the Exceptions of Mr. Keill*, London.
1711 *An Historical Preface to primitive Christianity, reviv'd with an Appendix containing an Account of the Author's Prosecution at, and Banishment from the University of Cambridge*, London.
1714 *The Cause of the Deluge demonstrated*, London.
1737 *A New Theory of the Earth, from its Original, to the Consummation of All Things*, London.
1753 *Memoirs of the Life and Writings of Mr. William Whiston . . . Written by Himself*, London.
1978 *A New Theory of the Earth* (reprint London 1696), New York: Arno Press.

Williams, Philip
1733 *Some Observations Addressed to the Author of the Letter to Dr. Waterland* (C. Middleton) . . . *In which from his own Words and Reasoning against the Author of Christianity as Old as the Creation, it is plainly proved that this Letter, Defence and Remarks ought all be burned and the Author of them banished*, London.

Wits, Hermann (Witsius)
1696 *Aegyptiaca*, Amstelodami.

Woodward, John
1695 *An Essay toward a Natural History of the Earth: and Terrestrial Bodies, especially Minerals . . .*, London.
1713 *An Account of some Roman Urns and Other Antiquities lately digg'd up near Bishop-Gate. In a Letter to Sir Christopher Wren*, London.
1714 *Naturalis historia telluris illustrata et aucta. Una cum eiusdem defensione praesertim contra nuperas obiectiones Camerarii . . . accedit methodica et . . . instituta fossilium in classes distributio*, Londini.

1723 *Remarks upon the Antient and Present State of London, occasion'd by some Roman Urns, Coins and Other Antiquities lately discover'd*, London.
1726 *The Natural History of the Earth Illustrated, Inlarged, and Defended. Written originaly in Latin and now first made English by Benj. Holloway*, London.
1728 *Fossils of all Kinds, digested into a Method*, London.

Wotton, William
1697 *A Vindication of an Abstract of an Italian Book concerning Marine Bodies which are found Petrified in several Places at Land written by Agostino Scilla, and abridged in The Philosophical Transactions for the months of January and February 1695–96* (see Arbuthnot 1697, 65–84).
1713 *Discourse concerning the Confusion of Languages at Babel*, London.

Wright, Thomas of Durham
1742 *Clavis caelestis, Being the Explication of a Diagram entituled a Synopsis of the Universe or the Visible World Epitomized*, London.
1750 *An Original Theory or New Hypothesis of the Universe, founded upon the Laws of Nature*, London.
1968 *Second or Singular Thoughts upon the Theory of the Universe*, edited from the manuscript by M. A. Hoskin, London.

STUDIES

Aarsleff, H.
1967 *The Study of Language in England: 1780–1860*, Princeton, New Jersey: Princeton University Press.

Adams, F. D.
1954 *The Birth and Development of the Geological Sciences*, New York: Dover.

Adams, H. P.
1935 *The Life and Writings of G. B. Vico*, London: Allen and Unwin.

Allen, D. C.
1963 *The Legend of Noah: Renaissance Rationalism in Art, Science and Letters*, Urbana: University of Illinois Press.

Apel, K. O.
1975 *L'idea di lingua nella tradizione dell'Umanesimo da Dante a Vico*, Bologna: Il Mulino.

Arrhenius, S.
1909 *The Life of the Universe as Conceived by Man from the Earliest Ages to the Present Time*, 2 vols., New York.

Aubin, R. A.
1934 "Grottoes, Geology and Gothic Revival," in *Studies in Philology*.

Badaloni, N.
1961 *Introduzione a G. B. Vico*, Milan: Feltrinelli.
1968a "Vico nell'ambito della filosofia europea," in *Omaggio a Vico*, Naples: Morano.
1968b *Antonio Conti: un abate libero pensatore fra Newton e Voltaire*, Milan: Feltrinelli.

Bailey, E. B.
1967 *James Hutton: The Founder of Modern Geology*, Amsterdam, London, and New York: Elsevier Publishing.

Baker, J. T.
1930 *An Historical and Critical Examination of English Space and Time Theories from Henry More to Bishop Berkeley*, Bronxville, New York.

Baldini, M.
1976 "Le riflessioni metodologiche di A. Vallisnieri," in *La medicina nei secoli* 3:547–59.

Bastian, H.
1959 *Und dann kam der Mensch*, Berlin: Ullstein A.S.

Bataillon, M.
1966 "L'unité du genre humain de P. Acosta à P. Clavigero," in *Mélanges à la mémoire de Jean Sarrailh* 1:75–95 Paris.

Belaval, Y.
1960 *Leibniz critique de Descartes*, Paris: Gallimard.

Bellone, E.
1978 *Le leggi della termodinamica da Boyle a Boltzmann*, Turin: Loescher.

Beringer, C. C.
1954 *Geschichte der Geologie und Geologisches Weltbildes*, Stuttgart.

Berlin, I.
1976 *Vico and Herder: Two Studies in the History of Ideas*, London: Hogarth Press.

Besterman, Theodore
1976 *Voltaire*, Chicago: University of Chicago Press.

Bligny, M.
1973 "Il mito del Diluvio universale nella coscienza europea del Seicento," *Rivista Storica Italiana* 85, 1:47–63.

Bloch, E.
1959 *Das Prinzip Hoffnung*, Frankfurt a.M.

Blum, H. F.
1962 *Time's Arrow and Evolution*, New York: Harper and Row.

Bock, K. E.
1956 *The Acceptance of Histories*, Berkeley and Los Angeles: University of California Press.

Bonanate, U.
1972 *Charles Blount: Libertinismo e deismo nel Seicento inglese*, Florence: La Nuova Italia.

Bontinck, F.
1962 *La lutte autour de la liturgie chinoise au XVIIe et XVIIIe siècle*, Louvain and Paris.

Borst, A.
1958–63 *Der Turmbau von Babel: Geschichte der Meinung über Ursprung und Vielfalt der Sprachen und Völker*, 6 vols., Stuttgart.

Brandon, S. G. F.
1951 *Time and Mankind: An Historical and Philosophical Study of Mankind's Attitude to the Phenomena of Change*, London: Hutchinson.

Brewster, T.
1928 *This Puzzling Planet*, Indianapolis.

Briatore, L.
1976 *Cronologia e techniche della misura del tempo*, Florence: Giunti-Barbera.

Brouwer, A.
1975 *Paleontologia generale: le testimonianze fossili della vita*, Milan: Est Mondadori.
Brunelli, V.
1977 "Religione e dottrina del linguaggio in Spinoza," *Verifiche* 6:755–88.
Bulferetti, L.
1952 "L'ipotesi vichiana dell'erramento ferino," *Annali della Facoltà di Lettere-filosofia e Magistero dell'Università di Cagliari* 19.
Burchfield, J. D.
1975 *Lord Kelvin and the Age of the Earth*, New York: Science History Publications.
Burke, J. B.
1966 *Origins of the Science of Crystals*, Berkeley and Los Angeles: University of California Press.
Busco, P.
1924 *Les cosmogonies modernes et la théorie de la connaissance*, Paris.
Busson, H.
1948 *La religion des classiques*, Paris: Presses Universitaires.
Campodonico, A.
1978 *Filosofia dell'esperienza ed epistemologia della fede in R. Boyle*, Florence: Le Monnier.
Canguilhem, G.
1968 *Etudes d'histoire et philosophie des sciences*, Paris: Vrin.
Cannon, W. F.
1960 "The Uniformitarian-Catastrophist Debate," *Isis* 51:38–55.
Cantelli, G.
1971 *Vico e Bayle: premesse per un confronto*, Naples: Guida.
1972 "Mito e storia in J. Leclerc, Tournemine e Fontenelle," *Rivista critica di storia della filosofia* 3:269–86 and 4:285–400.
1974 "Nicola Fréret: tradizione religiosa e allegoria nell'interpretazione dei miti pagani," *Rivista critica di storia della filosofia* 3–4:1–40.
1978 "Pitture messicane, caratteri cinesi e immagini sacre: alle fonti delle teorie linguistiche di Vico e di Warburton," *Studi filosofici*, Siena: Università degli Studi di Siena.
Carroll, R. T.
1975 *The Common Sense Philosophy of Religion of Bishop E. Stillingfleet*, The Hague: Martinus Nijhoff.
Casini, P.
1969 *L'universo macchina*, Bari: Laterza.
1973 *Introduzione all'Illuminismo*, Bari: Laterza.
Challinor, J.
1953–54 "The Early Progress of British Geology," *Annals of Science* 9:124–53; 10:1–19, 107-48.
Ch'en Shou-Yi
1935 "John Webb: A Forgotten Page of the Early History of Sinology in Europe," *The Chinese Social and Political Review* 19, 3:295–330.

Chilton, C. W.
1962 "The Epicurean Theory of the Origin of Language," *American Journal of Philology* 83, 2:161–63.
Chomsky, Noam
1966 *Cartesian Linguistics: A Chapter in the History of Rationalist Thought*, New York: Harper & Row.
Clagett, M., ed.
1962 *Critical Problems in the History of Science*, Madison: The University of Wisconsin Press.
Claparède, E.
1935 "Rousseau et l'origine du langage," *Annales de la Société J.J. Rousseau* 14:95–120.
Cloyd, E. L.
1972 *James Burnet, Lord Monboddo*, Oxford: Oxford University Press.
Colie, R. L.
1959 "Spinoza and the Early English Deists," *Journal of the History of Ideas* 20:43–46.
1963 "Spinoza in England: 1665–1730," *Proceedings of the American Philosophical Society* 107:184–86.
Collier, K. B.
1934 *The Cosmogonies of Our Fathers: Some Theories of the 17th and 18th Centuries*, New York: Columbia University Press.
Corsano, A.
1968 "Vico e la tradizione ermetica," in *Omaggio a Vico*, Naples: Morano.
Corsi, P.
1978 "Charles Lyell and the French Biological Thought," *Annali della Scuola Normale Superiore di Pisa*, Cl. Lettere, serie 3, 8 3:1253–85.
Cragg, G. R.
1964 *Reason and Authority in the Eighteenth Century*, Cambridge: Cambridge University Press.
1966 *From Puritanism to the Age of Reason: A Study of changes in Religious Thought within the Church of England, 1660 to 1700*, Cambridge: Cambridge University Press.
1975 *Freedom and Authority: A Study of English Thought in the Early Seventeenth Century*, Philadelphia: Westminster Press.
Croce, Benedetto
1911 *La filosofia de G. B. Vico*, Bari: Laterza.
1947–48 *Bibliografia vichiana, accresciuta e rielaborata da F. Nicolini*, 2 vols., Naples: Ricciardi.
1964 *The Philosophy of Giambattista Vico*, trans. R. G. Collingwood, New York: Russell & Russell.
Crowther, J. G.
1960 *Founders of British Science*, London.
Daniel, G.
1968 *L'idea della preistoria*, Florence: Sansoni.

Daubrée, A.
1880 "Descartes, l'un des fondateurs de la Cosmologie et de la géologie," *Journal des Savants*, March–April.
David, M. V.
1965 *Le débat sur les écritures et l'hiéroglyphe aux XVIIe et XVIIIe siècles*, . Paris.
Davies, G. L.
1964 "R. Hooke and his Conception of Earth History," *Proceedings of the Geological Association of London* 75:493–98.
1968 *The Earth in Decay: A History of British Geomorphology: 1578–1878*, London: Macdonald Technical and Scientific.
Dawson, R.
1967 *The Chinese Chameleon: An Analysis of European Conceptions of Chinese Civilisation*, London and New York: Oxford University Press.
Debus, A. G.
1965 *The English Paracelsians*, London: Oldbourne.
1977 *The Chemical Philosophy: Paracelsian Science and Medicine in the Sixteenth and Seventeenth Centuries*, 2 vols., New York: Science History Publications.
De Mas, E. et al.
1978 *Vico e l'instaurazione delle scienze: Diritto, linguistica, antropologia*, Lecce: Messapica.
De Mauro, T.
1968 "G. B. Vico dalla retorica allo storicismo linguistico," *La Cultura* 6:167–83.
1970 *Introduzione alla semantica*, Bari: Laterza.
Derrida, J.
1967 "La linguistique de Rousseau," *Revue Internationale de Philosophie* 21, 4:443–62.
1967 *De la grammatologie*, Paris: Editions de Minuit.
1976 *Of Grammatology*, trans. G. Chakrovorty Spivak, Baltimore: Johns Hopkins University Press.
Dillenberger, J.
1961 *Protestant Thought and Natural Science: An Historical Interpretation*, London.
Dini, A.
1979 "Le teorie preadamitiche e il libertinismo di La Peyrère," *Annali dell'Istituto di Filosofia della Facoltà di Lettere e Filosofia dell'Università di Firenze*, Florence: Olschki.
Dott, R. H., Jr.
1964 "James Hutton and the Concept of the Dynamic Earth," in Schneer 1969, 122–41.
Duchet, M.
1971 *Anthropologie et historie au siècle des lumières*, Paris: Maspero.
Dupront, A.
1930 *Pierre Daniel Huet et l'exegèse comparatiste au XVIIe siècle*, Paris.
Ehrard, J.
1964 *L'idée de nature en France dans la première moitié du XVIIIe siècle*, Paris: Albin Michel.

Eiseley, L.
1961 *Darwin's Century: Evolution and the Men Who Discovered It*, Garden City, N.Y.: Doubleday.
1972 *The Firmament of Time*, New York: Atheneum.
Eliade, Mircea
1954 *The Myth of the Eternal Return*, trans. Willard R. Trask, New York: Pantheon.
Ellegard, A.
1973 "Language, study of," in *Dictionary of the History of Ideas, 669*, New York: Scribner's.
Engel, C. E.
1950 *A History of Mountaineering in the Alps*, London: G. Allen and Unwin.
Etiemble, R.
1966 *Les Jésuites et la Chine, la querelle des rites (1552–1773)*, Paris: Juillard.
Europa Christiana
1978 *L'europa christiana nel rapporto con le altre culture nel secolo XVII*, Atti del Convegno di studio di Santa Margherita Ligure (19–21 maggio 1977).
Evans, A. W.
1932 *Warburton and the Warburtonians: A Study in some Eighteenth Century Controversies*, London.
Fano, G.
1973 *Origini e natura del linguaggio*, Turin: Einaudi.
Farrington, B.
1964 *The Philosophy of Francis Bacon: An Essay on Its Development from 1603 to 1609, with New Translations of Fundamental Texts*, Liverpool: Liverpool University Press.
Fleckenstein, J. O.
1949 *Scolastik, Barok, Exacte Wissenchaften*, Einsielden.
Formigari, L.
1970 *Linguistica ed empirismo nel Seicento inglese*, Bari: Laterza.
1972 *Linguistica e antropologia nel secondo Settecento*, Messina: La Libra.
Frankfurt, H. G., ed.
1972 *Leibniz: A Collection of Critical Essays*, New York: Anchor Books.
Frey, H. W.
1974 *The Eclipse of Biblical Narrative: A Study in 18th and 19th Century Hermeneutics*, New Haven: Yale University Press.
Funke, O.
1934 *Englische Sprachphilosophie im späteren 18 Jahrhundert*, Bern.
Furnish, V. P.
1974 "The Historical Criticism of the New Testament," *Bulletin of the J. Rylands University Library of Manchester* 56:336–70.
Garfinckie, N.
1955 "Science and Religion in England: 1790–1800," *Journal of the History of Ideas* 17:376–88.
Gargani, A.
1955 "Idea, mondo e linguaggio in Th. Hobbes e in J. Locke," *Annali della Scuola Normale di Pisa* 35:251–92.

Garin, E.
1934 "Bernard de Mandeville," *Civiltà Moderna*, 70–91.
1970 *Dal Rinascimento all'Illuminismo*, Pisa: Nistri-Lischi.
1971 "Divagazioni Cinesi," *Rivista Critica di Storia della Filosofia* 26:332–35.
1975 "Alla scoperta del 'diverso': i selvaggi americani e i saggi cinesi," in *Rinascite e rivoluzioni: Movimenti culturali dal XIV al XVIII secolo*, 327–62, Bari: Laterza.

Garulli, E.
1958 *Saggi su Spinoza*, Urbino: Argalia.

Gay, P.
1967 *The Enlightenment: An Interpretation*, 2 vols., London: Wildwood House.

Geikie, A.
1962 *The Founders of Geology*, New York: Dover.

Gerbi, A.
1955 *La disputa del Nuovo Mondo: Storia di una polemica: 1750–1900*, Milan and Naples: Ricciardi.

Geymonat, L. et al., eds.
1962 *Il metodo sperimentale in biologia da Vallisnieri ad oggi*, Padova.

Gillispie, C. C.
1959 *Genesis and Geology: The Impact of Scientific Discoveries upon Religious Beliefs in the Decades before Darwin*, New York: Harper.

Ginzel, F. K.
1906–14 *Handbuch der matematischen Chronologie*, 3 vols., Leipzig.

Giuntini, C.
1975 "Toland e Bruno: ermetismo 'rivoluzionario'?" *Rivista di Filosofia* 67, 2:199–235.
1979 *Panteismo e ideologia repubblicana: John Toland*, Bologna: Il Mulino.

Glass, B., et al., eds.
1968 *Forerunners of Darwin 1745–1859*, ed. B. Glass, O. Temkin, and W. L. Strauss, Jr., Baltimore: The Johns Hopkins Press.

Gliozzi, G.
1977 *Adamo e il Nuovo Mondo: La nascita dell'antropologia come ideologia coloniale dalle genealogie bibliche alle teorie razziali: 1500–1700*, Florence: La Nuova Italia.

Goretti, M.
1958 *Il paradosso Mandeville*, Florence.

Grafton, A. T.
1975 "Joseph Scaliger and Historical Chronology: The Rise and Fall of a Discipline," *History and Theory* 14, 2:156–85.

Greene, J. C.
1959 *The Death of Adam: Evolution and Its Impact on Western Thought*, Ames: Iowa State University Press.

Gregory, T.
1972 "Erudizione e ateismo nella cultura del Seicento: Il Theophrastus Redivivus," *Giornale Critico della Filosofia Italiana*, 194–240.

Grumel, V.
1958 *Traité d'études byzantines*, vol. 1: *La Chronologie*, Paris.
Gunther, R. W. T.
1930 "The Life and Works of Robert Hooke," in *Early Science in Oxford, 1923–1945*, 6, 131ff.
Guy, B.
1963 "The French Image of China before and after Voltaire," *Studies on Voltaire and the XVIIIth Century* 21:123–31.
Haber, F. C.
1966 *The Age of the World, Moses to Darwin*, Baltimore: The Johns Hopkins Press.
Hall, A. R.
1962 "The Scholar and the Craftsman in the Scientific Revolution," in Clagett 1962, 3–23.
1966 *Hooke's Micrographia 1665–1965: A Lecture in Commemoration of the Tercentenary delivered at the Middlesex Hospital Medical School*, London.
1972 "Science, Technology and Utopia in the Seventeenth Century," in Mathias 1972, 33–55.
Haller, E.
1940 *Die barocken Stilmerkmade in der englishen, lateinischen und deutschen Fassung von Dr. Thomas Burnet Theory of the Earth*, Bern.
Hampton, J.
1955 *N. A. Boulanger et la science de son temps*, Geneva.
Harnois, G.
1929 *Les théories du langage en France de 1660 à 1821*, Paris.
Hastie, W.
1900 *Kant's Cosmology*, Glasgow.
Haydn, H.
1950 *The Counter-Renaissance*, New York: The Grove Press.
Heinekamp, A.
1969 *Das Problem des Guten bei Leibniz*, Bonn: Bouvier u. Co. Verlag.
Heinemann, F. H.
1945 "Toland and Leibniz," *The Philosophical Review* 54:437–57.
1950–52 "J. Toland and the Age of Reason," *Archiv für Philosophie* 4:35–66.
Hill, C.
1971 *The Antichrist in Seventeenth Century England*, London.
Hintikka, J.
1972 "Leibniz on Plenitude, Relations and the 'Reign of Law,'" in Frankfurt 1972, 155–90.
Hodgen, M. T.
1964 *Early Anthropology in the Sixteenth and Seventeenth Centuries*, Philadelphia: University of Pennsylvania.
Hoelder, H.
1960 *Geologie und Paläontologie in Texten und ihrer Geschichte*, Freiburg and Munich: Alber.
Hooykaas, R.
1957 "The Parallel between the History of the Earth and the History of the Animal World," *Archives Int. d'Histoire des Sciences* 38:3–18.

1963 *The Principle of Uniformity in Geology, Biology and Theology*, Leiden: Brill.
1972 *Religion and the Rise of Modern Science*, Edinburgh and London.
1974 "La riforma protestante e la scienza," *Comunità* 28, n. 173:115–59.

Hoskin, M. A.
1963 *William Herschel and the Construction of the Heavens*, London.
1972 "Dark Skies and Fixed Stars," *Journal of the British Astronomical Association* 93.

Hunt, R.
1955 *The Place of Religion in the Science of R. Boyle*, Pittsburgh.

Iversen, E.
1961 *The Myth of Egypt and Its Hieroglyphs in European Tradition*, Copenhagen: Gec Gad Publishers.

Jacob, J. R.
1977 *Robert Boyle and the English Revolution*, New York: Burt Franklin.

Jacob, M. C.
1969 "Toland and the Newtonian Ideology," *Journal of the Warburg and Courtauld Institutes* 32:307–31.
1972 "The Church and the Formulation of the Newtonian World-View," *Journal of European Studies* 2:265–97.
1974 "Early Newtonianism," *History of Science* 12:142–46.
1976a "Millenarism and Science in the Late Seventeenth Century," *Journal of the History of Ideas* 37:335–41.
1976b *The Newtonians and the English Revolution*, Hassocks: The Harvester Press.
1977 "Newtonianism and the Origin of the Enlightenment: A Reassessment," *Eighteenth Century Studies* 2:1–25.

Jacob, M. C. and Guerlac, H.
1969 "Bentley, Newton and Providence: The Boyle Lectures Once More," *Journal of the History of Ideas* 30:307–18.

Jacob, M. C. and Lockwood, W. A.
1972 "Political Millenarism and Burnet's Sacred Theory," *Science Studies* 2:265–79.

Jaki, S. L.
1972 *The Milky Way: An Elusive Road for Science*, New York: Science History Publications.
1977 *Planets and Planetarians: A History of Theories of the Origin of Planetary Systems*, Edinburgh: Scottish Academic Press.
1978 "Lambert and the Watershed of Cosmology," *Scientia* 72:75–95.

Juliard, P.
1970 *Philosophies of Language in Eighteenth-Century France*, The Hague and Paris.

Kaye, F. B.
1924 "Mandeville on the Origins of Languages," *Modern Language Notes* 39:136–42.
1928 "The Influence of B. de Mandeville," *Studies in Philology* 19:83–108.

Keith, T.
1971 *Religion and the Decline of Magic: Studies in Popular Beliefs in Sixteenth and Seventeenth Century England*, London.

Kelly, S.
1969 "Theories of the Earth in Renaissance Cosmologies," in Schneer 1969, 214–25.

Keynes, G. L.
1960 *A Bibliography of Dr. R. Hooke*, Oxford.

Knight, W.
1900 *Lord Monboddo and Some of His Contemporaries*, London: John Murray.

Knowlson, J. R.
1965 "The Idea of Gesture as a Universal Language," *Journal of the History of Ideas* 4:495–508.
1975 *Universal Languages Schemes in England and France*, Toronto: Toronto University Press.

Koyré, Alexander
1957 *From the Closed World to the Infinite Universe*, New York: Harper.

Kraus, A.
1963 *Vernunft und Geschichte: Die Bedeutung der deutschen Akademien für die Entwicklung der Geistes-wissenschaft im späten 18 Jahrhundert*, Freiburg: Herder.

Kronick, D. A.
1962 *A History of Scientific and Technical Periodicals: The Origins and Development of the Scientific and Technical Press*, New York: The Scarecrow Press.

Krusch, B.
1938 *Studien zur christlich-mitteralterlichen Chronologie*, Berlin.

Kubler, G.
1977 "Vico e l'America precolombiana," *Bollettino del Centro di Studi Vichiani* 7:58–66.

Kubrin, D.
1967 "Newton and the Cyclical Cosmos: Providence and Mechanical Philosophy," *Journal of the History of Ideas* 28:425–46.

Kuehner, P.
1944 *Theories on the Origin and Formation of Language in XVIIIth Century France*, Philadelphia.

Kuhn, T.
1977 *The Essential Tension: Selected Studies in Scientific Tradition and Change*, Chicago and London: The University of Chicago Press.

Labanca, B.
1898 *G. B. Vico e i suoi critici cattolici*, Naples: Luigi Pierro.

Lach, D. F.
1941 "China and the Era of Enlightenment," *Journal of Modern History* 14:209–23.
1945 "Leibniz and China," *Journal of the History of Ideas* 6:436–55.
1965 *Asia in the Making of Europe*, Chicago: The University of Chicago Press.

Landucci, S.
1972 *I filosofi e i selvaggi: 1580–1780*, Bari: Laterza.

Launay, L. de
1905 *La science géologique, ses méthodes, ses résultats, ses problèmes, son histoire*, Paris.

Lenoble, R.
1954 "La géologie au milieu du XVIIe siècle," *Les conférences du Palais de la Découverte*, série D., no. 27, Paris.
Lepenies, W.
1978 *Das Ende der Naturgeschichte: Wandel kultureller Selbstwerständlichkeiten in den Wissenchaften des 18 und 19 Jahrhunderts*, Frankfurt a.M.: Suhrkamp.
Lerner, I. M. and Libby, W. J.
1968 *Heredity, Evolution, and Society*, San Francisco: W. H. Freeman
Levine, J. M.
1977 *Dr. Woodward's Shield: History, Science, and Satire in Augustan England*, Berkeley: University of California Press.
Lewis, P. W.
1957 *Time and Western Man*, Beacon Paperbacks.
Lough, J.
1968 *Essays on the Encyclopédie of Diderot and D'Alembert*, London: Oxford University Press.
Lovejoy, Arthur O.
1960 *The Great Chain of Being*, (Cambridge, Massachusetts: Harvard University Press, 1936) New York: Harper Torchbooks.
1968 "The Argument for Organic Evolution before the Origin of Species," in H. B. Glass et al., eds., *Forerunners of Darwin*, Baltimore: Johns Hopkins Press.
Lupoli, A.
1976 "La polemica tra Hobbes e Boyle," *Acme* 29, 3:309–54.
Mahudeau, P.
1915 "Un précurseur du polygénisme: Isaac de la Peyrère," *Revue Anthropologique* 25:21–26.
Mannarino, R.
1976 "Storia sacra e storia profana nel 'Triregno' di Giannone: il regno terreno," *Critica Storica* 13, 3:49–77
Manuel, F.
1959 *The Eighteenth Century Confronts the Gods*, Cambridge, Massachusetts: Harvard University Press (reprint 1967, New York: Atheneum).
1963 *Newton, Historian*, Cambridge: Belknap Press of the Harvard University Press.
1968 *A Portrait of Isaac Newton*, Cambridge: Belknap Press of the Harvard University Press.
1974 *The Religion of Isaac Newton*, Oxford.
Mastellone, S.
1965 *Pensiero politico e vita culturale a Napoli nella seconda metà del Seicento*, Florence.
Mather, K. F. and Mason, S. L.
1970 *A Source Book in Geology*, Cambridge, Massachusetts.
Mathias, P., ed.
1972 *Science and Society: 1600–1900*, Cambridge: Cambridge University Press.
Mayer, J.
1954 "Robinet, philosophe de la nature," *Revue des Sciences Humaines*, 295–309.

McKee, D. R.

1944 "Isaac de la Peyrère, a Precursor of Eighteenth-Century Critical Deists," *PMLA* 1:456–85.

Meek, R. L.

1976 *Social Science and the Ignoble Savage*, Cambridge: Cambridge University Press.

Mensch und Zeit

1952 "Mensch und Zeit," *Eranos-Jahrbuch* 20, Zurich: Rhein-Verlag.

Meyer, G.

1958 "Fontenelle and Late 17th Century Science," *The History of Ideas Newsletter* 4.

Meyer, H.

1951 *The Age of the World: A Chapter in the History of the Enlightenment*, Allentown, Pennsylvania: Muhlenburg College.

Milhaud, M.

1911 *Nouvelles études sur l'histoire de la pensée scientifique*, Paris.

Milic, L. T., ed.

1971 *The Modernity of the Eighteenth Century*, Studies in Eighteenth-Century Culture 1, Cleveland and London: The Press of Case Western Reserve University.

Momigliano, A.

1965 "La nuova storia romana di Vico," *Rivista storica italiana* 77:773–90.

1966 "Roman 'Bestioni' and Roman 'Eroi' in Vico's Scienza Nuova," in *Terzo contributo alla storia degli studi classici e del mondo antico*, 153–77, Rome (and *History and Theory* 5 [1966]:3–23).

More, R.

1962 *Man, Time and Fossils: The Story of Evolution*, London: Jonathan Cape.

Mornet, D.

1911 *Les Sciences de la nature en France au XVIIIe siècle*, Paris.

Mugnai, M.

1976 *Astrazione e realtà: Saggio su Leibniz*, Milan: Feltrinelli.

Munitz, M. K.

1957 *Theories of the Universe*, New York: The Free Press of Glencoe.

Naville, Pl.

1967 *D'Holbach et la philosophie scientifique au XVIIIe siècle*, Paris: Gallimard.

Nicolini, F.

1949 *La religiosità di G. B. Vico*, Bari: Laterza.

1949–50 *Commento storico alla seconda Scienza Nuova*, Rome: Edizioni di storia e letteratura.

1955 *Saggi vichiani*, Naples: Giannini.

Nicolson, M. H.

1959 *Mountain Gloom and Mountain Glory: The Development of the Aesthetics of the Infinite*, New York: The Norton Library.

1960a *Voyages to the Moon*, New York: Macmillan.

1960b *The Breaking of the Circle: Studies on the Effect of the "New Science" upon Seventeenth-Century Poetry*, New York: Columbia University Press.

1962 *Science and Imagination*, Ithaca, New York: Cornell University Press.

North, F. J.
1934 "The Anatomy of the Earth: A Seventeenth-Century Cosmogony," *Geological Magazine* 71:541–47, London.
Ogden, H. V. S.
1947 "Thomas Burnet's Telluris Theoria Sacra and Mountain Scenery," *English Literary History* 14:139–50.
Pacchi, A.
1973 *Cartesio in Inghilterra*, Bari: Laterza.
Pagliaro, A.
1961 *Altri saggi di critica semantica*, Florence: D'Anna.
1969 "Le origini del linguaggio secondo Vico," in *Campanella e Vico* (various authors), Quaderno 126 of *Problemi di cultura*, Rome: Accademia Nazionale dei Lincei.
Parkinson, G.
1969 "Language and Knowledge in Spinoza," *Inquiry* 12:15–40.
Pastine, D.
1971 "Le origini del poligenismo e Isaac Lapeyrère," in *Miscellanea Seicento* 1, Florence: Le Monnier.
1975 *Juan Caramuel: probabilismo ed enciclopedia*, Florence: La Nuova Italia.
1978 "Il problema teologico delle culture non cristiane," in *L'europa christiana* 1978, 3–23.
1979 *La nascita dell'idolatria: L'Oriente religioso di Attansius Kircher*, Florence: La Nuova Italia.
Patterson, L. D.
1949–50 "Hooke's Gravitation Theory and its Influence on Newton," *Isis* 40:327–41; 41:32–45.
Pécaut, C.
1951 L'oeuvre géologique de Leibniz," *Revue générale des Sciences*.
Philipp, W.
1957 *Das Werden der Aufklärung in theologiegeschichtlichen Sicht* 78–86 (*Kosmischer Nihilismus und nihilisticher Kosmos im 17 Jahrhundert*), Göttingen: Vandenhoeck und Ruprecht.
Phillips, O. M.
1978 *La geofisica*, Milan: Est Mondadori.
Pinot, V.
1932 *La Chine et la formation de l'esprit philosophique en France: 1640–1740*, Paris: Librarie Orientaliste Paul Gethner.
Piovani, P.
1969 "Vico e la filosofia senza natura," in *Campanella e Vico* (various authors), Quaderno 126 of "Problemi di cultura," 247–68, Rome: Accademia Nazionale dei Lincei.
Porter, R.
1973 "The Industrial Revolution and the Rise of the Science of Geology," in M. Teich and R. Young, eds., *Changing Perspectives in the History of Science*, 320–43, London: Heinemann.

Prandi, A.

1975 *Christianesimo offeso e difeso: Deismo e apologetica cristiana nel secondo Settecento*, Bologna: Il Mulino.

Primer, I., ed.

1975 *Mandeville Studies*, The Hague: Martinus Nijhoff.

Puech, H. C.

1952 "La Gnose et le temps," in *Mensch und Zeit* 1952, 76–85.

Raimondi, E.

1966 *Anatomie seicentesche*, Pisa: Nistri-Lischi.

Raven, C. E.

1942 *John Ray Naturalist: His Life and Works*, Cambridge.

1947 *English Naturalists from Neckham to Ray*, Cambridge.

Reale, M.

1970 "Vico e il problema della storia ebraica in una recente interpretazione," *La Cultura* 8, 81–107.

Redwood, J.

1974 "Charles Blount, Deism and English Free Thought," *Journal of the History of Ideas* 35:490–98.

Ricuperati, G.

1965 "Alle origini del 'Triregno': la 'Philosophia adamitico-noetica' di A. Costantino," *Rivista storica italiana*.

1970 *L'esperienza civile e religiosa di Pietro Giannone*, Milan and Naples: Ricciardi.

Righini-Bonelli, M. L. and Shea, W., eds.,

1975 *Reason, Experiment and Mysticism in the Scientific Revolution*, New York: Science History Publications.

Roger, J.

1953 "Un manuscrit inédit perdu et retrouvé: les Anecdotes de la nature de N. A. Boulanger," *Revue des Sciences Humaines*, July–September.

1962 Introduction to Buffon, *Les époques de la nature*, *Mémoires du Muséum National d'Histoire Naturelle*, Série C, Sciences de la Terre, 10, Paris.

1963 *Les sciences de la vie dans la pensée française du XVIII*[e] *siècle*, Paris.

1972 "Les conditions intellectuelles de l'apparition du transformisme," in Bourgeois, C. and de Roux, D., eds., *Epistémologie et marxisme*, Paris: Union générale des éditions.

Roller, D. H. D.

1971 *Perspectives in the History of Science and Technology*, Norman: University of Oklahoma Press.

Rossi, Paolo

1960 *Clavis universalis: arti mnemoniche e logica combinatoria da Lullo a Leibniz*, Milan and Naples: Ricciardi.

1968 *Francis Bacon, from Magic to Science*, trans. Sacha Rabinovitch, Chicago: University of Chicago Press.

1969 *Le sterminate antichità: studi vichiani*, Pisa: Nistri-Lischi.

1970 *Philosophy, Technology and the Arts in the Early Modern Era*, ed. and trans. Salvator Attanasio, New York: Harper Torchbooks.

1971a *Aspetti della rivoluzione scientifica*, Naples: Morano.

1971b *I filosofi e le macchine: 1400–1700*, Milan: Feltrinelli.

1974a *Francesco Bacone, dalla magia alla scienza*, Turin: Einaudi.
1974b *Storia e filosofia: saggi sulla storiografia filosofica*, Turin: Einaudi.
1977 *Immagini della scienza*, Rome: Editori Riuniti.
Routhier, P.
1969 *Essai critique sur les méthodes de la géologie*, Paris: Masson.
Rudwick, M. J. S.
1967 "A Critique of Uniformitarian Geology: A letter from W. C. Conybeare to Charles Lyell," *Proceedings of the American Philosophical Society* 111:272–87.
1976 *The Meaning of Fossils: Episodes in the History of Paleontology*, New York: Neal Watson.
Salmon, V.
1966 "Language-Planning in Seventeenth-Century England: Its Context and Aims," in Bazell, C. E. and Catford, J., eds., *In Memory of J. R. Firth*, London: Longman's.
Salvucci, P.
1972 *Adam Furguson: sociologia e filosofia politica*, Urbino: Argalia.
Sarton, G.
1914–19 "La synthèse géologique de 1775 à 1918," *Isis* 2:359–94.
Savan, D.
1958 "Spinoza and Language," *Philosophical Review* 57:212–25.
Savelli, R.
1961 "L'opera biologica di Vallisnieri," *Physis* 1961, 4.
Scherz, G.
1956 *Vom Wege Niels Stensens: Beiträge zur seiner naturwissenchaftlichen Entwicklung*, Copenhagen.
1958 Ed., "Nicolaus Steno and his Indice," in *N. Steno's Life and Works*, 9–86, Capenhagen: Munksgaard.
1969 Preface to Steno, *Geological Papers*, Odense University Press.
Schmitt, C. B.
1970 "Prisca Theologia e Philosophia Perennis: due temi del Rinascimento italiano e la loro fortuna," in *Il pensiero italiano del Rinascimento e il tempo nostro*, Florence: Olschki.
Schneer, C. J.
1954 "The Rise of Historical Geology in the Seventeenth Century, *Isis* 45:256–68.
1969 *Towards a History of Geology*, Cambridge, Massachusetts.
Schofield, R. E.
1970 *Mechanism and Materialism: British Natural Philosophy in an Age of Reason*, Princeton: Princeton University Press.
1971a "The Counter-Reformation in Eighteenth-Century Science," in Roller 1971, 39–54.
1971b "What Is Modern in the Eighteenth Century: Not Science," in *MILIC*, 1971.
Sciama, D. W.
1965 *L'unità dell'universo*, Turin: Einaudi.
Scott, W., ed.
1924–36 *Hermetica*, vol. 2, Oxford: Clarendon Press.

Scribano, M. E.
1978a "I 'Pensieri liberi' di Mandeville: una polemica antideista," *Atti e Memorie dell'Accademia Toscana di Scienze e Lettere La Colombaria* 43:139–99.
1978b "La presenza di Bayle nell'opera di Mandeville," *Giornale critico della filosofia italiana*.
Sina, M.
1976 *L'avvento della ragione: 'Reason' e 'Above Reason' dal razionalismo teologico inglese ad deismo*, Milan: Vita e Pensiero.
1978 *Vico e Le Clerc tra filosofia e filologia*, Naples: Guida.
Slotkin, J. S.
1965 *Readings in Early Anthropology*, London: Methuen.
Solinas, G.
1967 *Il microscopio e le metafisiche: Epigenesi e preesistenza da Cartesio a Kant*, Milan: Feltrinelli.
1973 *La 'Protogaea' di Leibniz ai margini della rivoluzione Scientifica*, Cagliari: Pubblicazioni dell'Istituto di Filosofia della Facoltà di Lettere e Filosofia dell'Università.
Spallanzani, M. F.
1977 "Esperienza e natura in A. Vallisnieri," *Contributi*, Biblioteca Municipale "A. Panizzi," Reggio Emilia, no. 2:5–36.
Spink, J. S.
1960 *French Free-Thought from Gassendi to Voltaire*, London: The Athlone Press.
Starobinski, J.
1967 "Rousseau et l'origine des langues," in H. and F. Schalk, eds., *Europäische Aufklärung, H. Dieckmann zum 60 Gebrustag*, 281–300, Munich and Allach: W. Fink Verlag.
Steinmann, J.
1960 *Richard Simon et l'exegèse biblique*, Paris: Desclée de Brouwer.
Stephen, L.
1962 *History of English Thought in the Eighteenth Century (1876)*, 2 vols., New York: Harcourt Brace.
Tait, P. G.
1876 *Lectures on some Recent Advances in Physical Science*, London: Macmillan.
Talluri, B.
1951 "Cinquant'anni di critica intorno al pensiero di Mandeville," *Studi Senesi* 43.
Taton, R.
1964 *Enseignement et diffusion des sciences in France au XVIIIe siècle*, Paris: Hermann.
Taylor, E. G. R.
1948 "The English Worldmakers of the Seventeenth Century and Their Influence on the Earth Sciences," *Geographical Review* 38:104–12.
1950 "The Origin of Continents and Oceans: A Seventeenth-Century Controversy," *Geographic Journal* 116:193–98.
Tega, W.
1971 "Meccanicismo e scienze della vita nel tardo Settecento," *Rivista di Filosofia* 62, 2:155–76.

Thorndike, L.
1947–58 *History of Magic and Experimental Science*, 8 vols., New York: Columbia University Press.
Timpanaro, S.
1972 "Friedrich Schlegel e gli inizi della linguistica indoeuropea in Germania," *Critica Storica* 9, 1:72–105.
1973 "Il contrasto tra i fratelli Schlegel e Franz Bopp sulla struttura e la genesi delle lingue indoeuropee," *Critica Storica* 10, 4:553–90.
Tindal, F.
1968 *L'homme sauvage: homo ferus et homo sylvestris, de l'animal à l'homme*, Paris: Payot.
Toulmin, S. and Goodfield, J.
1965 *The Discovery of Time*, London: Hutchinson.
Trengrove, L.
1966 "Newton's Theological Views," *Annals of Science* 22:277–94.
Troiani, L.
1977 *Commento storico al Contro Apione di Giuseppe*, Pisa: Giardini.
Tulloch, J.
1874 *Rational Theology and Christian Philosophy in England in the Seventeenth Century*, Edinburgh and London.
Tuveson, E. L.
1950 "Swift and the World-Makers," *Journal of the History of Ideas* 11:54–74.
1951 "Space, Deity and Natural Sublime," *Modern Language Quarterly*.
1964 *Millennium and Utopia: A Study of the Background of the Idea of Progress*, New York: Harper Torchbooks.
Van Kley, E. J.
1971 "Europe's Discovery of China and the Writing of World History," *The American Historical Review* 76:358–85.
Vartanian, A.
1953 *Diderot and Descartes: A Study of Scientific Naturalism in the Enlightenment*, Princeton: Princeton University Press.
Venturi, F.
1947 *L'antichità svelata e l'idea di progresso in N. Boulanger*, Bari: Laterza.
1958 "Postille inedite di Voltaire ad alcune opere di N. A. Boulanger e del barone d'Holbach," *Studi Francesi* 2:231–40.
Vernière, P.
1954 *Spinoza et la pensée française avant la Révolution*, Paris: PUF.
Verra, V.
1958 "Ragione, linguaggio e filosofia in Herder," *Filosofia* 2:221–59.
1966 *Mito, rivelazione e filosofia in J. G. Herder e nel suo tempo*, Milan: Marzorati.
Verri, A.
1970 *Origine delle lingue e civiltà in Rousseau*. In appendix, Rousseau, *Essai sur l'origine des langues*, Ravenna: Longo.
1975 *Lord Monboddo: dalla metafisico all'antropologia*, Ravenna: Longo.
1978 "Vico e Rousseau, filosofi del linguaggio," in De Mas et al., 1978, 141–162.

Wade, J.
1967 *The Clandestine Organisation and Diffusion of Philosophic Ideas in France from 1700 to 1720*, New York: Octagon Books.

Walker, D. P.
1964 *The Decline of Hell: XVIIIth Century Discussions of Eternal Torment*, Chicago.
1972 *The Ancient Theology: Studies in Christian Platonism from the Fifteenth to the Eighteenth Century*, London: Duckworth.

Watson, R. A.
1966 *The Downfall of Cartesianism, 1673–1712: A Study of Epistemological Issues in Late 17th Century Cartesianism*, The Hague: Martinus Nijhof.

Webster, C.
1975 *The Great Instauration: Science and Medicine Reform, 1620–1660*, London: Duckworth.

Wendt, H.
1970 *Before the Deluge*, London.

Westfall, R. S.
1958 *Science and Religion in Seventeenth-Century England*, New Haven: Yale University Press.

Willey, B.
1969 *The English Eighteenth Century Background: Studies in the Idea of Nature in the Thought of the Period*, (London: Chatto and Windus, 1940), New York: Columbia University Press.

Woodward, H. B.
1911 *History of Geology*, London.

Yates, F. A.
1964 *Giordano Bruno and the Hermetic Tradition*, London: Routledge and Kegan Paul.

Young, J. D.
1959 "Mandeville, a Popularizer of Hobbes," *Modern Language Notes* 74, 1:10–13.

Zittel, K. A.
1899 *Geschichte der Geologie und Paläontologie bis Ende des 19 Jahrhunderts*, Munich and Leipzig: Oldenbourg.
1901 *History of Geology and Palaeontology*, trans. M. Ogilvie, London: Gordon.

Zoli, S.
1972 "Le polemiche sulla Cina nella cultura storica, filosofica, letteraria italiana della prima metà del Settecento," *Archivio Storico Italiano*, 409–67.
1973 *La Cina e la cultura italiana dal Cinquecento al Settecento*, Bologna: Patron.
1974 "Il mito settecentesco della Cina in Europa e la moderna storiografia," *Nuova Rivista Storica*.
1978 "La Cina nella cultura europea del Seicento," in *L'europa christiana*, 1978, 87–174.

Index